21999
ECKEHART
AF784511

HANS REICHENBACH

SELECTED WRITINGS: 1909–1953

VOLUME ONE

VIENNA CIRCLE COLLECTION

Editorial Committee

HENK L. MULDER, *University of Amsterdam, Amsterdam, The Netherlands*

ROBERT S. COHEN, *Boston University, Boston, Mass., U.S.A.*

BRIAN MCGUINNESS, *The Queen's College, Oxford, England*

Editorial Advisory Board

ALFRED J. AYER, *New College, Oxford, England*

†Y. BAR-HILLEL, *The Hebrew University, Jerusalem, Israel*

ALBERT E. BLUMBERG, *Rutgers University, New Brunswick, N.J., U.S.A.*

HASKELL B. CURRY, *Pennsylvania State University, Pa., U.S.A.*

HERBERT FEIGL, *University of Minnesota, Minneapolis, Minn., U.S.A.*

ERWIN N. HIEBERT, *Harvard University, Cambridge, Mass., U.S.A.*

JAAKKO HINTIKKA, *Academy of Finland, Helsinki, Finland and Stanford University, Stanford, Calif. and Florida State University, Tallahassee, Fla., U.S.A.*

†VIKTOR KRAFT, *Vienna, Austria*

KARL MENGER, *Illinois Institute of Technology, Chicago, Ill., U.S.A.*

GABRIËL NUCHELMANS, *University of Leyden, Leyden, The Netherlands*

ANTHONY M. QUINTON, *New College, Oxford, England*

J. F. STAAL, *University of California, Berkeley, Calif., U.S.A.*

VOLUME 4

EDITOR: ROBERT S. COHEN

Photograph © August Scherl, Berlin

HANS REICHENBACH (Berlin, September 1931)

HANS REICHENBACH

SELECTED WRITINGS
1909-1953

VOLUME ONE

Principal Translations by

ELIZABETH HUGHES SCHNEEWIND

Edited by

MARIA REICHENBACH *and* ROBERT S. COHEN

With a Selection of Biographical and Autobiographical Sketches

D. REIDEL PUBLISHING COMPANY

DORDRECHT : HOLLAND / BOSTON : U.S.A.

LONDON : ENGLAND

Library of Congress Cataloging in Publication Data

CIP

Reichenbach, Hans, 1891–1953.
Selected writings, 1909–1953.

(Vienna circle collection ; v. 4–)
"Bibliography of writings of Hans Reichenbach": v. 1, p.
Includes index.
1. Philosophy–Addresses, essays, lectures. 2. Science–Philosophy–Addresses, essays, lectures. 3. Physics–Philosophy–Addresses, essays, lectures. 4. Social problems–Addresses, essays, lectures. I. Reichenbach, Maria. II. Cohen, Robert Sonné. III. Series.
B29.R423 1978 191 78–18446
ISBN 90-277-0291-8 (vol. 1) ISBN 90-277-0292-6 (vol. 1) pbk.
ISBN 90-277-0909-2 (vol. 2) ISBN 90-277-0910-6 (vol. 2)
ISBN 90-277-0892-4 (set) ISBN 90-277-0893-2 (set)

Principal translations from the German by Elizabeth Hughes Schneewind,
further translations from the German by Sheldon Gilman (Memories item 7),
by Gisela Lincoln (Memories items 1, 3, and 4),
by Maria Reichenbach (Articles 38, 42 and 43).
Translations edited by M.R. and R.S.C.

Published by D. Reidel Publishing Company,
P.O. Box 17, Dordrecht, Holland

Sold and distributed in the U.S.A., Canada, and Mexico
by D. Reidel Publishing Company, Inc.
Lincoln Building, 160 Old Derby Street, Hingham,
Mass. 02043, U.S.A.

All Rights Reserved
Copyright © 1978 by D. Reidel Publishing Company, Dordrecht, Holland
and copyrightholders as specified on appropriate pages within.
No part of the material protected by this copyright notice may be reproduced or
utilized in any form or by any means, electronic or mechanical,
including photocopying, recording or by any informational storage and
retrieval system, without written permission from the copyright owner

Printed in The Netherlands

TABLE OF CONTENTS

CONTENTS OF VOLUME TWO

PREFACE

These two volumes form a full portrait of Hans Reichenbach, from the school boy and university student to the maturing and creative scholar, who was as well an immensely devoted teacher and a gifted popular writer and speaker on science and philosophy.

We selected the articles for several reasons. Many of them have not previously been available in English; many are out of print, either in English or in German; some, especially the early ones, have been little known, and deal with subject-matters other than philosophy of science. The genesis and evolution of Reichenbach's ideas appeared to be of deep interest, and so we included papers from four decades, despite occasional redundancy. We were, for example, pleased to include his extensive review article from the encyclopedic *Handbuch der Physik* of 1929 on 'The Aims and Methods of Physical Knowledge', written at a time of creative collaboration between Reichenbach's Berlin group and the Vienna Circle of Schlick and Carnap. Reichenbach was a pioneer, opening new pathways to the solution of age-old problems in many fields: space, time, causality, induction and probability – philosophical analysis and interpretation of classical physics, relativity and quantum physics – logic, language, ethics, scientific explanation and methodology, critical appreciation and reconstruction of past metaphysical thinkers and scientists from Plato to Leibniz and Kant. Indeed, his own philosophical journey was initiated by his passage from Kant to anti-Kant.

To our pleasure, we have been able to assemble memoirs of Reichenbach from family, friends, students and colleagues of his all-too-brief life. They write from varying perspectives; they see him as a fellow-student, a brother, father, uncle, husband, a political leader, teacher, colleague, a human being with a rich and varied character . . . no philosophic star-gazer, not unworldly, but a man of flesh-and-blood, with a warm-hearted nature, practical enthusiasm, stubborn to some but to others the spirit of tolerance and permissive ease. We have been able to include some autobiographical papers too, written at turning-points of life.

These English-language materials can be seen as a supplement to the German edition of Hans Reichenbach's *Collected Works*, now under publication and comprising about 80% of his writings. For convenience we list the contents of the German volumes in our Bibliography [1977a]. To our knowledge the Bibliography is complete, but we would be grateful if readers would send any missing items, or references, particularly translations, to one of us. A full copy of the Reichenbach intellectual estate, including correspondence, drafts, manuscripts and other papers, lecture and other notes, will eventually be deposited at the University of Pittsburgh Library, where students and scholars will be able to work on scientific and historical understanding of this beloved and intelligent philosopher.

January 1978

MARIA REICHENBACH
ROBERT S. COHEN

PUBLISHER'S NOTE

The publishers of this volume wish to thank the following publishers for permission to include translations or reprints of the respective articles. They also would like to thank the suppliers of the illustrations mentioned in the captions.

Full details can be found in the Bibliography of Writings of Hans Reichenbach at the end of the book.

Part I

art. 1, 3, 5	-	Max Steinebach, Munich
" 2	-	*Das Monistische Jahrhundert,* W. Ostwald, ed.
" 4	-	*Berliner Freistudentischen Blätter*

Part II

art. 9, 10	-	*Neue Zürcher Zeitung*
" 11	-	*Hamburger Fremdenblatt*
" 12	-	*Kunstwart*
" 13	-	*Radio-Umschau* (Frankfurt)
" 14, 15, 17	-	*Berliner Tageblatt*
" 16	-	*Die Umschau* (Frankfurt)
" 18, 19, 20, 22, 23, 24, 26, 29	-	*Vossische Zeitung*
" 21	-	*Die Böttcherstrasse* (Bremen)
" 25	-	*Deutsche Allgemeine Zeitung*
" 27, 30	-	*Die Woche*
" 28	-	*Die Literarische Welt*

Part III

art. 31	-	*Symposion*
" 32	-	Drei-Masken Verlag, Berlin and Munich
" 33, 36, 37	-	Felix Meiner, Leipzig
" 34, 35, 39	-	Springer-Verlag, Berlin-Heidelberg-New York
" 38	-	Routledge & Kegan Paul, London
" 40	-	*Kantstudien*

Part IV

art. 42, 43	-	Routledge & Kegan Paul, London

HANS REICHENBACH: PRINCIPAL DATES

1891	Born on September 26th in Hamburg
1910–15	University studies in Berlin, Munich, Göttingen, Erlangen and the Technische Hochschule at Stuttgart
1915	Ph.D., Erlangen
1915–17	German soldier on the Russian front
1917–20	Radio engineer in the laboratory of a radio firm at Berlin
1920–26	*Privatdozent*, later Associate Professor, Technische Hochschule at Stuttgart
1926–33	Associate Professor of Philosophy of Physics, University of Berlin
1933	Dismissal from University of Berlin under Hitler
1933–38	Professor of Philosophy, University of Istanbul
1938–53	Professor of Philosophy, University of California, Los Angeles
1947	Visiting Professor at Columbia University, City College of New York, and the New School for Social Research
1947	President, American Philosophical Association, Pacific Division
1952	Lectures at the Institut Henri Poincaré, Paris
1953	Invitation to deliver the William James lectures at Harvard University
1953	Died on April 9th in Los Angeles

FAMILY TREE

Maria	Bernhard	Hans	Wendelina	Hermann
1887	1888–1977	9/26/91– 4/09/53 (Hamburg–Los Angeles)	1896	1898–1958

Parents:

Bruno Reichenbach, 2/04/1864 Hamburg – 1918 Hamburg. Jewish at birth, baptised in 1887. Bruno was the third of seven children.

Selma, née Menzel, 4/15/1863 in Reichenbach, Silesia – 1925, Milan, Italy. Evangelical Reformed Protestant.

Grandparents:

Bernhard Reichenbach, 1835 Ballenstedt (Harz) – 1887 Hamburg. Bernhard Reichenbach was a furrier who had a store in Hamburg.

Amalie, née Koppel, 1839 Hamburg – 1901 Hamburg.

Hermann Menzel, engineer, May 12, 1825 Breslau (Wrocław) – Nov. 10, 1910 Förstenberg (Oder).

Wendelina, née von Senden, Sept. 12, 1836 Emden – May 15, 1896 Tegel, near Berlin.

Maternal great-grandparents:

Parents of Hermann Menzel:

Joseph Menzel, manufacturer, born in Neustadt, Silesia – died 1834 Breslau.

Maria Magdalena Kluge, 1784 Breslau – 1845 Breslau. The Menzel family was Catholic but because of his marriage to Wendelina von Senden, Grandfather Menzel converted to the Evangelical Reformed church.

Parents of Wendelina von Senden:

Herman Friedrich Wilhelm v. Senden, pharmacist in Emden and District President; born 1797 Emden – died

1867 Emden. Married to his cousin.

Elisabeth Galama v. Senden, 1795 Emden – 1876 Emden.

Paternal great-grandparents are unknown.

Note: The von Senden family tree can be traced back to the Reformation when a certain abbot Bernhard von Senden joined Zwingli and became one of the founders of the Evangelical Reformed church. His descendants were largely doctors, pharmacists and clergymen. The von Senden pharmacy existed in Emden for many generations and until the twentieth century.

Hans Reichenbach: in his first marriage, married to Elisabeth, née Lingener.

Children: Elizabeth Austin, 05/19/1924, Stuttgart
Hans Galama, 06/09/1922, Stuttgart

in his second marriage, married to Maria, née Moll, Berlin.

MEMORIES OF HANS REICHENBACH

1. AUTOBIOGRAPHICAL SKETCHES FOR ACADEMIC PURPOSES

(1) *Berlin, 1932*

I was born on September 26, 1891 in Hamburg and went to school there until the age of seventeen. My father was a wholesale merchant in Hamburg; up to then, at least as far as I could find out, there had not been any scientists in my family, only some engineers. Originally, I myself wanted to become an engineer as well and from 1910 to 1911, for two semesters I studied civil engineering at the Technische Hochschule in Stuttgart. During this time, I first realized that my interests were predominantly theoretical; I then studied mathematics, physics and philosophy at the universities in Berlin, Munich and Göttingen. Among my teachers were Planck, Sommerfeld, Hilbert and Born. Unlike my studies of mathematics and physics, philosophy did not satisfy me at all because the kind of philosophy presented appeared to be inexact and too little connected with the natural sciences. The philosophers v. Aster, in Munich, and Cassirer, then at Berlin, however were exceptions; they showed understanding of the problems of natural philosophy and were encouraging and stimulating.

Among the philosophers, Kant interested me most. I studied his work very carefully; but otherwise, I was not too interested in historical philosophers; I read some of them with great respect, such as Descartes and Spinoza; yet, my own philosophical work always related directly to physical problems without consideration for historical connections.

As a high school student, I had already been interested in the kinetic theory of gases and, using mainly these particular problems as a starting point, I came to explore the question of the validity of the laws of probability for reality. This then became the topic of my dissertation [1915b]. I had to write this dissertation without assistance by any academic teachers because nobody there was interested in these questions. Although I had used Kantian philosophy to dress up the solution I offered for the problems, I am still convinced that the basic idea of this work is very essential; it is one of the foundations of my present conception of the problems of probability.

For this work, I received my Ph.D. on March 2nd, 1915 in Erlangen. In the meantime, the war broke out and I myself was a soldier for 2½ years. However, already at that time I viewed the war as a great tragedy and I have always felt since that scientific-minded people have the particular duty to fight the spirit which breeds such catastrophes for humanity.

Due to my work in the army radio troops [signal corps], I became involved with radio technology and during the last year of the war, after I was transferred from active duty because of a severe illness I had contracted at the Russian front, I began to work as an engineer for a Berlin firm specializing in radio technology (from 1917 until 1920). During this period, and in my capacity as physicist, I directed the loud-speaker laboratory of this firm. I also got married. Soon thereafter, my father died and for the time being I could not give up my engineering position because I had to earn a salary in order to provide for my wife and myself. Nevertheless, in my spare time I studied the theory of relativity; I attended Einstein's lectures at the University of Berlin; at that time, his audience was very small because Einstein's name had not yet become known to a wider public. The theory of relativity impressed me immensely and led me into a conflict with Kant's philosophy. Einstein's critique of the space-time-problem made me realize that Kant's *a priori* concept was indeed untenable. I recorded the result of this profound inner change in a small book entitled *Relativitätstheorie und Erkenntnis Apriori* [1920f]. Shortly after the completion of this work I was offered an opportunity to work at the Technische Hochschule in Stuttgart as an instructor. I stayed in Stuttgart from 1920 until 1926. My teaching activities there covered very different areas; I lectured about physical measurement technique, wireless telegraphy, theory of relativity, but also about philosophical topics. After several years, I was given a teaching position in philosophy and was furthermore appointed associate professor. In this position, I had to lecture about the history of philosophy and about natural philosophy. In 1926, I joined the University of Berlin where, thanks to Planck's efforts, I was appointed to a teaching position in natural philosophy. I still hold this position today. I declined an invitation to a chair in natural philosophy at the German University in Prague in 1930.

In Stuttgart, I continued my work on the theory of relativity which finally led to the construction of an axiomatic for the relativistic space-time-theory. In my eyes, the value of this axiomatic lies especially in that here the theory of relativity receives a philosophically correct justification; although Einstein, in the construction of the theory, had generally instinctively understood the philosophical element correctly, nevertheless conscious and systematic study from the point of view of the theory of epistemology was still missing. A decisive result of this axiomatic, in my mind, is the idea that the theory of relativity must be based upon a causal theory of time, and furthermore, upon the realization that physics requires special definitions of correlation. I was the first to systematically uncover and compile all necessary definitions of correlation of the space-time-theory. Furthermore, by using the construction of light geometry, I was able to separate merely definitional statements from those statements of the theory of relativity which make assertions about the contents of nature. Later I collected my various studies about the space-time-theory in a book entitled *Philosophie der Raum-Zeit-Lehre* [1928h].

"correspondence"

I continued to be fascinated by the problem of probability with which I had begun my scientific work. Although at that time, I had already established a close connection between the concept of probability and the concept of causality, I pursued my line of thought even further when, in the theory of relativistic time, I once again had come upon the problem of causality. Thus, in 1924, I wrote a treatise entitled 'Die Kausalstruktur der Welt und der Unterschied von Vergangenheit und Zukunft' [1925d] in which the generalization from the connections of causality in the world to a connection of probability was carried out in the hope of achieving in this way a determination of the direction of time. In this essay, I viewed determinism as a limes [limit, boundary]-statement; at the same time, I also demonstrated that a more general world can be imagined where the limes of probability cannot be brought arbitrarily close to certainty. And indeed, this generalization was later executed in quantum mechanics.

In Berlin I continued my work on probability. In 1929, I published a treatise about 'Stetige Wahrscheinlichkeitsfolgen' [1929l] where the mathematical theory of a probability connection is given which has a determinist character only in the temporally infinitesimal. I then developed a new basis for probability calculus which I published in 1932; in this 'Axiomatik der Wahrscheinlichkeitsrechnung' [1932f], this discipline is constructed in a special calculus which combines logical and mathematical formulas and operations in a unique way. Thereby, one succeeds in reducing the applied probability calculus to a single axiom, the axiom of induction. In the execution of this idea, one makes use of a thought which I had already presented in 1929 at the Prague meeting on epistemology, namely the thought of a special probability logic; for it turned out that the probability problem cannot be solved within the framework of strict logic. I see this as a very far-reaching result for the theory of physical-scientific knowledge since the concept of probability is of such fundamental importance for physics; for it can be shown that every physical statement is basically a probability statement.

I had already tried before to bring the school of natural philosophy closer together in respect to its organization. In these last years, I reached my goal through the *Berliner Gesellschaft für Empirische Philosophie*, and by the foundation of the journal *Erkenntnis*; there, works on 'exact' natural philosophy are published and at the same time intellectual exchange on an international level is sought. When I planned to found this journal, I addressed myself to the Vienna Circle with whose members I had been on friendly terms, and asked for their cooperation. As a representative of the Vienna Circle, Rudolf Carnap then became the co-editor of *Erkenntnis*. In collaboration with the Vienna Circle, we organized conferences on epistemology in Prague and Königsberg which were combined with local meetings held by natural scientists. The reports of these conferences where, among others, questions concerning causality and probability, and the logical basis of mathematics were discussed, can be found in *Erkenntnis*.

Presently, I am in the process of collecting my papers about the problem of probability in order to publish them as a larger book.

(2) *Istanbul, 1936*

Originally Kantian; later, under the influence of Einstein's theory of relativity, a break with the Kantian system. In my book *Relativitätstheorie und Erkenntnis apriori* [1920f] it is shown that the characteristics of time and space which Kant took to be *a priori*, are untenable and that they have been shown to be false by the theory of relativity; at the same time, the ground for the failure of Kant's plan of necessary presuppositions for knowledge is established: the agreement between nature and the human ability to know cannot be postulated; quite to the contrary, in principle, there exists the possibility of a contradiction between the human system of concepts and nature. Yet, in such a case, epistemology does not have to fail at all, for the human being possesses the ability to change his system of concepts and to adapt it to nature. Just this is what happened with the theory of relativity. In this work I then developed the program for a new philosophical method: in place of Kant's analysis of reason, the analysis of the sciences; this 'scientific-analytic method' expounded the most general presuppositions used by epistemology at a given time and examines them as to their validity.

This method then is applied to the analysis of the space-time-problem which is the topic of numerous papers and, later on, of two books, *Axiomatik der relativistischen Raum-Zeit-Lehre* [1924h] and *Philosophie der Raum-Zeit-*

Lehre [1928h]. In the first place, these books develop the separation of the definitional from the epistemological elements of a conceptual system. This is the first time that a complete system of all space–time definitions, the so-called *Zuordnungsdefinitionen* (coordinating definitions), is established; on the other hand, this is also the first time that the empirical contents of Einstein's theories is set forth with precision, and formulated in axioms (here the term 'axiom' refers to a statement that is used as a presupposition for the theory, yet that is in itself empirically controllable). Then, the causal theory of time is developed according to which the concept of time is reduced to the concept of causality; since, on the other hand, measurement of space is also reduced to the measurement of time, space and time are therefore shown to be the 'causal structure of the world'. But as the axioms are experiential statements, the laws of space and time express the most general qualities of the world; in this sense, space and time possess reality; and the theory according to which space and time are forms of knowledge of the knowing subject, i.e., the human being (Kant, Poincaré's conventionalism), is proven to be a catastrophic error. In addition, I presented a critique of the problem of appearance, and demonstrated that, in principle, non-Euclidian spaces can also be concretely imagined.

In two further papers, I presented comprehensive representations of the physical world-picture: firstly, regarding epistemological method, in 'Ziele und Wege der physikalischen Erkenntnis' [1929g], and secondly, regarding contents, in the popular-scientific work *Atom und Kosmos* [1930h]. In this latter book, I demonstrated that the conceptual world of everyday life as well as that of earlier physics originated from the medium size dimensions to which man belongs because of his natural size; on the large scale (astronomy) and on the small scale (atoms), this system of concepts fails, and a new one must be created. In this, then, is the accomplishment of modern physics.

All these investigations have shown that contrary to Kant's belief, synthetic judgments *a priori* do not exist; what Kant called synthetic judgments *a priori* are in fact very general experiential statements which possess rather limited validity and which in part have been proven to be of merely approximative validity to our world. But one problem remains still unsolved which since then has caused greatest difficulties to philosophy; and moreover no consistent empiricism can be developed as long as it remains unsolved: that is the problem of induction. Since Hume's splendid critique, this problem dominates all epistemology and, now that the solution suggested by Kant has been proven untenable, one had to find another.

These questions led me to the problem of probability. For a conclusion based on inductions is in fact a conclusion based on probability. Thus, I frequently interrupted other work in order to work on the problem of probability. In my Ph.D. thesis [1915b], I had already dealt with the problem of the application of probability calculus. Although this work was still written from the Kantian standpoint, it nevertheless represents an important contribution to the problem of probability because I showed that the so-called principle of the missing cause is in fact untenable and can be eliminated. I published other papers in this direction in scientific journals, primarily in *Erkenntnis*; I should mention a work of 1925 about 'Die Kausalstruktur der Welt und der Unterschied von Vergangenheit und Zukunft' [1925d]. There, I predicted the possibility of just that kind of development of the causality concept which indeed, later on, entered into quantum mechanics. These investigations, however, required a further mathematical development of probability calculus; such a work then was published in 1932 in 'Axiomatik der Wahrscheinlichkeitsrechnung' [1932f]. Here, I developed a combination of probability calculus and logistic and then proceeded to demonstrate that the problem of probability can be reduced to the problem of induction alone. On the other hand, I was able to point the way to a possible solution of the problem of induction in connection with the problem of probability. These ideas then led to an expansion of logic, namely to the logic of probability which was published for the first time in [1932g]. In 1935 in my book *Wahrscheinlichkeitslehre* [1935h], I published a comprehensive representation of the entire problem area.

The decisive idea is that we can never assert that inductive conclusions are true or false but always only that they are more or less probable. This factual situation corresponds to a logic where the two values of truth, namely true or false, are replaced by a continual scale of truth values whose borderline cases, 0 and 1, mean false or true. But this logic is also the logic of all scientific statements. For, as all scientific statements are statements about the future, one can merely assert their probability. Thus, this logic of probability also offers a solution to the problem of statements about the future which until now could not be solved by symbolic logic (e.g., in Carnap's work). The logic of probability, however, can fully cover scientific thinking. In this framework, the problem of induction can be solved by demonstrating that the induction rule is a necessary presupposition of statements about the future. Of course, this presupposition is not sufficient and therefore success cannot be guaranteed for the inductive method; one can only say: when we can make statements about the future at all, then we shall find them by using the inductive method. And this is a completely satisfactory justification for the inductive conclusion.

Not before this solution of the induction problem was found could the development of logistic empiricism be completed since it disputes the existence of synthetic judgments *a priori*. There exist only facts and their tautological transformations and there is no principle which reaches further than them, and which could supply us with knowledge. But by applying the principle of induction, we organize the facts in such a way that they supply us with the most advantageous predictions, i.e., those predictions which – if predictions are possible at all – are successful in the majority of cases. Predictions are 'bets', i.e., we bet on the future in the same sense as one bets on a horse in a horse-race. Science is the discovery of the system of the most advantageous bets.

In addition to my scientific work, I have also been involved in organizational activities, mostly under the joint auspices of the 'Berlin' and 'Vienna' Circles; among others, publication of *Erkenntnis*, organization of congresses, etc. My little book of 1931, *Ziele und Wege der heutigen Naturphilosophie* [1931j] was programmatic in character.

(3) *Los Angeles, 1951* [written in the third person; only a portion is given, the remainder duplicating earlier statements – Ed.]

Turning back to physics proper, he undertook an analysis of quantum mechanics and succeeded in giving a philosophical solution for this much discussed part of modern physics. He showed that the difficulties concerning the question whether the elements of matter are particles or waves can be solved by the introduction of a three-valued logic, in which a category called 'indeterminate' stands between the truth values 'true' and 'false'. These ideas were presented in his book *Philosophic Foundations of Quantum Mechanics* [1944b].

Summarizing his studies in symbolic logic, he wrote a book *Elements of Symbolic Logic* [1947c] in the form of a text book for students. But this book contains at the same time a great deal of research in logic. He applied the methods of symbolic logic to the study of conversational language and devised a new kind of grammar, which promises to have much practical use. These problems were brought to his attention by the particular circumstances of his personal life: his travels compelled him to teach in various languages, and he saw that mathematical logic can be a help for understanding the structure of languages.

He is now working on a comprehensive exposition of his ideas on philosophy, including a criticism of traditional philosophic systems, which is addressed to a wider public and written in a non-technical form. The book is scheduled to appear in the fall. [*The Rise of Scientific Philosophy*, 1951a].

2. MEMORIES OF WENDELI ERNÉ, HANS REICHENBACH'S SISTER

Once upon a time . . . .

I must, of course, start my tales of old Hamburg and our youth with the escapade of Bernhard and Hans in the park. Hans must have been still quite young, as one can judge from the words he used, and Bernhard probably too, not yet old enough to go to school.

Bernhard felt an urgent need for which both boys deemed a visit to a public structure erected for this purpose indispensable. But they did not have any money and this deficiency left Bernhard, the older one, completely speechless. Hans took over. In front of the little house he turned to the woman in the white apron who was waiting there for her guests and said: "scuse us, scuse us, my brother has to." They found understanding for their lack of money and were allowed to enter.

When they reemerged and left in a hurry, 'Hänschen' turned around once more and called consolingly to the woman who was gazing after them: "But he couldn't even!"

In the following story he was not much older. Hänschen threw a mighty tantrum and was so naughty that our Mama lost her patience. She had been a teacher before her marriage; she led him gently but firmly to a corner of the room and said: "If you can't behave decently, you must stay by yourself for a while."

Hänschen broke into an ear-splitting howl. Quite abruptly, however, he stopped. Hänschen turned around to the family and said indignantly: "You always say that I am spoiled; but I don't see that at all!"

My early memories concern also his occasional sudden fits of rage which he tried to fight with all his might and – as I can attest – successfully when he was a little older and already in school. His will-power impressed me very much. If I remember correctly, Hans was fourteen years old, since Bernhard was still at home, when each of the brothers got a big wart, Bernhard on the back of his hand and Hans on his forearm. Bernhard treated himself, I don't know on whose advice, with a very strong ointment, evidently too radically, for he had to wear a bandage as a consequence.

Hans had his own method of treatment. "Warts can be burned off" he said, got a candle from his night-table, lit it, put the part of his arm with the wart into the flame and waited until it was cauterized black. I did not look at the clock how long or short a time it took, but I still remember how I, his little sister, watched with respectful shudder and in breathless suspense.

Against me Hans was very rarely angry or mad. But once even I had to run for cover.

Before going to bed Hans regularly put a glass of water on his night-table. This happened when the three of us were still sharing one big bedroom that was divided by a reddish curtain between me and the boys.

One evening it so happened that I could not withstand an attractive temptation: Always falling asleep late, I was still awake when I heard him enter. I got out of bed, saw how he thirstily took the glass, put my head through the curtain and called: "I have spit into it and that serves you right!" And to do that to Hans who was always so fussy, as we children said. Seeing how his rage was just beginning to erupt I could still, barely avoiding him, flit to the door and shout: "It ain't true, it ain't true!"

And it was not true.

Five years older than I, Hans was in those times always ready to console me and to calm me down; he was my mentor, the highest authority. Unfortunately, I was a "valerian child" [needing sedatives], lying awake in the evening I don't know how long, a victim of my fantasies and therefore often full of anxiety.

How often did Hans when he went to bed sit at mine and talk to me in a low, soothing voice, taking all my troubles seriously.

When I now suffer sometimes, especially from nightly old-age anxiety, I think: if there were an individual after-life after death – Hans would still come to my bed once in a while and I would hear again his low calm voice.

A clear memory of Hans, the student, awakens in me while I am writing:

We went to the first "Juryfree Ball". Tonio [my husband] and I had come with a number of colleagues from the Moissi Acting School, and Bernhard and Hans had arrived, too.

Very soon we, i.e., our student actor circle, noticed that my Hans was asking the prettiest girls to dance with him, apparently successful in conversation and dancing, a fact that my colleagues disapproved of. One of them, a boy from Cologne, a nice fellow and a somewhat close friend of Tonio's and mine, suddenly emerged from the crowd, right in front of us, purple with rage and with genuine tears in his eyes, calling to me in desperation: "Your brother has stolen my girl, she is dancing only with him, she is really teaching him the steps, I have to fight him in a duel tomorrow morning!"

I explained to him gently that my brother, Hans Reichenbach, would never fight a duel with anybody in the world; he, my colleague, could go to sleep tonight without being afraid of tomorrow.

That happened in Berlin, 1913.

3. AT THE END OF SCHOOL DAYS: A LOOK BACKWARD AND A LOOK FORWARD (1909)

For twelve years – two thirds of my life – I've been sitting at a school-desk learning and learning. There were so many new things to absorb that I entirely lacked the opportunity to create anything on my own. All the more joyfully therefore do I anticipate the day when I shall be free to do what I like.

But what is the origin of my drive to deal with the sciences in my own way? Why do I now claim this ability? Because I have studied for twelve years. And it is probably quite justified now, before the *Abitur* [final high-school exam], to look back once more at my school-years; after all, up to now school has been the most important element of my life.

All the things I came to know in those twelve years! In the lower grades, the tools for dealing with the sciences. Reading, writing, arithmetic, orthography, and grammar. In addition, the first beginnings of real knowledge, natural history, local history and geography, biblical history. It is strange that, in the lower grades, one considers these subjects as ends in themselves; for that reason only do they become interesting. And exactly for that reason must they be taught at a young age; later on, one would find instruction in reading and writing so boring that one could not learn it at all anymore. As the mind becomes broader, more mature, and more seriously searching, instruction as well must become deeper, otherwise it appears tedious and goes by uselessly. But since minds in a large class are all too different, instruction necessarily becomes often tiring for the brighter students of the class; at the same time, they lose much valuable time. Certainly one could get to the *Abitur* in much less than twelve years if classes were smaller, or, even better, if in these twelve years one could learn much more.

In the intermediate grades, one learns foreign languages, German, geography and history. How overjoyed I was when I could express myself in a foreign language for the first time! What strange fascination lies in the ability to say something in a new and different way! How nifty the fifth-grader feels as against the lower grades when he can say (in English): "I can speak English!" Interest in foreign languages grows with the difficulty of the material; it is a pity, however, that in most cases, it grows more rapidly than class instruction can keep up with, and therefore leads to many a boring class-period.

The greatest joy, however, I derived from lessons in mathematics and the natural sciences. I had always had a particular interest in this field; but how much greater and nicer did these subjects appear when I learned the method for penetrating them more deeply. Whenever I came to know a new area, I was convinced that before me lay an apex of science beyond which one could not reach, and where I would have preferred to linger. And every time, I was amazed that one can get much, much further, so that everything that preceded it appears insignificant. And it was only due to this experience that the desire to continue research in the sciences stirred in me, for now I know that they do not have an absolute apex.

The greatest upheaval in my school-life came after I passed the final exam of the *Realschule* (non-classical secondary school). In the three upper-class years, one learns more than in all of the nine lower grades put together. The lessons became much more difficult and accelerated because the class was smaller and the weaker students had been eliminated. Only here did I get the right conception of the sciences. Only here did I learn to speak French and English fluently, and, above all, to grasp the spirit of these languages. Only here did I become intimately acquainted with German literature. All that which in legendary times inspired the minds of the wild Teutons, which, in medieval times the belligerent knight chose as his ideal, and which modern times have chosen as their goal, – all that I came to know. Great men spoke to me through their works, Goethe and Schiller became familiar examples of noble men to me. The thoughts of the great philosophers of all times were explained to me; I got an idea of the monumental intellectual struggles of humanity.

The exact natural sciences in particular sparked anew my old preference for mathematico-logical thinking. For what I came to know here surpassed by far all earlier knowledge. Only now could I understand mathematics in all its greatness; only since I have known of analytic geometry, differential and integral calculus, have I understood why mathematics is of such enormous importance for the technical sciences. In chemistry and physics classes I learned that true science does not consist of the knowledge of a certain amount of facts and numbers but of the inner appreciation for the great interconnectedness of nature. The kinetic theory about which I wrote a paper made me particularly aware of organized, logical thinking which must lead to a true understanding of nature; it also led me to perceive the close connection that exists between chemistry, physics and mathematics. Biology, finally, acquainted me with living organisms, frequently by means of experiments which I carried out myself. To biology I owe a deeper appreciation for the medical arts.

And now, shortly before the *Abitur*, which will finally end my school-years, I am once again supposed to account for all that school has meant to me. To be sure, I have learned much, very much in school, but I also have lost a lot. School does not end with the instruction period at 2 p.m., but because of the homework it takes up the better part of the afternoon. The worst of it is that this homework is precisely prescribed. Whether I felt like it or not, I had to prepare first a chapter for French, then solve a cubic equation, then work on a German essay, then learn a piece of world history, etc. Hereby, the mind is constantly tossed to and fro and finally tires; at the end of my school-years I no longer possess the same mental freshness I had when I entered the first grade as a small boy.

I must criticize this pressure which is constantly exerted upon the mind. Wouldn't it be much better if, at home, the student could do the kind of work he really likes? The pressure weighed heavily on me, especially because it frequently robbed me of any possibility to occupy myself with things which interested me. In school, only knowledge is being taught, the arts are not considered at all. In this respect, even the efforts by some teachers who are not permitted to stray too far from the prescribed curriculum, cannot be of much help. But it is even worse that school robs one of the time to study the arts independently. I have never been able to systematically study painting and the graphic arts, although I would have liked to do so very much. The same applies to literature. School stimulated an interest in it without leaving me with the possibility to penetrate it more deeply. But school also robs me of the time to thoroughly concern myself with fundamental moral and religious problems which I want to work out after my interest in them has been stimulated in class. The afternoon has to be used exclusively for learning, for the absorption of new knowledge; the little time that remains is needed urgently for walks or sports in order to fortify the exhausted mind.

Remembering this mental pressure leads especially to joyful anticipation of the day I say goodbye to school. I must confess that in the sciences I have learned very much in school; wouldn't it be better, though, if I could say that I had made the same progress in the arts and in ethics?

After Easter, I shall leave for the university. For my field of studies I chose my favorite subjects. But since I also seek to apply in practice all that I have learned I want to become an engineer. Here, I hope to be able to deal with theory as well as with practice. But above all, I intend to catch up on all I missed in school. I want to improve myself in the arts and in ethics. I don't want to become an engineer who can do nothing but calculate.

(20 YRS. OLD)

4. LETTER FROM REICHENBACH TO HIS FOUR YEARS OLDER BROTHER BERNHARD

Charlottenburg, November 20, 1911
Friedbergstr.16

Dear Bernhard,

There is no other science so much given to controversy and dissension, in which views are splintered in so many ways and are so diametrically opposed, than philosophy. Mathematics marches absolutely securely, fortified by its method; the natural sciences test their assertions unequivocally by means of experiments; the humanities check themselves by observations and historical research – philosophy alone, which should range above all other sciences, is exposed without protection to the free play of contentious opinions. One might compare this situation to the antagonistic views of economics held by the various political parties; however, these quarrels have their reasons ultimately in philosophical, especially ethical, differences.

Come to think of it, this is very strange since philosophy possesses in its method, logical thinking, the most univocal means for advancement. This would be correct if philosophy were not to occupy the outer limits of our dimension of thinking. In epistemology it is charged with investigating the limits of thinking itself; thus the most intricate problems arise when the laws of thinking are to be applied to the laws of thinking themselves. This means dissecting and analyzing until one reaches the ultimate foundations which can no longer be proved, but which, nevertheless, must be checked with regard to their correctness. On the other hand, philosophy is supposed to establish the fundamental principles for human conduct; here it is even more difficult for philosophy to select adequately the first principles upon which it can progressively build. Everywhere emotion and tradition play an important role. When can it be said of an assertion that it has been derived purely intellectually rather than having arisen from unconscious prejudice? The emotions, on the other hand, have a perfect right to exist. Yet how can I decide what is good and what is bad in them; according to what principle should I choose my values?

The third task of philosophy is to develop a certain conception of reality. Do events happen purposefully, i.e., teleologically, or causally, i.e., according to causal connections? Can the world be explained in terms of one principle, monistically, or in terms of several principles? Does something exist behind the things? And what is man's place in the universe, is he free or is he, too, merely a link in the chain of events?

With these latter questions philosophy touches areas to which the natural sciences have already made some important contributions. It is in this domain that prejudices are most strongly rooted – that is the reason why here disputations become most outspoken, although this does not mean that the differences are always rigorously argued.

As far back as we can trace human history, however incompletely, mankind has been interested in all the questions I have sketched briefly above. And there were always some towering minds among them who tilled especially deep furrows in the realm of thought and who carried large masses along with them, even though frequently in opposite directions. These giant pillars have always been of special interest; their contemporaries found support in them, future generations regarded them as models. It is a justified demand of liberal education to expect a cultured person to have a close acquaintance with these geniuses.

Therefore, you need a so-called general history of philosophy. It is extremely difficult to recommend a specific book to you. Most of such works are very detailed and written in sober, scientific style; they are more like reference books, written for the purpose of consulting specific chapters rather than for the perusal of the whole. Only someone with special interests will study them in their entirety. Above all, the greatest danger is the fact that all such presentations are more or less *critical*, seen from the viewpoint of the individual author. The beginner who cannot yet survey the material is in danger of wholly losing his own judgment. The great works are relatively hard to understand for a person not yet used to philosophical terminology and method. In addition, there are small, brief expositions 'for the educated person' written by decent high school teachers and similarly harmless people; what I have seen of them is completely useless and contains nothing but textbook-type articles that do not say anything. The purpose is not to learn the years of death or the famous sayings of the great philosophers; the whole study is valuable only in so far as one penetrates their mode of thinking and recognizes with them that great truths are involved in their sayings. If one hears, for instance, that Thales said water is the ultimate constituent of all things, one is dismayed and exclaims "Gosh! How did the guy ever get this crazy notion?" But once a more thoroughgoing investigation shows how Thales saw change and flux in all happenings, how as a child of the seafaring Ionians, he admired the sea in its mystical darkness as the most enigmatic and powerful force of nature, one begins to understand that there may perhaps be more to this saying than is immediately obvious.

I have scanned about fifteen different works for your purpose and I shall recommend one which I do not even like very much because of the philosophical outlook of its author. It is the work of Rudolf Eucken in Jena: *Die Lebensanschauungen der grossen Denker*. Eucken is very much the idealist and as far as I am concerned he lacks an orientation toward the natural sciences. Yet there does not exist anything better for your requirements; the others are too detailed and often also too difficult. Eucken writes very clearly without many foreign words and in a captivating style; he is pleasant to read. If you want to study some chapters [of the history of philosophy] more thoroughly, you can do so in works such as Wilhelm Windelband, *Geschichte der Philosophie*, or Kuno Fischer, *Geschichte der Philosophie* (a dozen volumes), or Friedrich Überweg, *Grundriss der Philosophie* (four volumes, very objective, but very pedestrian and scientific) or *Klassiker der Philosophie*, edited by Frommann.

Friedrich Paulsen wrote a good *Einleitung in die Philosophie*; but it is a very thick book. You might look into it some time and read a few chapters. Excellent, but very detailed and treating only certain philosophers, is Friedrich Albert Lange, *Geschichte des Materialismus* (two volumes). I shall indicate some chapters in them to you later.

It would, of course, also be appropriate to read some works in the original; Nietzsche, for instance, is fascinating and writes in a scintillating style. I shall write you again concerning this matter. For the time being I recommend only the work of Eucken together with Paulsen's *Einleitung*. I am sure you will be able to find them in a library.

I would like to write you at greater length, but I better hurry and get to bed; otherwise I won't be able tomorrow to pay enough attention in Planck's seminar. I am very busy; I live only in integrals and equations of state.

The reviews about you were excellent; one of them seemed to try to discover the character actor in you. And that would certainly be right.

My cordial regards to you.
Yours,
Hans

5. FROM A LETTER OF BERNHARD REICHENBACH TO MARIA REICHENBACH (1975)

In contrast to myself, Hermann and Wendeli, Hans was considered to resemble the Menzel family more closely. And indeed there is much to be said for it, as for example his early appreciation for technology and mathematics; well, Mother's father and brother were engineers – and when little Hans, then six or seven years old, was asked what he wanted to become later in life he answered "technician" – "and what are you now" – "quarter-technician", – an answer that became part of the family treasure of quotations. By the way, unlike me, Hermann and Wendeli, Hans had none of the external Jewish features [*Züge*] ; in his temperament however – he could flare up very quickly – he was very much the son of his father. Like me, he was equally close to Mother and Father. For both, their daily lives were completely dedicated to their children. . . . the speech that Hans delivered at the funeral ceremony for our father in its emotion – while always directed at the dead father – shows religious feelings – certainly something quite unusual for Hans.

6. MEMORIES OF ILSE REICHENBACH, HANS REICHENBACH'S SISTER-IN-LAW

My husband [Bernhard Reichenbach] and I loved Hans very much and admired him for his integrity, his intellectual powers, his energy, enthusiasm and warm heart. There were some people who thought, he had been too ambitious: Of course he was delighted when he had been asked to give the William James lectures in Harvard. I still have the letter he wrote about it. Of course he wanted to be fulfilled, he had so much to give. He was so sure of his mission that he could even afford to be modest. But never did he seek popularity at the expense of others. He had great interest in all our personal affairs; I just found a letter from his sister Wendeli in which she thanks him for his *empathy* with her present troubles. And my husband and I were so grateful to him when he proved his generosity by helping us with the school fees which we then could not afford for our son. He was so warmhearted in personal relationships. This is my personal experience.

My husband often talked about his early youth: He was two years older than Hans, but though Hans had been always on top of his class and my husband never, he never was jealous, because he loved him. Hans was so versatile, he also could do everything in the home and my husband remembers, whenever he tried to do something, his mother would say: "Let Hans do it." (In consequence it was only in later years that Bernhard, for example, could drive a nail into the wall correctly.) Hans always had been a natural 'authority' at home. And though we had been separated from Hans, we in England, he in Turkey and America, our friendship with him never weakened.

When I was a student, a few years younger than Hans, also having studied science, mathematics and philosophy, I took part with others in a small private circle, which Hans had organized, where we read Kant. And I still remember how Hans helped us to clear things up and how interesting it was and how superior he was to all of us.

About his family in the early years I remember this: He always then was interested in his children, and I remember that he wanted to celebrate *his* birthday by asking his children to invite *their* friends. The relation with his wife, Elisabeth, was somewhat complex. As a matter of fact he wanted the child that Elisabeth had from her first marriage living with him, and Elisabeth did not want that. We never could understand this. (Much later, when Elisabeth visited us in England in 1958, after having been psychoanalyzed, she told me: I now cannot understand why I refused.)

Elisabeth was a great personality in her own right, with her own interest especially in education, and she did not share Hans' interest. Hans however often was so completely absorbed by his work that for days he was so remote from her. She suffered from that, she told us, and Hans also must have suffered that she could not share his work. And soon there were other emotional difficulties. But they stayed together; both were pedagogically interested in their children and personally in their problems. Later, when we met Hans and Mia [Maria Reichenbach, Hans Reichenbach's second wife] in Switzerland in 1952 Hans said: We stayed together because of Hitler. The marriage ended with great bitterness on both sides, that was shown in the way they acted, with great resentment.

There was an episode about which I can only tell what I heard from Hans, his daughter and his son Hans Galama who stayed with us in 1950. Hans Galama was a highly gifted young man, brilliant at school and also interested in his father's subjects. But Hans also had an artistic leaning. After having finished school, he refused to study. Apparently Hans' father was very disappointed and both seemed to be very stubborn. Anyway this led to an estrangement between them which never healed. We felt that Hans did not understand his son's refusal to study. He foresaw how difficult his son's life would become (and in hindsight he was right). But apparently Hans [Reichenbach] was using the wrong method. We, my husband and I, were always fond of Hans Galama, having sympathy with his very complicated, very difficult personality. In Switzerland we told Hans that perhaps he had not done the right thing with Hans Galama. He listened to us but did not agree. Later that summer in 1952 Hans and Mia visited us in London. I never forgot that afternoon: Hans had helped me with the washing up and we talked, and when the conversation came to Hans Galama I saw tears in Hans' eyes. Never had I seen this.

7. MEMORIES OF UNCLE HANS: NINO ERNÉ

My mother [Wendeli], not I, ought to write about Hans Reichenbach. She was a wealth of anecdotes and, although she possessed a good deal of poetic fantasy, her childhood memories were not only true in a deeper sense, but also factually accurate. Hans was not her favorite brother, for she demonstrated an unusually strong attachment for all three brothers, even when the ocean or 'merely' a canal lay between them and her. Bernhard was the lone brother 'she still had' and when he died the entire bulwark of her existence, the foundation of her origin and deepest past, literally collapsed beneath her. She died soon thereafter.

I meant to say: Hans was not her 'favorite' but the one on whom she depended most for her inner and, whenever possible, physical support. He was, for her, the rock in the midst of a storm-lashed existence. She shared most of her interests with Bernhard: first, the theater; later, a political activism which continued until he fled to England and she warned the last of his comrades, who had remained behind in a German city, of the Gestapo about to close in on them. In her heart, Hermann would forever remain the younger brother, the 'little one.' Of all her brothers and sisters, Hermann was, without doubt, the most fragile, unstable and psychically vulnerable.

Hans, however, was her 'support'. Our far-flung family was reunited once more in 1924 at the foot of Monte Rosa in Campertogno. It was on this occasion that my mother broke her foot while climbing a hill. I do not remember who was present; in any case, she told me subsequently that Hans carried her on his shoulders down to the valley below and back to camp. As far as she was concerned, 'only' Hans could have done that. No one else, not even her own husband, made her feel so safe. It was this sense of security she felt in Hans' presence which caused her to acquiesce in this form of 'transport', as if it were the most natural thing in the world.

As a child, she slept for a time in the same room with Hans and Hermann. Hans, the eldest, used to say goodnight to the two younger ones. Then, with the quiet methodology which he possessed even as a child, he would place a glass of water on his night table. This done, he could then undress in darkness when he wished to go to bed.

My grandmother tells the following ancedote. Once, when Mother Selma lay ill or perhaps simply exhausted in childbed, a term still used in those days, Hans prevailed upon the maid to give him 60 Pfennigs, a debt he promised to work off. Hans could not have been more than four or five years old as he rushed off with the money clenched in his fist. Upon his return, he carefully constructed a miniature rhinoceros family from wood, glass or some such material, all the while his face radiant with expression. He placed his creation at his mother's bedside, hoping it would comfort and strengthen her.

As regards his sister Wendeli, in those days we called her Lina, Hans assumed almost paternal duties. Lina's father, Bruno Reichenbach, worked every day until late in the evening in the office of his grain importing company. Only on Sunday afternoon did he allow himself a few free hours, at which time he would go to his chess club for relaxation. And yet, based upon what his children say, he must have been an extremely affectionate and anxiously concerned father. Of all the family, it was he who most enjoyed the Christian Christmas tree during the holiday season, despite his Jewish origin. His Aryan wife and children had already liberated themselves from all religious ties, in so far as they had ever possessed any.

It was Hans who 'explained' the world to his younger sister. He remained calm and patient when confronted with all the questions with which four to six years old children drive their parents to distraction. Even as a child he must have possessed a distinct pedagogical talent and an extreme sense of responsibility. There is 'historical' proof that on certain occasions he even felt responsible for his older brother Bernhard. Given the difference in age, it was generally Bernhard who contrived their joint adventures and played the leader. On one such adventure, they purposely got themselves locked in the zoological garden, so that they might spend the night with the animals. Whoever knew the brothers can readily imagine that Bernhard had been motivated by a temperamental desire for adventure. As regards the quiet and reflective Hans, the dominant impulse must have been a kind of nascent 'scientific interest', the researcher's urge to discover how, for example, a *genuine* rhinoceros family behaves during the night.

But somehow or other, they were discovered by the keeper and brought home – I do not know whether or not they suffered the wrath of parental displeasure. My mother still remembers that Papá (the accent on the second 'a') had been much more excited than Mamá and had spent the whole evening wandering through the city looking for his sons. Without doubt, Hans had been led astray by following his brother's lead.

I retain only a hazy recollection of 'Uncle Hans' during that family reunion at Monte Rosa when he proved himself as an alpine stretcher-bearer. My memories of him become clearer only four years later, in 1928. It was then that we tried to reestablish ourselves in Germany again, after having spent several years in Italy. The move came about as a result of the fact that one day after school my sister and I proudly sang the Fascist hymn 'Giovinezza' at the supper table – whereupon my father, in one of his rare moments of resoluteness, decided to save his children from the State of the *Duce*. He did not suspect that we were leaping from an Italian frying pan into a German fire which was to break out a few years later under the Nazi regime. In those days, the German Republic seemed to be still half intact. My parents had another, equally important, reason for their decision to return to Germany. As she had done so often in her childhood, my mother turned to Hans for advice and help, because my chronic middle ear infection had not responded to treatment and the doctors in Milan had threatened, as a final resort, to chisel my head open! Hans, who was then a professor in Berlin, consulted his medical colleague, the Director of the Charité [Hospital], and the latter explained that I could be cured by other means which were not so barbarous, such as ultra-violet therapy. My mother went on ahead with us children, while my father had to attend to the dissolution of his business interests, his lease, and similar things. We had hardly arrived in Germany, my tonsils and adenoids had barely been removed – during the operation I had bitten through a surgical instrument while under sedation – when my father contracted an almost fatal case of meningitis. My mother had to take the next train back to Milan. Without the slightest hesitation, the brothers 'assumed' responsibility for the children: my oldest sister Eva went to Bernhard and Ilse in Büderich. I remained with Hans and Elisabeth in Berlin (somewhat later I stayed for a short time with Hermann and Lo who also lived in Zehlendorf, a suburb of Berlin). Many short scenes from those weeks remain in my mind – fleeting, flashing shreds of memories, but most of those would be out of place here. I saw Hans seldom, for he spent much of the day at the university. However, I know for certain that I felt more drawn to him, more 'secure' with him than with his wife of that time. This was due to the fact that I had developed an unusually strong sense of justice. I knew well that Hans was incapable of lying or being unjust to us children. I would not say the same for Elisabeth, who was supposed to be an excellent teacher in school. With her own children, however, she was rather helpless, as far as I could judge.

The atmosphere in Hans' study was unique. It was the only thing in the whole apartment which exerted a calming effect on me. Even when Hans was not there, I used to like to crawl, alone or with my cousin Hans Galama, under his large dark desk. I remember also how large the dark, scented, easy chairs in the study seemed to the eyes of a six year old. (I no longer have any recollections about the rest of the apartment – yet I can still see myself standing outside the corner house at 45 Schützenallee, and the little path that led from the front door to the garden gate.)

When Hans had time, he treated us to his own peculiar mixture of seriousness and humor. It was strange how, on the one hand, he could be so serious about world questions and problems, while, on the other hand, he would prevent us from taking these matters too seriously and, in the end, we could even laugh at them. Perhaps it would be more fitting for me to say 'I', since I do not know whether the two Reichenbach children shared my feelings. If I am not mistaken, they were more attached to their mother, which is quite normal for small sons and daughters.

High points of my visit were the performances of the 'Potato Comedy', and I think I can remember Hans fashioning puppet heads from potatoes and participating in the play itself – his hands above a curtain held between two sticks. In any case, I have retained the very distinct and definite association that he liked to call us children 'You swell heads' [*ihr Pomuchelsköpfe*], an expression whose exact significance has eluded me to this very day!

Another high point: the first drive 'around the block', in the tiny car which Hans purchased at that time [previously, he had had a motorcycle; his brother Hermann even owned a bike with a sidecar, shiny black with green stripes.] The new car was a Dixi, the smallest car on the market in those days. In spite of its size, we all managed to fit in. I shall always remember the sound of Uncle Hans' voice when he, in an almost childlike pride, said, "our Dixi"!

I did not see Hans any more in those decisive formative years because I had to leave Berlin again soon after. My sister and I were enrolled in a country boarding school at Solling where we remained for six years until we were expelled in 1935 for political reasons. When we returned to Berlin, he was already living with his family as an emigrant in Turkey.

It was not until a few years after the war, about 1952, that I saw Hans again, when he visited me and my young family. He came with his second wife, Maria, and I must confess that even today I can still taste the surprising, overwhelming joy with which I gazed at the couple. Hans was not only quiet, humorously serious, reliable and kind as in my childhood memories of him, but so relaxed, so free and easy – a hackneyed word will have to serve in this instance – so happy, as I had never before known him to be. He was not ashamed to flirt with his wife on the street – there is a snapshot of such a scene. I do not know whether Hans was teasing Mia, but both were smiling, she started to blush, while he shook his finger jokingly – I never knew what was going on and was careful not to ask any questions. But that was only a small part of the day we spent together; it passed so quickly, just as the sun will, at times, peek forth from between the clouds and then disappear behind them – what remained was a distinct impression of two happy people.

Unfortunately, I no longer remember much of our conversation; the essential thing, however, was the silent feeling of being together. I still remember with what profound, profound (I must repeat the word, sorry!) satisfaction he listened as I related to him the ungrudging and warm admiration my father had once expressed for him in 1944, the last year of the war: "Hans – yes – *he* made it." My father was referring not so much to the successful escape from the Nazis and the establishment of a new position, a livelihood, as to the realization of those potentialities dwelling within man which are so often buried by external and internal catastrophes or just plain obstacles.

It is to Hans that I, more or less, owe my preoccupation with Sigmund Freud which began soon thereafter. At that time, I was still laboring under the prejudice that Freud related everything exclusively to sex. Armed with blinders Freud had obstinately entrenched himself in a *cul-de-sac*. There was not sufficient time for me to ask Hans for a detailed lecture (free, of course, just as he had lectured to his students in Berlin!). Yet a calm and friendly remark by Hans, spoken with a disarming decisiveness, shook my doubt: "No, no. That is pure nonsense!" or: "No, that is not correct. . .". His words had so strong an effect on me that for the very first time I really began to read Freud correctly. For when Hans defended Freud with such determination. . . .

What impressed me most was his attempt to feel his way around in Germany again. With the same dispassionate objectivity with which he viewed himself and the world, and without the slightest hint or resentment, he concluded that it would be difficult, if not impossible, for him to teach or lecture to German students again. He had tried to establish contact with young students here, only to notice how accustomed he had become, in the interim, to the American system of teaching and to dealing with American students. He looked forward to his return to Los Angeles – not because he no longer wished to be a German, but because he had become, in the meantime, an American – yet without any patriotism or nationalism, feelings which had always been totally foreign to him. Moreover, he felt that Anglo-Saxon countries understood him better, that people there were more prepared to accept his brand of philosophy. In Germany, on the other hand, the old philosophical hero worship, albeit in a somewhat modified form, was still in vogue.

In a few precise sentences he conveyed to me that the Germans were more concerned with the history of philosophy than with philosophy *per se*. The past decade, in fact, has witnessed a gradual change in this field and his complete works are beginning to appear in Germany too.

The crowning climax to that day in 1952 was our trip to Starnberg Lake where Hans treated us to dinner. Although my family was not literally starving, the invitation to dinner was a unique treat for someone who awaited each milk bill for our child with increasing anxiety. This child, my son Marcel, sat with us at the table; he was, unfortunately, still too young to derive any future benefit from our conversation. He is, as far as I can detect, the only offspring of our family who has, more or less, continued in Hans Reichenbach's footsteps. As a twelve year old, he would wander during his holidays along Mediterranean shores and, like Archimedes, he would use his stick to draw in the sand; but by then, his great-uncle was beyond his reach. What joy these two could have provided each other! Marcel, now 30 years old and a member of the West German Scientific Council in Hannover. In the course of his research, Marcel repeatedly encounters the name, Hans Reichenbach. Yet, he is saddened by the thought that he will never be able to submit his dissertation to this great pioneer, who was not only his relative, but even more importantly, a kindred spirit.

Nevertheless, they did shake hands once, they did spend a day together, the sixty year old man and the five year old boy. Although I did not become a scientist, I will permit myself the poetic license of comparing their brief encounter to the passing of the baton in that imperceptible relay race which began in the past around 1900, and will continue into the future year 2000.

[*Mainz, 1978*] [trans. by Sheldon Gilman]

8. HANS' SPEECH AT THE FUNERAL OF HIS FATHER

Hamburg, October 18, 1918

Dear Father,
Standing yesterday in front of your deathbed I only then began to realize that you have left us. We just could not believe it, we still saw you so much alive before us – would we not suddenly hear your short, quick steps, would not your voice, so vivid and so firm, again draw our attention? Yet you were lying there so immobile, so infinitely inflexible; your tender hands were pale and rigid, and your dead brow lay there so cold. Very slowly I began to understand that you were not unhappy lying there. Your face showed that distant seriousness that is no longer aware of our feelings, that does no longer partake in our troubles, and that gives to our mourning the incomprehensible answer: I am not here anymore.

We shall sadly miss you. Your departure has left us with deep pain. For you were our beloved good father, concerned about us from morning till night; you were the secure pole in our circle from which all our existence started. We feel a great sorrow that you have left us so early, that you were unable to finish your work and that you could not enjoy the fruit of your labor.

For you it may not be so sad that you have died. You look so calm and peaceful. You are resting now from your endeavors, and your face is illuminated with that nobility that says: I am no longer part of your pleasure and your grief.

Dear Father, we have come together once more to say good-bye to you. Your children are standing now at your bier. And as you are lying there so pale and cold before us – all this life, all of your children as they are standing here realize that you were their creator. And your wife who made you very happy, our mother, stands among us and finds new life in our young families. Is it not beautiful, dear father, that your life has brought forth so many strong young people? I believe you deserve to be very content, dear father.

And now, as you are leaving us, a great love lies in your earnest features that tells us once more: I have loved you very much. May we say it to you once more? We have loved you very much, too.

Good night, dear father.

9. APHORISMS OF A DOCENT FORMALLY ADMITTED TO TEACH AT A UNIVERSITY (1924)

One is thirty-three years old. One has officially been declared to be mature enough to teach students though not to receive a salary for this occupation – it is regarded as presumptuous to even think of it.

Thus one sits like a wall flower and waits for the offer of a regular position. Supposedly it is bad manners to apply openly for one. For that purpose one has an uncle or a father-in-law who may assist the young talent at any time. Or one belongs to the circle of a powerful person. There are powerful men who have found chairs for all of their assistants. Others are less successful; for them it is difficult to find assistants.

The universities are like the theatrical concerns. Nobody wants an unknown person. But if somebody is famous, they fiercely compete for him. Almost all who received an offer during the last years immediately received several; others, however, wait in vain.

"Dear Colleague, your father is Jewish or Communist? I certainly do not want to complicate your academic career – but naturally I have to take the faculty into consideration."

"Dear Colleague, don't be so dissatisfied, you have a future ahead of you." "Possibly – but if I'll yet have to wait much longer, its present will be that much shorter."

If Leibniz would come back today – not *he* would get an offer for a teaching position, but the author who wrote a book about Leibniz.

Only in his forty-seventh year did Kant's colleagues regard him as mature enough to receive an official salary for his lectures. Yet his colleagues have been nourished by his mind for over a century.

One plays the role of assistant, i.e., one does at thirty-three what a twenty-three year old could do. As a consequence one receives a higher salary than the twenty-three year old because one has been formally admitted to the faculty as a qualified university teacher. At that, one receives an amount equivalent to the beginning salary of a twenty-six year old instructor at an intermediate high school.

A worker once told me: "If I had learned as much as you, I would not work for this salary." He did not know that one does not work *for* the salary but *in spite of* the salary.

A captain of industry once told me: "On my travels I always recognize university teachers from the fact that they go third class on trains, wear badly cut suits, and are accompanied by badly groomed women. Why do you put up with such salaries?" "You forget", I said, "that only that person will acquire a university position who can bear up under such conditions; the others quit long before."

It is certainly admirable if someone goes hungry until he becomes a full professor. But the quality of the sciences should not be made to rest on the assumption that they require this approach – one does not even know how many people with higher demands from life they lose.

H.R.

10. UNIVERSITY STUDENT: CARL LANDAUER

I have no firm recollection when and where I first met Hans Reichenbach; it may have been in Munich 1909, or in Berlin in 1911. It was almost certainly through our common activity in the *Freie Studentenschaft*, which had been founded as an organization of non-fraternity students, intended to remedy their underdog role within the student body, and had developed a liberal, democratic and socially progressive ideology; it also was a sort of self-help organization, because it attempted to supplement university courses by organizing its own lectures and discussion groups. Ours was one of several circles of personal friends that grew out of the common work for the organization.

To this circle belonged, aside from Hans and myself, a few others, some very closely, some in a looser fashion. Definitely a member of our 'inner circle' was a law student, Alfred Werner, one of the sharpest intellects I have ever known; after 1911 my future wife, Hilde Stein, then a student of Germanistics and history, also belonged to the core. More loosely connected with our circle was Herbert Weil, like myself a student of economics, a little bohemian in life style, much interested in some sports – I remember that he introduced Hans to rock climbing. Still others were farther on the outside: Philipp Loewenfeld, a student of law, whose father was one of the outstanding experts on social insurance and labor law, and Philipp was interested in the same fields; the brothers Kranold, Hermann and Albert, who were in economics; and we eventually developed some contacts with some of the leaders of the *Wandervögel* movement and groups related to it, about which more will be said later.

What drew us together, aside from the common work for the *Freie Studentenschaft*, is hard to say. One can never wholly explain how and why ties of friendship develop among some persons and not among others. In our case it was certainly not any conscious effort to form a circle whose interests and knowledge were complementary; just the same, we did form such a circle. Since interest in social science, history and law predominated among us, it was particularly important for our thinking that Hans represented a link to mathematics and natural science. Also, whereas Alfred Werner, and to a less extent myself, had included philosophy in our studies, we would not, without our innumerable talks with Hans, have been able to develop any kind of understanding for the new kind of philosophy in which physics was destined to play a key role.

We had not quite the same life style. Hans had a romantic streak; the rest of us were prevailingly rationalistic. Hans, I think, had been in the *Wandervögel* while in high school. This was an organization of young people, below college age, who were engaged in a sort of upheaval against the system of education at home and in school as it existed in Germany around 1900. But it was not an active revolt; it was, what one might call, a 'revolt by withdrawal'. The young people formed hiking groups; '*Wandervogel*' means bird of passage. By wandering through the country, camping, sometimes staying in villages with farmers, they loosened their contacts with their families and largely escaped from control by family and school. But this did not – at least not in the beginning – meet with disapproval by the older generation. Failing to understand that the movement had originated from antagonism to traditional ideas and forms of living, the adults regarded it as an organized attempt at physical invigoration through marching and primitive living, commendable and not out of line with German traditions.

But the movement developed its own culture, or, in present-day terms, its counterculture. The core of this was a mystique of fellowship; just being together, as young people, feeling close to each other, free from external restraint, seemed of decisive importance. Common enjoyment of nature, singing while wandering through the country or sitting around the campfire and accompanying the songs on the guitar, enhanced the feeling of comradeship. It was decidedly a culture of emotions and sentiment, not of intellect; there was a definite tendency to deprecate the intellectual culture to which the school tried to introduce the young people.

As the movement grew, it diversified. Although most of the groups were restricted to boys, some were coeducational. Some groups assumed a taint of nationalism, even antisemitism (although the Jewish youth groups did not remain untouched by the *Wandervogel* style); others were rather inclined toward pacifism. Here and there within the *Wandervogel* movement the idea cropped up that withdrawal from the existing civilization was not enough, that the cultural values of the *Wandervögel* must permeate society which ought to be reconstructed accordingly.

I had not known Hans in his *Wandervogel* days. It was quite clear, however, that he could not permanently be satisfied with the sentimental emotionalism of the *Wandervögel* nor with its antagonism to the intellect; Hans was a scholar even before he reached the level of mature scholarship, and therefore could not share in the denigration of brainpower and its works. So it is no wonder that he found his way to the unromantic, scholarly oriented *Freie Studentenschaft*. But, of course, man does not live by brainpower alone; it was an excellent thing that Hans brought to our circle some elements of his *Wandervogel* past.

The *Wandervogel* movement encountered its most severe crisis when its members reached college age; many of them entered the university. One cannot be a student in any field and despise the intellect. The university branch of the *Wandervogel* movement, the *Akademische Freischar*, could not master this dilemma. It tried to preserve the *Wandervogel* life style, but it did not succeed in expressing its cherished values in a philosophy, and it did not even seriously try.

We of the *Freie Studentenschaft* kept an attitude of friendly aloofness toward the *Freischar* people. We were aware of the differences that separated us; at the same time, like ourselves, the *Freischar* wanted freedom of individual development; it was as much opposed as we were to the kind of academic life which the fraternities had created, with their duelling, their compulsory drinking sessions and their social exclusiveness; in some sense, therefore, the *Freischar* was our ally. Again, it was of great help to us to have Hans in our ranks, who could assist the rest of us in understanding the spirit of the *Wandervogel* movement and its academic branch.

At one point in its process of growth the movement of *Wandervogel* youth felt the necessity of an attempt at unification – at least of an effort to find a formula that would express the common beliefs of the diverse groups. To this end a great meeting was called in 1913 to the *Hohe Meissner*, a mountain in the Hessian chain of hills. There a resolution was adopted stressing the will for the autonomy of youth, its right to shape its own life under its own responsibility. More important than this formula, which was too ambiguous to have much practical effect, was the echo which the meeting created in public opinion.

The older generation, or at least large parts of it, suddenly woke up to the fact that the youth movement was not simply a bunch of hiking clubs but an attempt to secede from traditional culture. Alarm spread and soon caused vicious attacks by conservatives. The wing of the movement which attracted the bitterest hostility was led by Gustav Wyneken, a remarkable man with whom Hans was in fairly close contact until the outbreak of World War I, when Wyneken, like many Germans, succumbed to the temptation of extreme nationalism.

A few years before the *Meissner* meeting, Wyneken had founded the *Freie Schulgemeinde Wickersdorf*, which was based on essential equality of rights of students and teachers in all matters concerning the school. This was very much in line with the spirit of the youth movement, and Wyneken and his supporters participated in the meeting at the *Hohe Meissner*. But there were differences. Wyneken wanted to train the intellect as well as to give youth a chance to develop its emotions and its artistic sense. Moreover, the members of the *Freie Schulgemeinde* or at least many of them, went farther in rejecting the life style of the older generation than most of those of the *Wandervögel*, or they were perhaps only more open about it; some of their expressions seemed to demand sexual freedom and thereby aroused horror in a part of public opinion; the whole youth movement seemed to be tainted with licentiousness.

In the face of these attacks, the leaders of the youth movement felt rather helpless. They lacked the skill in debating issues, in defending a position in speech and print, that we had acquired in the *Freie Studentenschaft*. Feeling helpless, they did what even good and honest people have often done from weakness: they separated themselves from those in their group who seemed to be most vulnerable to the enemy's arrows. They drew a thick line between themselves and the Wickersdorfers; as it appeared to us, they threw Wyneken to the wolves.

We tried to do our best to help the youth movement in defending itself and to heal the rift between Wyneken and the rest. More than ever, it was a great help in these efforts that Hans knew how to appeal to the youth movement. Whereas the relations of most of us to Wyneken now became cordial, the other leaders accepted our assistance only grudgingly: they could not rid themselves of the suspicion that we wanted to draw them into politics, and they had no understanding of the fact that, if one is politically attacked, one cannot stave off the attack except through a political struggle. It was a complex situation with ramifications of fundamental significance; but the outbreak of the first World War, only a year after the Meissner meeting, put an end to the development.

I think it must have been in 1917 that I again met Hans in Berlin, when he was either on leave from the army or had been discharged, and I had moved to Berlin from Kiel, after a short military service, and was working in the war food administration. We were together quite a bit, exchanging views on the war and on the coming peace – our positions were very similar. We also engaged in an undertaking that would have greatly benefitted at least my career, and perhaps even influenced that of Hans, if we had been able to finish it: we tried to find a mathematical approach to the fundamental problems of economics. At that time few economists made much use of mathematics; it was not before the 1930's that the mathematization of economic theory became widespread, and not until after World War II that it became general. It was my fault that our efforts came to naught: being the economist on our two-men team, it would have been part of my task to study the few but significant previous attempts in the field of mathematical economics. I failed to do that and as a consequence we discovered too late that what we had intended as an original endeavor amounted merely to the tracing of a few steps which others had taken long before us. Later I had to pay the penalty: I have been severely handicapped in my own work by my lack of mathematical knowledge.

For personal reasons, I and my family had to move to Munich in the fall of

1917, and we did not return to Berlin until 1926. I do not know how active my correspondence with Hans was in this period, since I have no letters anymore, nor do I have a clear recollection. Even after we returned to Berlin, our meetings with Hans were infrequent, although our relations remained as cordial as ever. But I held two full-sized jobs: teaching at the College of Commerce and acting as managing editor of an economic weekly, *Der deutsche Volkswirt.* Hans must in these years have been very busy with developing that unique combination of mathematics, physics and philosophy which is the essence of his life work. Moreover, after 1930, the thought of the impending political disaster absorbed time and energy. I remember vividly, however, our last meeting in Germany, shortly before Hans left for Istanbul and while I was also making plans for teaching abroad – I am not sure whether I had already received my invitation to Berkeley. Of course, we did not know at that time whether we would ever see each other again, and certainly not that we would meet in California.

We did meet in California a few times, before or during World War II, but not as often as we might have. For both of us the appointments in Los Angeles and Berkeley were strokes of great good luck, but the difficulties were in the beginning great for me, in spite of the friendliness and understanding with which I was received by my colleagues, and probably no less so for Hans. The academic set-up and the ideas of what a professor should do or not do were quite different from those in German institutions, and it took time and effort to fit into the new environment. Moreover, my time – and I presume it was the same with Hans – was mortgaged with responding to the cries for help in emigration that reached us from friends in Germany who were persecuted for political or racial reasons. All this explains why we found it difficult to cover the distance from Berkeley to Los Angeles and vice versa more often in order to meet; afterwards, one realizes that one should have made more of an effort – such repentance over sins of omission is part of almost every human life.

February 1976 Berkeley, Calif.

[Ed. note: See our introductory note to Part I, p. 00. A fully documented account of the German youth movements at the time of Reichenbach's own activities as a university student may be found in Ulrich Linse, 'Hochschulrevolution: Zur Ideologie und Praxis sozialistischer Studentengruppen während der deutschen Revolutionszeit', *Archiv für Sozialgeschichte* **14** (1974) pp. 1–114. Further discussion and analysis may be found in Hans-Harald Müller, *Intellektueller Linksradikalismus in der Weimarer Republik. Seine Entstehung, Geschichte und Literatur – dargestellt am Beispiel der Berliner Gründer-*

gruppe der Kommunistischen Arbeiter-Partei Deutschlands (Scriptor Verlag, Kronberg/Taunus, 1977), and in the forthcoming work of U. Linse on the 'Entschiedene Jugend 1918–1921', the left-wing of the German youth movement, *Die Freideutsche Jugend*.]

11. UNIVERSITY STUDENT: HILDE LANDAUER

I became acquainted with Hans Reichenbach some time in 1912. It may have been within the frame of the *Freie Studentenschaft* but in any event in a way which was quite characteristic of Hans and of the style that had become natural to part of our student generation. One day, in the big and beautiful center hall of the University of Munich, a young man with a very round head, accentuated by well groomed, slightly wavy hair that added to the roundness of the head, came up to me, saying: "Last night, you said something quite intelligent in the discussion, we might take a walk together in the *Englische Garten*." And this we did, and it became the first of many walks and of many discussions, personal ones and talks about our activities within the student movement. We covered quite a wide area; I want to select two topics.

One centered around the function of the study of philosophy and the question of the form in which it should be offered to the beginner. We agreed that the big lecture course of four weekly hours in history of philosophy did little to help the newcomer in obtaining an understanding of the importance of the philosophical approach for his studies and his life. Hans was in favor of a beginner's seminar starting with methodical questions of a given system and demonstrating step by step how questions had been asked and answers been tried under given conditions, thereby taking in what has become the general problem of integrated social science.

The second topic hotly discussed between Hans and myself – as well as in our groups of the *Freie Studentenschaft* – concerned education in the specific sense of 'family or institutions?' We favored different sides, although our discussion was carried on in the most friendly terms. I felt that the initial role of the nuclear family in bringing up an infant enabled and even destined it to be a potential source of mutual assistance, enjoyment and enrichment in the relationship of the generations; Hans, possibly on the basis of personal experiences, was inclined to emphasize the shackling influence of the family and visualized the institution as a tool for liberating the personality. It is a pity that we did not meet often enough and long enough in later life to take up the old topic in view of our life experiences.

For many months of our student life in Munich in 1912 and 1913 we met almost daily. This happened at the office of the *Freie Studentenschaft*, in the big center hall of the university, in the Café Glasl at certain times of the day, at evening meetings about student affairs, and most regularly at noon dinners. Many of the student meetings centered on the problem: in what way guidelines of the Catholic church worked or did not work as an impediment to research. These discussions developed on a rather impressively high level. The representative of the Catholic group, Karl Nischler, as well as those of the opposite view, Alfred Werner, Carl Landauer, Albert Kranold and Hans Reichenbach, debated the issues without losing sight of the fact that we had to live together on a basis of human solidarity.

The noon dinners came about in this way. I lived in a small boarding house, three rented rooms and a more widely open dining room. The owners, Mr. and Mrs. Huith, of Swiss and Bavarian parentage, were limited in their outlook on life, but these limitations were made acceptable by a natural warmth and capacity of friendship. They also met other members of my family. When my mother and I rented the room for me in the fall of 1911, Mrs. Huith finally asked my mother: "May your daughter have male visitors with your consent?" My mother answered "yes" and the Huiths accepted her judgment. Many of the male visitors ended up becoming dinner guests. Thus we had a lively table with Alfred Werner, Hans Reichenbach, Hans Arneke, later a casualty in the war, Erich Trost, a highly capable bookkeeper, accountant and budgeteer of the *Freie Studentenschaft*, and occasional joiners. After the dinner the party often adjourned to my room which for not quite obvious reasons was designated as the Café of the Purple Monkey and indulged in whatever topic came to mind, and sometimes the discussions ended in a walk in the *Englische Garten*. I have never ceased to look back at this time as distinctly productive and closely related to later attitudes, decisions and enrichments.

February, 1976

12. MEMORIES OF HANS REICHENBACH, 1928 AND LATER: SIDNEY HOOK

My memories of Hans Reichenbach are warm and affectionate. I met him during the fall and winter of 1928–1929 during my Guggenheim year in Germany when I attended intermittently his seminars in the philosophy of science and the theory of probability. His courses, if I recall correctly, were not offered under the aegis of the Department of Philosophy but that of Physics. Both the manner and substance of his courses differed profoundly from that of the regnant philosophical schools. In an article on 'A Personal

Impression of Contemporary German Philosophy' in *The Journal of Philosophy*, 1930 (the first part of which was subsequently translated and published in the first number of *Erkenntnis*) I gave an account of these differences. Reichenbach was a rigorous logical empiricist for whom most of the metaphysical disquisitions of his colleagues in philosophy were not cognitively significant, and he never hesitated in saying so although he tried to avoid giving needless offence in the manner of his expression. Conventional German philosophers who were acquainted with his work, and there were few of them, would brusquely dismiss him as a philosopher.

Not only did I find Reichenbach's philosophy congenial to my own pragmatic naturalism, his way of teaching was more like that to which I was accustomed. He had an engaging, simple, and direct manner both in lecturing and discussion, always invited questions and criticisms, welcomed give and take, was patient, invariably well-prepared, and had a strong sense for the cumulative development of the subject matter.

I got to know him personally when we both dodged some meeting of the Kant Gesellschaft and found ourselves at the same table of a *Bierstube*. We talked late into the night, and then he rode off on a motorcycle. After that I visited him several times at his home; and we would often adjourn for a drink after meetings of the group he had organized around him. Among the members of that group I recall Hempel, Helmer, Grelling and Dubislav.

Reichenbach had never heard of Charles Peirce and was astonished and a little incredulous of the account I gave him of Peirce's frequency theory of probability and his doctrine of 'leading principles'. It was only after he had received a copy of Peirce's selected essays, edited by Morris R. Cohen under the title *Chance, Love, and Logic*, which I sent him immediately upon my return to the United States, that he acknowledged the work of his distinguished forerunner. I was heartened to find that he seemed to be in agreement with the naturalistic theory of the *a priori* I had developed in my book on *The Metaphysics of Pragmatism*. On this issue he was quite critical of Carnap whom I had not yet read, and asserted that there were Platonic elements in Carnap's position. The only philosophical issue on which we differed was whether ethical judgments had any cognitive significance and could be declared 'true' or 'false'. This always remained a point of difference between us.

I recall also a matter of mutual shock at the time. He was startled when I asserted that the theory of psychoanalysis, especially in its Freudian formulation, was unscientific, and I was startled when he insisted that it was. It seemed clear to me that on his own criteria of scientific verification that psychoanalysis was no more scientific than Christian science. In passing, I should note the curious fact that all the logical empiricists or positivists I have known were quite vehement in defending the scientific validity of Freud's basic views – something which in my obtuseness I could never square with their professed philosophy of science.

We became even friendlier when Reichenbach discovered that I had strong socialist views. He had never met an American socialist before and seemed as surprised to learn that there were American socialists as some proto-Nazi students at Munich had been when I told them that there was a Jewish proletariat in the United States. I then learned that Reichenbach had been head of the German Socialist Student Union and had played a very active role. He regaled me with stories about events that anteceded the First World War.

Some of my friends who have differed with Reichenbach over whether all intelligible concepts of probability can be interpreted in terms of the frequency theory have asserted that at crucial points in the argument, he would restate his position with great stubbornness and without any recognition of the lacuna in it. They believe that he was devoid of a sense of humor. I cannot agree with this. Like most philosophers he had a strong sense of vanity but I always found him possessed of a quiet sense of humor reflected in a benign twinkle.

I recall in this connection a story told me by Paul Tillich at a meeting of the New York Philosophy Club. "I once asked Hans Reichenbach to hold up his hand when he thought I was talking nonsense in a short discourse I was giving on Being. As soon as I uttered the first sentence his hand shot up and he kept it there until a few minutes before I finished. Afterwards I told him I was happy that he understood at least something of what I said. 'I didn't', he replied. 'But your hand came down', I interjected. 'Oh! I just got tired of holding up my hand!' he said."

Before I left Germany in the early Fall of 1929 I predicted that the Nazis' nationalistic appeal would win a large following. Reichenbach like most Socialists of the time was skeptical of it. I was happy to learn therefore that at Hitler's advent to power Reichenbach had gone to teach at Istanbul, and in the course of our correspondence I learned that he was planning before long to lecture in Turkish. But he was not very happy in his Turkish exile and I resolved to bring him to the United States. I wrote to Einstein asking for a letter of evaluation and support of Reichenbach's work. Einstein replied with an enthusiastic recommendation of Reichenbach's qualifications and of his contributions to the philosophy of science. On the strength of it I went to Chancellor Harry Woodburn Chase and received permission, despite the straitened economic situation of New York University, to cable Reichenbach an offer of a full Professorship. Both the history of Hans Reichenbach and that of the Philosophy Department of NYU would have been profoundly different if events had turned out as planned. To our great disappointment, Hans cabled back that the Turks had refused to let him go. He had signed a five year contract to teach at Istanbul, and the Turkish Department of Education refused flatly and adamantly to release him. As a gesture of retaliation, Hans abandoned his study of Turkish.

By the time the five year contract was up, Donald Piatt had already snatched Hans for the University of California at Los Angeles. I saw little of him because of the distance but always enjoyed my meetings with him and Maria. One of our most memorable reunions was at the meeting of the American Philosophical Association at which he participated with C. I. Lewis and Nelson Goodman in discussing the character of judgments of immediate perception. Hans gave his talk in flawless English without consulting a note. He was as fluent in French as well.

Other contributors will assess his specific contributions to the field of philosophy. I content myself here in offering these memories of an important thinker in the life of our time, a dedicated defender of intellectual and cultural freedom, an opponent of all varieties of fanaticism, and a man of great moral courage.

January 1976

P.S. [I remember the story of Tillich differently. It took place at the house of a psychoanalyst in Los Angeles. Tillich had asked to meet Hans and it was arranged. Tillich said (and I was present): "Professor Reichenbach, would you please hold up your hand every time I say something meaningless." Hans replied with a very charming smile: "I am afraid, Professor Tillich, that hand would never come down." M.R.]

13. A YOUNG UNIVERSITY TEACHER

[from a letter of Carl Hempel to Maria Reichenbach, March 21, 1976]

When Hans Reichenbach joined the faculty of the University of Berlin, I was an advanced student there; while majoring in mathematics and physics, I had also taken various courses in philosophy, including a seminar on Kant's *Critique of Pure Reason*. In an early conversation with Reichenbach – I think it was during a break in his course, and in startled response to ideas he had just set forth on causality and determinism – I told him that I regarded the principle of causality as true *a priori* and found it impossible even to imagine that it might be false. With a characteristic smile, and an aside on how limited one's imagination could be, Reichenbach remarked that I seemed to have swallowed Kant's conception hook, line, and sinker, but that he was confident I would change my mind on consideration of the arguments he would be presenting. I did change my mind on this subject and many another, and Reichenbach's new ways of viewing and tackling philosophical issues exerted a strong influence on the development of my thinking.

The very occurrence of that early conversation reflected Reichenbach's open and refreshingly informal way with students. At the time, many professors at German universities were practically unapproachable to their students, especially beginners. They delivered their lectures, but were not accessible to questions or objections. In some philosophical courses, I had the sense that we students were handed down *ex cathedra* a system of profound ideas which we were expected to try to absorb, but not to question. Sometimes, in rewriting in some detail my shorthand lecture notes, I had a feeling of performing a ritual dance in words whose meaning was frustratingly elusive.

Not so at all in Reichenbach's classes – and, happily, in those of some other teachers I remember with similar gratitude. Reichenbach conveyed the conviction that the problems under consideration were important and interesting, that they could and should be stated clearly, and that any proposed solution must be formulated in precise language and supported by clear, objective arguments. He was receptive to requests for clarification, and to objections, concerning his own views; and he encouraged the students in his seminars to offer their ideas on how to tackle open problems that were encountered in the discussion. Thus we had a sense of being engaged in a common effort at philosophical exploration and of being potential contributors to progress in the field.

The exchange of ideas often continued at informal after-seminar gatherings in a nearby cafe, where the conversation tended to turn also to scientific, social, education, and political subjects, in all of which Reichenbach took a highly informed and contagiously intense interest.

14. A PROFESSOR IN TURKEY, 1936: MEMORIES OF MATILD KAMBER

It was a most rewarding and exciting experience for me to meet Dr. Maria Reichenbach, the philosopher-wife of the late philosopher-scientist, Professor Hans Reichenbach, at a philosophical convention in Los Angeles in March, 1975. When I introduced myself to her as one of the former students of Professor Reichenbach from the University of Istanbul, she asked me if I could write my memories of her husband. Of course, I was greatly pleased; nothing could have satisfied me more than the opportunity to express my indebtedness for the guidance his genius had provided the students and philosophers of Turkey.

My memories of Professor Reichenbach take me back to the year 1936, the period when Ata Türk had started his great educational reform by inviting a group of German professors to teach at the University of Istanbul. Among these renowned intellectuals, Professor Reichenbach had come from Berlin in 1933 to be the head of the Philosophy Department at the University of Istanbul. At the time I happened to be a student of Philosophy and attended the three classes which he was teaching: History of Philosophy, Epistemology, and Symbolic Logic, which was then called 'logistics'. I cannot find words to describe how much we enjoyed these classes and how much of his philosophy was imparted to us in a very simple and lucid manner. I shall not attempt to discuss the main philosophical problems which were the subject of the courses, because I believe that philosophers all over the world are well acquainted with Reichenbach's philosophy. Instead, I shall try to describe his activities as the head of the Philosophy Department, thinking that this will reveal the valuable service that this great philosopher-scientist has rendered, not only to the history of philosophy, but also to the field of education.

Today everyone in teaching is acquainted with the new trend of interdisciplinary teaching. Many colleges and universities and, in America, junior colleges are experimenting with this kind of teaching. According to a report of the Center for Educational Research and Innovations in 1970, interdisciplinary teaching does not have a long history, but was practiced in Germany and France only thirty years ago. But forty years ago Professor Reichenbach introduced this type of teaching at the University of Istanbul.

In an attempt to show Professor Reichenbach's concept of knowledge and education, I shall try to describe how he organized the Philosophy Department at that time. Like many European universities, the University of Istanbul was divided into two faculties: The Faculty of Letters and the Faculty of Science. Philosophy belonged to the Faculty of Letters, yet Professor Reichenbach put philosophy into a unique position. Like regular science students, philosophy students had to attend classes in the Faculty of Science. To be more specific, they had to study two theoretical sciences and one experimental one each semester for four years. So they studied mathematics, biology, physics, medicine, chemistry, physiology, and genetics in addition to the different humanity courses such as literature, sociology, psychology, history, etc. In a way this could be classified as multi-disciplinary and cross-disciplinary teaching, yet the seminars which were an adjunct represented an interdisciplinary activity. I should like to say a few words about how these seminars were conducted. Professor Reichenbach called them Collegiums [for faculty], yet the philosophy students were asked to attend. All the professors from the different departments were present. There were physicists, biologists, mathematicians, physicians, and dentists, and each philosophical problem was discussed from different points of view.

Unfortunately, this ideal situation did not last many years, for as soon as Professor Reichenbach moved to the United States, the Philosophy Department reverted to the old system without any interdisciplinary activity. Even today it is not at the University of Istanbul, but at the University of Smyrna and the Hacitte Tepe in Ankara where interdisciplinary teaching is practiced.

Besides Professor Reichenbach's great popularity as a philosophy teacher and educator, he also created a new image of teacher–student relationships. Most of the students did not know English or French and thus had no direct way of communicating with him, yet one would never see him alone at lunch or walking around the university campus. Everyone gathered around him wherever he appeared and learned from him (by way of translation) outside the classroom as much as they did inside. He would speak to us about the books he had written or was writing, and was humble enough to ask our opinions on specific subjects. He informed us of current philosophical problems about which he was corresponding with other philosophers and scientists.

One incident which impressed me greatly was a telegram I received from him one midnight. When he was writing his book *Experience and Prediction*, he had asked me to type the manuscript. At that time my home was very far from his home on the Asiatic side of Istanbul. His midnight telegram, an unusual experience which upset my entire family, concerned a change of one of the examples in his manuscript. As I remember, he wanted to change the word 'sheep' to 'goat'. The timing and significance of his request seemed strange, but he was so enthusiastic and devoted to his work that it never occurred to him that he would see me the following morning.

Another peculiarity of the great yet very human man was his unique approach to the use of foreign words. Our lecture classes were in German, and we had interpreters who translated these lectures into Turkish. Usually Professor Reichenbach did not write out his lectures as his interpreter was quite proficient in German. On several occasions I remember his catching a mistake of translation, knowing that the interpreter had not translated what he had said, and this he would detect not only by examining the facial expressions of the students but also by recognizing a Turkish word with which he was familiar and which he thought might contradict what he had said in his lecture.

Another unusual incident that I remember concerns his approach to language. One day he was supposed to give a lecture in English and, after completing a very impressive address, he told us that he had used only 1500 different words in English. When I asked him how he could manage with so few words, he said that he had learned those English words which, according to a certain calculation which he himself had made, were the words that enabled him to lecture on that particular problem with no difficulty. Before he left for the United States, he had improved his English and French and could also make very intelligent guesses about what we were saying in Turkish.

As a concluding remark, I want to mention that in many periodicals and in the *Annals* of the University of Istanbul the years 1933–1938 are referred to as the 'Golden Age' of the university. These were the years when Professor Reichenbach laid down the principles of interdisciplinary teaching by establishing close ties between the various departments of the University of Istanbul and creating a new approach to teaching.

March 1975

Columbia Junior College
Columbia, California

15. CONCERNING REICHENBACH'S APPOINTMENT TO THE UNIVERSITY OF CALIFORNIA AT LOS ANGELES: CHARLES MORRIS

I first met Reichenbach at the Prague Preparatory Conference in 1934. My subsequent contacts with him centered around two projects: the publication of his book *Experience and Prediction*, and his appointment as Professor of Philosophy at the University of California in Los Angeles.

Reichenbach had written *Experience and Prediction* in English, and surprisingly good English for a German writer. He asked me, however, to go carefully over the manuscript before it was submitted to the University of Chicago Press. The book was accepted for publication in 1937, and published in 1938.

The problem of a teaching appointment for Reichenbach in the United States was an urgent one, since if he had not received an appointment by October 1938 he would have had to sign a five year extension of his contract in Istanbul, Turkey – where he had found refuge from Nazi Germany.

To this end I corresponded with philosophers at the University of Illinois, Northwestern University, New York University, the University of California at Berkeley, and the University of California at Los Angeles. An attempt to get an offer for Reichenbach at the University of Chicago was rejected by the Administration.

It was at the University of California at Los Angeles that matters finally took a favourable turn. I had asked Einstein, Ernest Nagel, Victor Lenzen, and others to write their opinions of Reichenbach, and I turned these letters over to Donald A. Piatt, then Chairman of the Department of Philosophy at UCLA. President Robert G. Sproul offered Reichenbach the position of Professor of Philosophy at UCLA, and it was accepted by him early in 1938. His teaching began in the Fall of that year, and continued until death in 1953.

[From a note prepared for Herbert Feigl in 1967]

16. MEMORIES OF HANS REICHENBACH: RUDOLF CARNAP

[In his 'Intellectual Autobiography', Carnap wrote of his good friend Reichenbach at several places. Some excerpts follow. The full essay appeared as 'Carnap's Intellectual Autobiography' in *The Philosophy of Rudolf Carnap*, P. A. Schilpp (ed.), *The Library of Living Philosophers*, Vol. XI, 1963, La Salle, Illinois and London, pp. 3–84. – R.S.C. and M.R.]

Among all those who worked in Germany in a similar direction in philosophy and in the foundations of science, Hans Reichenbach was the one whose philosophical outlook was nearest to mine. He was at that time Instructor of Physics at the Technological Institute in Stuttgart. Both of us came from physics and had the same interest in its philosophical foundations, and especially in the methodological problems created by Einstein's theory of relativity. Furthermore, we had a common interest in the theory of knowledge and in logic. At first we communicated only by correspondence. It was not until March 1923, that we met at a small conference in Erlangen, which we organized with a few others who were likewise working in the field of symbolic logic and its use for the development of a scientific philosophy. Among the participants were Heinrich Behmann, Paul Hertz, and Kurt Lewin. There were addresses on pure logic, e.g., a new symbolism, the decision problem, relational structures, and on applied logic, e.g., the relation between physical objects and sense-data, a theory of knowledge without metaphysics, a comparative theory of sciences, the topology of time, and the use of the axiomatic method in physics. Our points of view were often quite divergent, and the debates were very vivid and sometimes heated. Nevertheless, there was a common basic attitude and the common aim of developing a sound and exact method in philosophy. We were gratified to realize that there was a considerable number of men in Germany who worked toward this same aim. The Erlangen Conference may be regarded as the small but significant initial step in the movement of a scientific philosophy in Germany.

After the Erlangen Conference I met Reichenbach frequently. Each of us, when hitting upon new ideas, regarded the other as the best critic. Since Reichenbach remained in close contact with physics through his teaching and research, whereas I concentrated more on other fields, I often asked him for explanations in recent developments, for example, in quantum-mechanics. His explanations were always excellent in bringing out the main points with great clarity. I sometimes explained to him developments in special problems of logic or the logical foundations of mathematics (pp. 14–15).

Among philosophical groups at other places which were close to our philosophical conceptions in the Vienna Circle, I wish to mention especially Reichenbach's circle in Berlin and the Warsaw philosophical group. Reichenbach had been teaching at the University of Berlin since 1926 and had gathered a group of people with similar philosophical interests, which later developed into the Society for Empirical Philosophy. Among the active participants in the discussions were Walter Dubislav, Kurt Grelling, and Carl G. Hempel, and also some scientists with philosophical interest, among them the psychologists Kurt Lewin and Wolfgang Köhler. I sometimes went to Berlin, gave talks in the Society and in Reichenbach's seminar, and had intensive private discussions with Reichenbach and his friends (pp. 29–30).

As early as 1923, when Reichenbach and I announced our plans for the Erlangen Conference, we found that many people were interested in our type of philosophical thinking. Reichenbach suggested the idea of a new periodical as a forum for our kind of philosophy. However, it was only years later that his efforts were successful. Our periodical, *Erkenntnis*, began to appear in 1930. It published not only articles but also reports on conferences and congresses of our movement, beginning with the report of the First Conference on the Epistemology of the Exact Sciences, which took place in Prague in September 1929 (p. 32).

17. MEMORIES OF HANS REICHENBACH: HERBERT FEIGL

[In his essay on 'The Wiener Kreis in America' Herbert Feigl wrote briefly of Reichenbach in America. The full essay appeared in *The Intellectual Migration, Europe and America 1930–1960*, Donald Fleming and Bernard Bailyn (eds.), Harvard University Press, Cambridge, Mass., 1969, pp. 630–675. – R.S.C. and M.R.]

Charles Morris was . . . instrumental in bringing Hans Reichenbach to the United States. Reichenbach had to leave Germany because of the increasing menace of the Nazi regime. He went first to Istanbul (Richard von Mises was also there at that time) where he taught first in French, and later even in Turkish. In 1938, the year in which Reichenbach's important book *Experience and Prediction* was published by the University of Chicago Press (Morris had arranged for this), he began teaching at the University of California at Los Angeles, where he taught until his untimely death in 1953. He soon established himself as one of the leading philosophers of science in America. He was a most productive scholar; he worked in many areas, and published several important books. Reichenbach was also a brilliant speaker and lecturer, beloved by his students. He and Carnap and I met occasionally at specially arranged conferences, or – as in 1951 – at a combined meeting of the American Philosophical Association and the American Association for the Advancement of Science at Bryn Mawr and Philadelphia (p. 648).

. . . After Schlick's death, Carnap and Reichenbach were in my estimation the leading exponents of our philosophical approach (p. 672).

18. RECOLLECTIONS OF HANS REICHENBACH: ERNEST NAGEL

When I first met Hans Reichenbach in 1934 at the Eighth International Congress of Philosophy held in Prague, I was already heavily in his debt for the instruction I had received from a number of his publications. I had studied with enormous profit his *Philosophie der Raum-Zeit-Lehre* shortly after the book appeared in 1928, and had read with excitement what were to me his illuminating contributions, especially those dealing with probability and induction, to the periodical *Erkenntnis* (which he and Rudolf Carnap had been editing since 1930). Here is a philosopher after my own heart, I remember thinking, who takes seriously the classical conception of the task of philosophy to provide a solidly based interpretation of science and its logic, and with whom I would dearly love to study. The award of a Guggenheim Fellowship, which obligated me to be in Europe during the academic year 1934–5, offered the opportunity to realize this desire; and I looked forward eagerly to talks with Reichenbach, first in Prague and eventually at greater length elsewhere. Since his *Wahrscheinlichkeitslehre* had not been published at the time of the Congress, I was particularly interested in discussing with him whether statements about the probability of general hypotheses (or theories) can be explicated in terms of the frequency conception of probability to which I knew he subscribed. But an international congress is hardly an occasion for much private discussion, especially with a man as much in demand as was Reichenbach. In the one conversation he was able to have with me he had time for little more than to tell me that he thought the notation Keynes used in his *Treatise* was misleading, and that his own forthcoming book on the subject would establish the adequacy of the frequency notion of probability for dealing with inductive reasoning. Reichenbach had resigned his university post in Berlin, and was on his way to fill the professorial chair he had been offered in Turkey. He urged me warmly to visit him in Ankara, but my limited finances made it impossible for me to accept his invitation.

I saw Reichenbach again only after he joined the Philosophy Department of the University of California at Los Angeles. My recollections of him are still vivid as a participant in a number of philosophical meetings and conferences at which I was also present, but especially as a colleague when he was Visiting Professor in my Department at Columbia. His books testify to his impressive mastery of modern physical theory. But it was a mastery that was also evident in his conversations as well as in his lectures. It was obvious that he believed that a firm grounding in fundamental physical theory and in the mathematics relevant to it is essential equipment for a competent pursuit of philosophy. On the other hand, he rejected the view held by some of his friends that a thorough education in science makes training in philosophy unnecessary, and that professional philosophy can no longer play a significant role in a modern university or a rationally ordered society. In his opinion, professional scientists are as a rule neither trained nor competent to articulate the structure of physical theories, to analyze the meanings and interrelations of scientific concepts, or to make explicit the logic of scientific method. These are tasks, he believed, that need to be undertaken for the sake of the continued vigor and growth of science itself, but which fall into the province of philosophy. In consequence, he was not apologetic for being a 'mere' philosopher, nor did he doubt that the sort of constructive analysis of science in which he was engaged contributes to the advance of our knowledge of nature. He recognized that a high degree of technical specialization is unavoidable in a philosophy so conceived; but he did not think that such specialization is self-justifying, and he had little interest in subtleties or so-called 'ontological' problems that have no bearing on the pursuit of science. As he conceived and practiced philosophy, it is not a trivial enterprise, for its ultimate aim is the clarification of the conclusions and the method of scientific inquiries into the nature of things.

Reichenbach was a remarkably fluent and soft-spoken speaker, and an admirably clear and cogently organized lecturer. He was singularly successful in communicating his ideas to professional audiences as well as in the classroom, and was remarkably resourceful in his use of illustrations to make evident the import of the ideas he was presenting. He was patient with students, quickly won their respect and often their affection, and succeeded in convincing them that philosophy is not an idle indoor game, but is a crucially important though difficult undertaking, directed to the clarification of the findings and the conduct of a major human enterprise – certainly these things were so when he was at Columbia.

Philosophical conferences and colloquia were invariably enlivened by Reichenbach's presence. I do not recall his ever *reading* a paper on such occasions (as is the custom in the U.S.), even though he was on the program to deliver one. He had the ability (shared by many Central European scholars) to state what he had to say in coherent and well-constructed prose ready to be sent to the printer, usually without recourse to any notes; and the virtuosity of the performance contributed to the effectiveness of the argument. When he was not a scheduled speaker at the meeting he was attending, it was his wont to take a front seat as near as possible to the reader of the paper and the chairman of the meeting, presumably because his own hearing was poor and he needed to be in close proximity to the speaker. But I suspect that in addition to this reason he occupied a front seat in order to be quickly recognized by the chairman when discussion of the paper was in order.

Reichenbach's comments on papers presented at such gatherings were unfailingly pertinent and precise. However, it was often possible to anticipate what he would say, especially when he had already dealt with the problem under discussion in one of his publications, for he was completely confident that he had the definitive answer to such problems. Indeed, he seemed to find it just about impossible to believe that anyone who had read his books carefully and really understood what he was saying could disagree with his resolution of the issue under discussion. At any rate, my conversations with him during his visit to Columbia were uniformly pleasant and interesting, but sometimes also frustrating. It happened more than once that when I raised with him what seemed to me a serious difficulty in the position he had taken on some question – for example, in his view that statements assigning some degree of probability to a theory in the light of the available evidence can be explicated in terms of his frequency construal of probability – he would in reply simply read to me a familiar passage in one of his books.

Reichenbach was genuinely friendly and democratic in his overt behaviour toward colleagues and students, and in my experience with him he was always the soul of courtesy. However, he did exhibit what I thought was a certain condescension, doubtless unintended and unconscious, in his assessments of, and attitudes toward, American scientists as well as American philosophers. He apparently never quite fully outgrew what was once a common European conviction that unlike physical scientists trained in Germany, for example, American physicists had only a superficial understanding of contemporary physical theory – at any rate, I heard him talk in this vein on more than one occasion. But to balance this example of Reichenbach's one-sided high esteem for the accomplishments of European science, it should be noted that he did less than adequate justice to the achievements in the philosophy of science of European thinkers other than himself. In his *The Rise of Scientific Philosophy*, in which he purported to describe the development during the present century of a 'scientific' (or positivistically oriented) philosophy of science, the only contributions to the solution of major problems in this branch of inquiry that are mentioned are his own.

However, these are foibles which do not diminish Reichenbach's stature as a creative philosophical analyst and as an important figure in the endless struggle for philosophical enlightenment.

August 1977 — Columbia University

19. UCLA: DONALD KALISH

It is both an honor and a pleasure to have the opportunity to comment briefly on my association at The University of California, Los Angeles with Professor Hans Reichenbach. This association began when I was considered for an appointment at UCLA. What might have been for me, and too often is, an anxiety generating experience turned out to be an extremely enjoyable and philosophical evening – the pleasant memory of which is still strongly with me. Hans and Maria Reichenbach had invited the Philosophy Department's faculty and graduate students to their home to meet me. After socializing in a most hospitable and completely informal atmosphere, I presented some ideas I had on meaning and truth. My presentation was followed by a discussion of those topics, guided by Hans in such a manner that I felt that I was already one of the group and not a job-applicant being examined or cross-examined by the group. That evening has served as a model for our Department in its recruiting activities over the past two decades, but Hans himself provided an essential ingredient that could not be duplicated.

Years later, as department chairperson (1964–1970), I had to play a key role in the recruitment of new, especially junior, faculty. One of the first questions that would be posed to me upon making a job-offer to a young applicant was: "Will I have an office of my own?." In contrast, what seemed most attractive to me when first employed by UCLA was the fact that I would share an office with Hans Reichenbach. Listening to Hans talk with students, and just taking advantage of his presence, I had a continuing educational experience that certainly equalled and significantly augmented the best of my graduate student training. Hans seemed to have almost unlimited patience with even the slowest students. Once, I asked him to account for his patience with students, and his willingness to explain points [to them] many times and in many different ways, in contrast to his slight impatience at times with some of his peers. His response to me was that the latter in contrast to the former clearly had the ability to understand him easily and quickly. Hans not only patiently gave time individually to students who approached him during his office hours, but he had an extraordinary ability for classroom teaching of beginning freshmen as well as advanced undergraduates and graduate students. Indeed from Hans' untimely death until today we have not offered lower division courses in inductive logic or the philosophy of science, for no one seems capable of doing a respectable job of presenting these technical subjects to students who have no background in logic, mathematics, or science. Hans did more than a respectable job; he did it superbly.

Hans and I taught the Department's beginning course in symbolic logic in alternative semesters. He had recently published his *Elements of Symbolic Logic*, and W. V. Quine's *Methods of Logic* had just become available. Of course, we discussed the question of suitable textbooks for the course, and Hans immediately made me a gift of an autographed copy of his book. But I had become strongly attached to Quine's methods of natural deduction, which he had set forth in syllabi printed by the Harvard bookstore before the publication of *Methods*; and hence I decided to use Quine's *Methods* rather than Reichenbach's *Elements*. Not only was this decision accepted most generously, but Hans always strongly encouraged a variety of approaches to the study of logic. From the number of *my* now younger colleagues who use the textbook that I authored with the late Professor Richard Montague, I sometimes wonder whether I have lived up to the excellent example of generosity set for me by Hans Reichenbach.

It is almost impossible to believe that two decades have passed since Hans was teaching with us at UCLA, for the development of our department has followed so closely the direction he set for it. Indeed, if he could only return to us now, he would recognize us immediately as *his* department, and be more at home in it now than he was when he undertook to make it a department of scientific philosophy.

January 2, 1976

20. UCLA: PAUL WIENPAHL

Hans Reichenbach came to the Department of Philosophy at UCLA in the Fall of 1938 when I was commencing my second year of graduate work. He came to us from Istanbul via the Alps, where on the journey he had a heart attack. Partly because of this Donald Piatt, Chairman of the Department, asked me to meet Professor Reichenbach at the Los Angeles Union Station and drive him to an apartment in Santa Monica. I dreaded and could hardly wait for that day to arrive. I was a beginner and he was the Department's great new arrival. The jaunt turned into a pleasure because the man, whom I eventually came to call 'Hans', was so easy to be with.

At that time the Department at UCLA was not large: Donald Piatt, John Elof Boodin, Hugh Miller, D. C. Williams (who left in a year for Harvard), and Isabel Hungerland (then Creed). The Dean of the College, Rieber, and the Provost of the campus, E. C. Moore also taught courses in philosophy. There were, as I recall, five regular graduate students, including Hans Meyerhoff.

Because of his attack Reichenbach required an office on the ground floor of Royce Hall, and for a time I drove him to and from the campus. Thus I got to know him reasonably well rather more quickly than I would have otherwise. Several things impressed me. The first, I suppose, was his rotund shape. Everything about him either was or seemed round. Yet I never thought of him as fat. He was muscular and possessed of great energy. This gave rise to a second impression. His doctor had advised extreme care in physical exertion. Reichenbach took this advice by doing everything with a calm composure that sometimes set my teeth on edge. I could not see how a man of such obvious energy could walk and move in other ways with the obvious deliberation that he did. A third impression was of extreme cheerfulness and a good deal of laughter.

The arrival of Miss Neyire Adil Arda from Turkey gave me another view of Reichenbach. Here was a student who had followed him almost half way around the world to pursue her studies with him. I gradually realized that he was a man who attracted disciples and enjoyed doing so. Within a short time he asked me to translate *Die Wahrscheinlichkeitslehre*. It was a task at which I finally failed because I could not get the Germanisms out of my English, but in giving it to me he attracted me to him.

In 1940, Russell joined the department, bringing with him from Chicago Abraham Kaplan and Norman Dalkey. If matters had tended to get above my head with Reichenbach and his scientific empiricism, weaned as I was on pragmatism, the thing now seemed hopeless. The use of symbols and formal systems threatened to swamp the graduate students, except, of course, for the two new-comers. It was during that year that I then saw another side of Reichenbach. He was a superb teacher. In his seminar on space and time he was able to make the issues clear to students who had nowhere near the mathematical background normally required for this. I could understand how he could write *Atom and Cosmos*.

Then Maria, later to become Mrs. Reichenbach, came to the campus. There would be an occasional visit by the Carnaps. We graduate students got a new sense of how to do philosophy. We became deeply involved with logical positivism and scientific empiricism, and we found that they were in a way simply a more systematic version of Dewey's instrumentalism. Russell left after only a year, and it was Hans Reichenbach who became chiefly responsible for this new stamp on our thinking. Our earlier teachers had done a good job, but he helped us to get a little 'elegance' into what we did. He was not my dissertation-director, Piatt was, but he was on the committee and the result, inadequate as it may be, markedly bears the stamp of his way of thinking.

Making these changes was not easy, I mean from a completely American way of doing philosophy to the more or less European way of doing it. Reichenbach's *élan* helped. The man was so full of energy, hopes, drive, and good will that he drew more than pushed us along with him. I remember his often laughing about things. One story comes to mind. Otto Klemperer had conducted a concert in Royce Hall. In addition to his music we talked of his stature. Reichenbach thought this great in several ways. It turned out that he had known the Klemperers in Germany. In fact at one point he tutored the future composer's sisters in mathematics. They were apparently as large as their brother, and Reichenbach was gleeful about what this did to his own short stature.

The army and other things got to me in February of 1942. Returning to teach for a year in 1946–47, I became a colleague of Hans Reichenbach – and of Abe Kaplan's, who also returned to the fold that year. Hans was now fully recovered, as far as we could tell, from the attack in 1938. His office was in the upper reaches of Royce, a tall building without elevators. Busy as I was with my first year of teaching, I audited his year-course in logic for lower-division students. The point of this was not merely to brush up on formal logic after five years' absence from it. It was also to observe an excellent teacher of the subject. The class was large for a course in those days, and huge for one in logic (130 students, if I remember correctly). This is a testimony to Hans' reputation on the campus. I learned a great deal about teaching logic, however well or poorly I may have subsequently done it. During many of the years that I did, *Elements of Symbolic Logic* was the text. It was one of the ways in which Hans stayed close to me after I left UCLA and could no longer see him daily.

February 6, 1972

University of California,
Santa Barbara

21. UCLA: NORMAN DALKEY

A disquieting feature of distant memories of people is that they don't form neat rounded wholes. The images come back in a pastiche, a montage of intertwined impressions. Although I knew Hans Reichenbach for many years as student, friend and colleague, the strongest memories are not of a sharply etched figure such as one might find in a history of philosophy, but rather a sequence (not necessarily in temporal order) of vignettes . . . .

Quite unforgettable is the opening session of Hans' seminar on probability in the Fall of 1939 at UCLA. I had gone to UCLA as a third year graduate student after two years at the University of Chicago. Hans' ebullience contrasted sharply with the grave persistence of Carnap and the intellectual urbanity of Charles Morris. Hans clearly loved his subject. A discovery or a new insight, were occasions for delight, as well as for scholarly satisfaction. Hans taught his courses (even undergraduate logic courses) in a free-wheeling stream-of-discovery mode, following the dictates of his latest investigations, with second priority to standard content. Some of his students were not altogether happy with this intensely personalized subject matter. At the time, I thought it was the only way to fly.

Mixed with the technical textures – illata and posits and probability meanings – there was a strong flavor of mid-European culture. I recall sessions at the Sonja Henie outdoor skating rink in Westwood, long since usurped by a Sears outlet (and later by a parking lot) with a simulated Swiss Chalet for a service area, and Hans cutting precise figure eights on the ice like a defrocked Santa Claus – straight out of a Viennese Christmas Card. While gliding he explained the exquisite relationship between the center of gravity and the edge of the skate – if one wanted to keep his anatomy unbruised – and the ubiquity of the principle of conservation of momentum, such as the problem it posed for stopping on the ice, or in the opportunity to accelerate a swirl by drawing the arms close to the body . . . .

Hans was aware of science all around him. As our relationship grew from student to friend, he took me and other favored students on trips, often to the San Gabriel mountains. There he would take pictures with his Zeiss Ikon camera – as often as not stereo pictures of the mountains and trees with a wide base, 100 or more feet apart. It required careful alignment, and a steady tripod. But viewed under a stereoscope, slopes, which in ordinary stereo pictures would still look flat and part of the background, would spring out of the border of the prints and stand out in airy three-D.

Continual viewing of the world through the prism of physics was not always pleasing. Hans was very well acquainted with the night sky. And the stars and fuzzy patches of light were not only a visual treat, but a reminder of the curvature of space and the conventionality of temporal simultaneity. I spent the summer of '40 with Bertrand Russell and his family at Fallen Leaf Lake, adjacent to Lake Tahoe. The clear mountain air, and the wide expanse of the lake created a brilliant natural planetarium. Hans visited on several weekends, and the first night or so his virtuosity in delineating the constellations and creating the conviction that β-cygni – wavily viewed through binoculars – indeed looked like a double star, was a welcome enhancement of the evening. But as it became clear that Hans felt the lectures were a necessary component of *every* evening, darker feelings stirred in the Russell household. Before his next visit, they carefully coached young Conrad, then about four or five and cognitively mature far beyond his years, on the location and special features of a small, obscure constellation, I believe it was Scutum Sobieski (the shield of Sobieski, in the vicinity of Saggitarius). When Hans returned and began his evening discourse on astronomy, young Conrad enthusiastically broke in to point out Scutum (it's not easy to find) and recounted at length its arcane properties. It so happened that Hans was not up on Scutum: subsequent evenings were devoted to other entertainments.

Hans, more than any other of the logical positivists, felt that philosophy – or at least the sensible part of philosophy – had utility in all walks of life, and not just in a rational reconstruction of theoretical understanding. Language could be reformed nearer to the precision of symbolic logic. The mess of grammatical cases and tenses could be replaced with clear syntactical forms and a precise representation of time. The theory of probability could clarify notions of evidence and reasonable doubt in jurisprudence. Decisions could be improved with concepts like expected values and variance. The time was not ripe for this sort of thinking prior to World War II. I am certain Hans would have been delighted to see the increasing applications of decision theory and probabilistic analysis in government and industry. But he probably would have been irked by the currently favored subjectivist interpretation of probabilities.

Like most of us, Hans lived in a world filtered through his own conceptual framework. It was difficult for him to absorb an idea until he had translated it into his language. The trait irritated some colleagues who, in the same spirit, felt that some of their ideas were clearer in *their* language. In many ways, though, this was a strength. The translation was a test of the adequacy of his own conceptual apparatus, and it assured that whatever passed the test found a role in his thinking.

At times, the trait took a physical form. Hans, at the time I knew him, was somewhat hard of hearing, and he wore a hearing aid, the kind where the battery was carried in a small separate holder worn under his jacket. He was not above surreptitiously turning down the volume when someone was speaking in an alien universe of discourse. I recall vividly a walk I took with him across campus one day. I was expounding an idea that had come to me the previous evening (I've forgotten the topic). He walked quietly beside me as I described my insight, nodding occasionally as a cue for me to continue. We paused at the point where our ways diverged. I noticed his hand move beneath his jacket, and then he announced, "Dalkey, I was just thinking . . .". And then he described the idea I had been discussing, but formulated in a quite different idiom – one where its consequences could be seen directly.

World War II cast an extended shadow forward. Long before Pearl Harbor, Hans was convinced that the United States would be drawn into the slaughter. He counseled me somberly – philosophy would not be prized in the trenches. He recommended acquiring a second profession, one that would have martial relevance, yet make use of my mind. I finished my thesis on language structures in '42 while struggling to parlay an undergraduate course in calculus into some competence as a meteorologist.

The war derailed my career in philosophy. The next close association with Hans came when I joined the Rand Corporation as a mathematician in 1948. Rand in those days was a remarkable place where philosophers mingled freely with 'hard' scientists. The problem of trading off intangible values with some all-too-tangible facts of international life, was a live issue in a world grown immensely dangerous. Hans was prepared to beat some of his conceptual scalpels into decision tools. He became a consultant to Rand and initiated several studies. One concerned the problem of combining evidence, another the aggregation of values (now a distinct field called multi-dimensional criteria) and a third on decisions under risk. Like much that was being explored in those days, Hans' initial efforts did not convert the Air Force brass to a more rational outlook. But a number of his seminal ideas persisted. I can identify several in my own work on group judgment.

Hans thought that philosophers would probably make lousy kings. But he also thought that the conceptual frameworks devised by philosophers – the good ones, that is – could make as big a difference in the world of affairs as the more visible advances in the physical sciences. In fact, I don't think he separated these two realms in his own thinking. The same frame of mind that allowed Coulomb to perceive the central force law for electrical attraction in the wanderings of a pith ball could enable a clear-headed observer to see the rule of multiplication for joint probabilities in the hunches of a decision-maker. That abstract impression of Hans as a man who felt the unity of things in the intellectual nets we use to capture them is probably my most enduring recollection.

1977 Los Angeles, California

22. UCLA: HERMANN F. SCHOTT

Sometime in 1939 or 1940 I came across an article in *The New Republic*, or it may have been in *The Nation*, commenting upon the fact that Hans Reichenbach would be lecturing at UCLA. I had been puzzled by the claim of some Marxists that certain historical events must necessarily occur in the future. Although I had a B.S. in chemistry and a Ph.D. in biochemistry and had carried out some scientific research, I was unsure of the limits to knowledge obtainable by deduction from a seemingly consistent theory. I read Korzybski and was pursuing his extensive bibliography which took me through Piaget's studies of realist and rationalist errors in children and Carnap's *Logical Syntax* from which I learned the elements of symbolic logic. Reichenbach was on my reading list, so I made arrangements to attend his courses and purchased a copy of *Wahrscheinlichkeitslehre*. I worked my way through the volume, drawing geometrical analogues of the theorems, one of which I found to be in error (the derivation had tacitly assumed two sets to be independent and also mutually exclusive). I derived the correct theorem, went to Reichenbach's office, introduced myself and told him I had found an error in one of his theorems. "That is impossible!" he exclaimed telling me that they had all been carefully checked. However, he looked at my notes, admitted that I was correct, thanked me and later acknowledged my contribution in the introductory pages of his English *Theory of Probability*.

I sat in on most of Reichenbach's courses and seminars after that as well as those of other professors in the Philosophy Department, including Wesley Robson, Abraham Kaplan and Bertrand Russell (who also gave a personally conducted tour through *Principia* for a handful of students with the background and interest). I was impressed by Reichenbach's rigor and clarity. The systematic, step by step presentation and the clear, colorful and often humorous illustrations made his lectures easy to follow and their content easy to remember. In his inductive logic class he used Sherlock Holmes and other detective stories as well as actual murder cases.

I was fortunate in being welcomed into the Philosophy Department circle and even joined the staff temporarily, during a war-caused personnel shortage, to teach some logic courses. Thus I came to know Reichenbach better. I was intrigued by his sense of humor and his love of life. He liked the outdoors and was a first rate figure skater. He once remarked that he sometimes wished he had taken up skating as a profession with physics and philosophy as hobbies. He did combine philosophy with his hobbies; thus he used his skill as an amateur photographer to prepare stereoscopic slide pairs taken with lens separations of up to 200 times the normal interocular distance, giving cities and mountains the perspective of table models. He used such slides to demonstrate that visual perception is not just a matter of retinal images. When he lectured on the effect of the discovery of non-euclidean geometries in discrediting the synthetic *a priori*, he used the ideas of spherical, plane and pseudospherical surfaces as illustrative analogies of the different possible kinds of space. I made him a wooden pseudosphere on my lathe which he then used regularly for these lectures.

He once said to me, "I have something to tell you, but you must keep it confidential." I agreed. "Do you know what confidential means?" he asked. "Of course," I replied, "I must not tell anybody else". "Oh no" he said "You may tell anyone you please, but you must tell them that it is confidential." What he then told me I should like to reveal, confidentially of course, but I can no longer recall it. He had a great store of fruitful sources. Once he assigned me an article by Helmholtz. I was excited and enlightened to be made to realize that physical quantities such as weight, length, resistance and conductivity, can be treated arithmetically just in so far as there are methods of measurement which can be coordinated to the axioms of arithmetic.

In summary, Hans Reichenbach was a clear thinker who could explain well, partly, perhaps, because he liked people and understood their needs. He had a sense of humor and a sense of drama that helped establish him as a favorite lecturer. He had a joy of living and an open manner which made him an excellent companion.

March 21, 1977 Los Angeles, California

23. A BLIND STUDENT RECALLS HANS REICHENBACH: H. G. BURNS

When I showed my Braille copy of his book *Atom and Cosmos* to Dr. Hans Reichenbach in 1941 at UCLA, he was both curious and delighted over the method by which I would study his fine survey of the implications of modern physics for present-day philosophy. Our informal discussion in his office led to the psychological problem of creating for the sense of touch the visual illusion of three dimensions on a two-dimensional surface. Blind from age five, I had little experience with depth perception and no direct knowledge of the techniques used to create illusions of depth on plane surfaces. Dr. Reichenbach seemed as eager to impart this simple bit of information to me as he was to present the fascinations of the history of light theory in his wonderfully organized classroom lectures. Here was a real pleasure of the mind for him: the explanation and illustration of a phenomenon that could not be directly experienced by his blind student. How could he convey the methods for three-dimensionalizing a two-dimensional drawing?

He seemed so pleased that I had taken the special trouble to ask trained Braille copyists to transcribe his book in order to take his class that he took extra time in our one-to-one teaching relationship to show me how a two-dimensional drawing of a square becomes a cube through the use of parallelograms at the edges of the square. It was not his patience and generosity that impressed me as much as the eagerness, excitement and curiosity with which he took on the task of teaching something new.

When later I became an associate professor of psychology at Los Angeles City College in 1949, my informal acquaintance with Hans Reichenbach continued when he visited our campus. Occasionally he picked up his wife, Dr. Maria Reichenbach, after her day's teaching in our Philosophy Department, where she successfully demonstrated the influence of modern science on philosophy.

A few years after Hans Reichenbach's death, I requested a Braille transcription of his book *The Rise of Scientific Philosophy*, using it for reference material in my psychology courses. Maria Reichenbach and I arranged for its recording on disks provided by Recordings for the Blind, a volunteer agency in Hollywood, California. I am pleased and greatly honored by the opportunity to have brought Hans Reichenbach's teachings to numerous other blind students.

There is a bitter-sweet flavor to my memory of the UCLA philosophy course with Dr. Reichenbach. It was designed for the nonspecialist in physics and advanced philosophy, categories in which I fear I hold very high qualifications. It is with a twinge of regret that I recall thirty-six years later that I got an 'A' at the midterm – but a 'B' in the course.

December 18, 1977 Los Angeles, California

24. RECOLLECTIONS OF HANS REICHENBACH: DAVID BRUNSWICK

I have some definite recollections of Professor Hans Reichenbach from the last of the 1930's or beginning of the 1940's at meetings of the Psychoanalytic Study Group of Los Angeles. A Professor of Philosophy at that time at the University of California at Los Angeles, he was well known to our President and leader, Dr. Ernst Simmel, whom we had invited from Berlin. Dr. Simmel invited Professor Reichenbach to the Study Group and got him to give us several lectures.

I remember Hans Reichenbach as a devotee and expert in a realistic scientific philosophy, which was the only type of philosophy which had ever had any interest for me since my days at college, where I heard it in my first course in Psychology with Professor Edwin B. Holt, an early behaviorist psychologist and one of the few realist philosophers of that time.

Reichenbach was a friendly, very pleasant and interested and interesting man, who gave us talks about scientific philosophy and philosophy of science in a very clear and forthright way, always welcoming questions and discussion. This was a welcome addition to the discussions and deliberations of our Psychoanalytic Study Group, contributing a broader context and background for the science of psychoanalysis.

I am grateful to Professor Reichenbach's widow, Mrs. Maria Reichenbach, for reminding me of a part of those pleasant times of the early development of psychoanalytic activity in Los Angeles and of Hans Reichenbach, a fine intelligent human being.

1977 Los Angeles, California

25. UCLA, 1945–1950: CYNTHIA SCHUSTER

The first time I graded tests for Hans Reichenbach, I was ambivalently flattered by his reliance on my judgment. He never checked my work. I wondered whether to interpret his behavior as complete trust in me or as indifference to students of elementary logic. Only after the grades had been recorded in the registrar's office, did he ask to see the grade sheets. Fearing that he might want to plot a distribution curve, I prepared to argue that problem-solving tests should yield an upside-down distribution curve. He was way ahead of me, asking, before he looked at the record: "Did it come out right, with relatively few C's?" Then he stared at the sheets for five minutes, and declared, with a twinkle: "We are measuring something. Very nice correlations: D, C–, C–, D+; A–, A, B+, A; D–, F, F, F; B, B–, B+, B. Yes we are undoubtedly measuring something. *What* we are measuring, we do not know!"

I recall the meeting of Reichenbach's seminar in the Philosophy of Mathematics which took place the week after his article on ethics had appeared in *The Nation* ('Philosophy: Speculation or Science', *Nation*, Jan. 4, 1947, Vol. 164, No. 1, pp. 20–22). Eight or ten graduate students in his large office were wrestling with the analysis of the parallel axiom when the door burst open and Professor Hugh Miller stalked in, slapping his copy of *The Nation*, and nearly shouting: "Hans, you don't mean this! You can't believe this! Look: In 1933 when Hitler dismissed you from the University of Berlin and from your position in the educational radio programs, you *knew* he was wrong and you *knew* you were right. So how can you pretend to believe this stuff about no knowledge in ethics?"

With practiced, perhaps infuriating, calm Reichenbach responded: "What you do not understand, Hugh, is that I do believe what I have said there. I did not like Hitler's aims; he did not like mine. He had power. I got out and fortunately was able to come to a place where most people's desires and aims are close to my own. But I do not see how I could claim to *know* that Hitler was wrong or that I was right. Such claims to knowledge do not stand up to the criteria by which I define 'knowledge'."

Muttering something to the effect that it was hopeless, Professor Miller left in a huff, and Reichenbach herded us back to our muttons with: "Some lines of thought will not meet in one generation, but the lines we are concerned with in the parallel axiom are happily a simpler kind of line."

I once repeated to Hans a conversation I had had with a visiting professor (Hunter Mead) who had enquired about the conclusions entailed by non-cognitivism in ethics. The question was whether it entailed any task for the betterment of the human condition. Confident that I would receive approval, I reported my remarks to the effect that in his philosophical work Reichenbach was exclusively concerned with logic and epistemology, not with social reform. To my surprise, he was angry; "But no! That is not true. The whole movement of scientific philosophy is a *crusade.* Is it not clear that only by ending the dogmatism of irresponsible claims to know moral truth, that only by clarity and integrity in epistemology, people can attain tolerance and get along with one another? Don't be misled by the frequency with which others mention their concern for mankind and the infrequency with which I use such words. I dare say I care as much about the future condition of man as anyone you'll ever meet, and what I'm doing aims as directly at social consequences as the programs of those who call themselves 'social reformers'."

Reichenbach's relations with his colleagues at times left much to be desired. His formal Herr Professor manner was, rightly or wrongly, interpreted by some as arrogance. One episode taught me how stupid top-ranking intellectuals can become when preconceptions guide their perceptions. A department meeting was scheduled to discuss Hans's request that the comprehensive examination in the history of philosophy be made either easier or optional. We graduate students awaited the outcome with baited Ph.D.'s, knowing that Hans proposed to prove by a calculation from the number of minor names in the index of a history of philosophy that at least 300,000 questions could be constructed of the detailed kind customarily asked on that history exam.

After the meeting Hans told me that he believed he had convinced the others. At any rate, he said, no one had even attempted to answer his arguments. Within the next few days two of the other professors took me aside to tell me what they thought of the Great Exponent of Reason and Tolerance. They were seething. What they had to say boiled down to: "He spoke his piece. We listened. We were ready with counter arguments. But when he had finished the so-and-so switched off his hearing aid!" By the second time I heard this, I had guessed what had occurred, and was ready with the question: "What makes you think he switched it *off*, rather than *on*? You know, he always turns it off before making a speech, for his own voice comes in too loud."

Reichenbach's English was extraordinarily elegant, correct, idiomatic, forceful. Those of us who listened to him for years found ourselves correcting our sloppy American enunciation and diction. Once, however, under the stress of suppressed anger, he used a turn of phrase which has become a handy family expression in several households. The occasion was a meeting of the American Philosophical Association; he chaired a session during which some egregiously stupid papers about his epistemology were read. His face became rigid with a pseudo-saintly expression, and for some time he honored equally confused comments from the audience, maintaining a saccharine courtesy. Finally he looked at the floor, began to walk back and forth, and said: "*I shall now make myself a remark.*" . . . Later, when he asked me whether I had learned anything from that session, I answered: "Yes, I learned that when you are really mad you make like a saint," at which he chuckled and answered: "What else could I do?"

It does good things to a dinner-table conversation to open with: "I shall now make myself a remark."

Reichenbach's students used to discuss how it was possible for him to convey the impression of keenness of intellect and wit, and of personal dignity as strongly as he did, when all the details of his bodily equipment suggested the opposite: short, almost rotund, stubby hands and feet, round face, snub nose, thick glasses, false teeth, a hearing aid, and a thin high-pitched speaking voice. At beer busts ingenious theories were proposed concerning mind-over-matter, the power of an ice-skater's muscles, and the effects reading his books had on our perception of his person. The best 'explanation' was an analogy: Just as Socrates spoke with authority on the superiority of the pleasures of the mind over sensuous pleasures *because* his face evinced carnal lust, so Reichenbach spoke with authority against miracles *because* he performed one whenever we looked at him and saw a physique which was not there.

February 10, 1975

Department of Philosophy
University of Montana

26. UCLA, 1949: W. BRUCE TAYLOR

I traveled from Minneapolis to Los Angeles to study with Hans Reichenbach in the fall of 1949. Herbert Feigl had recommended him as the leading exponent of the logical analysis of physical concepts and of scientific philosophy in general. My master's thesis had been on the subject of time so that it was a natural choice for me.

As with his other students, he took a personal interest in me, inquiring as to my intellectual interests and my arrangements at the university. This was especially true for the new students, who arrived from all over the United States and abroad to study with him. For me, he arranged a lunch together where we discussed many of my interests. At this time he also insisted that I not pay tuition, and facilitated this by promising (and later arranging) a teaching assistantship for me.

Another student, from New Zealand, described the 'piper' which was the basis for his earlier degree in philosophy. Hearing this, Reichenbach arranged his hands in the manner of a flute player and asked if he meant this kind of a piper. He was razor sharp on all matters concerning natural languages, as is well-known from his logic text.

Another student, one from Harvard, had a number of opinions about the work of Quine as an analyst of science. It became clear in the discussion that Reichenbach had in mind *de facto* physical science as the starting point of analysis whereas Quine appeared much more inclined to linguistic speculation about science – at least as described by the graduate student. The upshot was the statement by Reichenbach that many logicians simply do not understand science. The student, a good logician himself, resented this statement for some time, but after a semester or two of Reichenbach's in-depth treatment of relativity theory and quantum mechanics, he accepted this verdict also.

The ebb and flow of the interaction with students centered not only in his seminars, but also in his undergraduate classes and in discussions at his home which many of us attended. The seminars were well attended, and by students of logic, mathematics, and physics. At the beginning of the semester, topics were chosen and presented, one per seminar meeting. His comments on our papers were generally to emphasize positive points, and always the important ones. He asked once, why an especially critical survey of logical concepts had placed such a great emphasis on the open-ended, incomplete character of the functional calculus while neglecting to mention the completed, more closed character of the propositional calculus. He also was not afraid to use colorful language to communicate with us, asking about all those 'little holes' that can appear between certain linear sets of elements.

One seminar was especially eventful when it was learned that a janitor in the building had asked to present his ideas on relativity to Professor Reichenbach. He (Reichenbach) explained that Einstein once refused an audience with such a fellow and was threatened. So we sat very quietly and asked polite, simple questions when the janitor explained how matter 'unwinds'. This meant that electrons rotating around a moving nucleus would by themselves travel in helical-like paths, and thus 'unwind' the atom. After talking for a while, the man was thanked, left quietly, and was not heard from again.

It was in his classes that a consistent and very systematic picture of his scientific philosophy was presented. All the natural sciences were represented, often with a good amount of straight exposition of basic principles prior to, or in conjunction with, the logical analysis. The most extensive was probably Einstein's relativity theory, with special attention to its relation to the history of formal geometry. He would often use a black-board eraser as a 'meter stick' to show how special definitions (coordinative definitions) were needed to establish the equivalence of the length of the same meter stick as he carried it from one part of the class room to another.

He would occasionally pick up a heckler in one of his classes when the subject came close to traditional religious beliefs. One student argued at some length that Darwin's concepts of chance variation and natural selection were either logically inconsistent or non-explanatory. This was of course resisted by Reichenbach. And in 1953, shortly after he died, one philosophy student who was also a priest insisted that Reichenbach must have embraced the Church in the final hours of his life. No way.

At his home, intellectual and social stimulation were combined for students and friends, with the evening centering around some recent work of his. It was here that I first encountered, in depth, his work on the logical analysis of nomological relationships. Two interesting personal reactions to this were those of some British students. One of them wrote home to a former professor to say he thought Reichenbach now believed in real metaphysical connections. The other who had an epistemological bent, said he did not believe in any scientific truths at all and hence could not accept nomological statements. Both missed the point probably because they did not understand the method and approach in the work of Reichenbach. That is, the scientific results came first, and then the logical analysis came as the creative philosophical contribution, no more and no less.

Finally, life in professor Reichenbach's circle of students had a strong element of high tradition and excitement about it. He passed on to us his esteem for Einstein and other greats. He arranged for us to attend the lectures of von Neumann at nearby Cal. Tech., and introduced us to him one-at-a-time after the lecture. He discussed his development of the foundations of physics with Teller, and with Feynman, and either arranged for us to attend or told us the results later. The students themselves also provided their contributions to the quality of the 'Reichenbach Circle' and have demonstrated in the subsequent years by their own attainments, in many walks of life, the high quality of students attracted to the man. It was one of the most exciting and rewarding periods of my life.

July, 1976 Chevy Chase, MD.

27. 1950: DONALD A. WELLS

I first met Hans in 1950 when we invited him to Washington State University as a lecturer. We later came to know him and Mia intimately when I went to UCLA on a post-doctoral Ford grant, with the explicit purpose of studying with him. Two aspects of his personality remain vivid.

The first was the seriousness with which he approached the teaching of undergraduates. Unlike many stellar academicians, he seemed to prefer the contact with the beginner in philosophy. He zealously guarded his privilege of teaching introductory courses, several of which I sat in on to observe his teaching style. In order to keep contact at the same time with the graduate students he carried an inordinate teaching load with enthusiasm.

The second recollection concerns his style with advanced classes. I can still picture him in his classes on 'Time' and one where he used *Experience and Prediction*, erect before the class, with an occasional short pace, hearing aid turned off, peering through his thick lenses at some eternal object. He spoke precisely without any notes as if he were reciting. The lectures were systematic and clear. At the end of the lecture, he would turn on his hearing aid, seemingly adjust his vision to objects in the room and entertain questions. Students soon learned that this would not be the high point of the day. A typical dialogue went as follows: *Student*: "I don't understand why a principle of absolute falsification is not tenable." *Hans*: "Have you read the text?" *Student*: "Yes." *Hans*: "But did you read page 88?" *Student*: "No." *Hans*: "Read those pages and your question will be answered." Should the student have said, "Yes, I read those pages," Hans would reply, "Then why are you asking the question."

September 18, 1975

Department of Philosophy
University of Hawaii at Hilo

28. UCLA, 1951–53: RUTH ANNA PUTNAM

I have tried to write something about Hans; I am not terribly happy with it because I think that I have failed to convey how alive he was. What makes remembering so difficult emotionally is just that he comes to life again, and then, of course, the loss also comes back. Still, I think it is important that young philosophers who have only his books get some feeling for the kind of man he was . . . .

In the Fall of 1951, Reichenbach taught, as he did every Fall, a tremendously popular course entitled 'Inductive Logic', a course intended for what we called 'lower division students'. It was, in fact, much more than a course in inductive logic, it was an introduction to scientific philosophy, to the probability theory of meaning, to epistemology, etc. Although it was a popular course, it was very serious and very exciting. Reichenbach made that course;

after his death it rapidly lost students and was soon abandoned. In retrospect, what made the course so valuable to those of us who took it was the fact that he taught us not merely philosophical doctrines but a way of looking at the world. He told once, in that class, of teaching someone to drive. In those days of the two lane highways, the great terror of novice drivers was to have to pass a car by moving into the lane of oncoming traffic. Reichenbach said that his particular pupil claimed that whenever he attempted this a car would be approaching from the opposite direction. Reichenbach said that he encouraged his pupil to keep a record, to prove to himself scientifically, statistically, that it was not so. It was the very triviality of the example, the fact that it made sense to each of us, which drove home the point that one should find out the facts rather than let one's view of the world be distorted by one's fears and hopes. Along the same lines, I remember clearly his discussion of the gambler's fallacy, while we must all act on the best available posit, the person who commits the gambler's fallacy does precisely the opposite.

In the Fall of 1952 Reichenbach taught his celebrated course in the Philosophy of Space and Time. Those of us who had the good fortune to attend this course at one time or another will never forget his 'conducted tour through a non-Euclidean world'. We all had the feeling that we were actually there in that world in which one wandered from one White Horse Tavern to the next. When Reichenbach taught, his whole being was involved in the teaching. He was an excellent teacher, of course; but he managed to do something which went beyond that. This was an undergraduate course and the material covered was not anything on which he was currently working; nevertheless he managed to give us the sense of participating in intellectual discovery. He did that partly by asking the right kinds of questions, questions which would force us to try to solve a problem before he presented the solution. He did it also by telling stories, that is to say by making the situation in which the question would arise vividly real; what would one say, for example, if one were to encounter one's younger self, and what would that self reply?

Thinking back now to these two courses as well as to the course on probability which he gave in the Spring of 1953, it seems to me that he taught us something more than Reichenbach's philosophy, important and exciting as that was, he taught us courage. On the one hand, the intellectual courage to be clear, and on the other hand the courage to accept the fact of having to live with uncertainty. For Reichenbach, human knowledge was the result of cooperation; only if one was sufficiently clear so that one's mistakes could be recognized did one participate in this cooperative effort. And since certainty is unattainable and one must act, there is no point in bewailing the fact that one has to act on the basis of mere probabilities.

I should like now to turn to more personal memories in an effort to show what is to me an equally important aspect of Reichenbach, his extraordinary ability to discover other people's needs and to respond to them. I introduced myself to Reichenbach in the Fall of 1951, and in that brief first conversation he discovered that I was working my way through school and had a hard time making ends meet. Immediately, he offered me the job of typing the manuscript of *The Direction of Time*, which he was then in the process of writing. Although I was, in fact, a very slow typist, I accepted of course. The arrangement was that I would pick up and deliver the work in his office at lunch time. This developed quickly into my sharing his lunch, or rather he brought two lunches instead of one. This kindness helped not only my budgetary problems but it provided me with the marvellous opportunity to ask him questions, either about the lecture I had just heard or about the part of the manuscript I had just typed. His patience and his great pedagogical gifts must have been sorely tried, yet I never had the sense that he found these sessions any less enjoyable than I did. I never saw him impatient. Once I raised an objection to his justification of induction. He did not reply then; but two days later he returned to the subject and led me to see that if one has the best possible justification (he used other examples as well), then one could not complain that one didn't have 'perfection'.

During one of the terms I was enrolled in a very bad physics course. The lectures were boring, the textbook was bad, and, worst of all, the class met at lunch time. I rarely went to class and was doing rather badly. When I told him that I expected a 'C' in physics, he immediately offered to tutor me. But when I confessed that I rarely went to class because it prevented me from eating lunch with him, Reichenbach decided that I did not need tutoring, which was, of course, what I had hoped he would decide. I did not want to waste our lunches on learning optics. Sometime during the first year, he invited me to attend a colloquium restricted to graduate students and faculty. When I came to the room, people were milling about and chatting in small groups and I felt utterly intimidated. Reichenbach noticed me and came across the room to offer me a cigarette. Another time, Reichenbach spoke to the philosophy club about his recent trip to Europe and mentioned that he had visited Russell. Someone asked him whether he had discussed his philosophical differences with Russell. Reichenbach replied, "No, Russell is so old and frail, I didn't want to excite him."

On a lighter note, there was the time when a janitor came to Reichenbach's office to tell him that he, the janitor, had refuted Einstein's theory of relativity. Reichenbach listened patiently to a long explanation of the supposed refutation, and he also discovered that the man had been a janitor at Princeton. Finally, Reichenbach asked whether the janitor had ever presented his refutation to Einstein. The man replied, "Yes, but it is very hard for a man to admit that his life's work has been in vain." When the man left, it was obvious that Reichenbach had made him feel good. And that, perhaps, is the last and most important thing to be said. One always left Reichenbach feeling good, there was never a sense of failure to communicate, never a sense of being 'dismissed', and always the belief that there would be a next time.

[From a letter to Maria Reichenbach: . . . I never observed Hans with anyone except myself or with you. When you were there, then one could feel his love for you. Of course, when he was lecturing, then he was totally involved in that; . . . .]

February, 1976 Wellesley College

29. MEMORIES OF HANS REICHENBACH: FRANK LEROI

I called Hans Reichenbach 'Pep'. Pep was my stepfather. I loved him very much. He died, when I was fourteen.

Pep gave me an unusual set of attitudes with regard to my environment. He was such a competent person in physics, mathematics, and philosophy. His competence also included being highly verbal and being at ease with people. He set these examples before me, and I assumed that such knowledge and competence was normal for all human beings. As a consequence of his example I have been motivated to be interested and conversant with many more fields of knowledge than most people are inclined to get involved in. His example led me to believe that I could become competent in any field I chose.

Pep fostered an attitude in people by setting an example. He was happy when he was thinking about something. 'Thinking about something' means focusing your mind and energy on some task or idea. This happiness or excitement he often imparted to others. In concentrated thought one makes connections and discoveries which give one satisfaction. Pep was happy when he was solving problems. These problems were philosophical as well as day to day. Together we solved a lot of problems, too. We worked on the design and building of hi-fi components, designed and built furniture, and did experimental photography.

As a Junior High School student I wanted to be in the presence of Pep a great deal. He was a man who was not easily accessible to me, because of his concentrated efforts at writing. He had a den at home and when the door was closed he was off limits. What I did do in order to gain his company was walk three miles every Monday from my school to his office at UCLA where he held a weekly graduate seminar. I would arrive in the middle of the seminar at around 4:15, enter without knocking and go directly to a large comfortable chair which was reserved for me; I sat there and watched the graduate students perform for Pep. I did not understand a great deal of what went on, but I enjoyed seeing Pep preside over the graduate students in a fashion which drew the graduate students out. He had a manner which succored communication; it was as if he would provide a crucible in which ideas could be discussed in a relaxed and concerned fashion.

The real reward was the private time I had with my stepfather when we drove home. We would leave the graduate students and secretary of the Philosophy Department on the third floor of Royce Hall and walk down the stairs to the parking lot. We would talk about physics, electronics, or photography from the time that we started down the stairs until we got home a half hour later.

In my conversations with Pep he rarely set the agenda as to what we would talk about. The rare time he would dominate the conversation would be more as a parent than as an intellectual. He did not try to bring any particular ideas to me or put something in front of me that I 'was supposed to know'. What I think happened was that he set an example of interest in a wide variety of ideas. I simply gravitated to certain areas and we would discuss some of them.

As a youngster I wanted to pattern myself on Pep at times. This patterning took the form of being interested in the slide rule, mathematical tables, formulae, and seven-place logarithm tables with proportional parts. I often would pour for hours over these things trying to understand and interrelate them. His example of drawing together more aspects about a particular idea than most people do has been something which I have been able to assimilate into my own character. This has given me a depth and expanse in my own understanding and appreciation of the world around us that I would not have had without knowing Pep.

30. HANS REICHENBACH'S DEFINITIVE INFLUENCE ON ME: ADOLF GRÜNBAUM

In the late 1940's (1946–1950), when I was a graduate student in philosophy and physics at Yale, I lived for some time in a kind of Cooperative somewhat impishly called 'Steamview,' which Robin H. Cohen and Robert S. Cohen ran in Hamden, Connecticut. Being a serendipitous bibliophile, Bob Cohen used to comb book stores that carried out-of-print treasures, whenever he was in New York City. One such store – the Science Library Service located in the Stechert-Hafner Building on East 10th Street – was run by the philosophically literate Sam Orlinick, who sold me a precious copy of Carnap's neo-Husserlian Ph.D. thesis *Der Raum* from the 1922 volume of the *Kantstudien*. On one of his visits to Sam Orlinick, Bob Cohen bought a copy of Hans Reichenbach's *Philosophie der Raum-Zeit-Lehre* and showed it to me.

Its effect was uniquely electrifying: it made me feel that even if – *per impossibile* – I could achieve the vocal feats of a Beniamino Gigli or a Jussi Bjoerling, I would prefer to devote my life to struggling with the kinds of questions which Reichenbach treated so magisterially in that book. It was my good fortune that Carl Hempel, who had done his own Ph.D. thesis in Berlin on probability under Reichenbach's direction, was on the Yale faculty at the time. Hempel had had a course from Reichenbach in Berlin which was based on *Philosophie der Raum-Zeit-Lehre*. And Hempel was most encouraging to me when I came to him with the proposal to write my Ph.D. thesis on Zeno's paradoxes, once I outlined just how I thought Reichenbach's ideas on the physics of temporal order could be adapted to supply the *nervus probandi* which, in my view, had been lacking in Bertrand Russell's refutation of Zeno's paradoxes of motion. Since the *Philosophie der Raum-Zeit-Lehre* was not available in English translation in the late 1940's – the English translation by Maria Reichenbach and John Freund first appeared a decade later (1958) – I impetuously asked Paul Weiss to devote an entire issue of his *Review of Metaphysics* to a digest of the book's major ideas, which I proposed to write in English but never actually did.

The decisive manner in which Reichenbach shaped my subsequent work via the impact on me of his writings on space and time is detailed in Wesley Salmon's masterful comprehensive survey of Reichenbach's far-ranging major contributions: [published in the 1977 Reichenbach memorial issues of *Synthese* and in Maria Reichenbach's German translation as introduction to Reichenbach's *Gesammelte Werke*, published by Vieweg in Braunschweig]. Unfortunately, I never had the opportunity to study under Reichenbach in person.

But with Hempel's encouragement, I mustered the courage to write to Reichenbach at UCLA regarding my utilization of his ideas in my doctoral dissertation on Zeno's paradoxes. And I was aglow with joy – and indeed on the brink of getting a swelled head – when I saw Reichenbach's book on *The Direction of Time* and found that his two references to the literature on Zeno's paradoxes (p. 6, fn.1) consisted of Bertrand Russell's *Our Knowledge of the External World* and an excerpt from my Ph.D. thesis, published in the 1950 volume (Vol. 4) of *The Review of Metaphysics*.

As a result of my correspondence with him, Reichenbach generously and graciously invited me several times to join him and his wife Maria for lunch or other informal discussion when they came to the eastern United States for a professional meeting. One such occasion is especially memorable: It was at a Bryn Mawr meeting of the Eastern Division of the American Philosophical Association, immediately preceding a packed plenary session at which Reichenbach, C. I. Lewis and Nelson Goodman debated the issue of the incorrigibility of 'protocol' observations – C. I. Lewis' '*thin* given' – in the foundations of empirical knowledge. Reichenbach introduced me to Professor C. I. Lewis and asked me to join him and Lewis for lunch, which I did despite my trepidation at the thought of finding myself alone at the table in the presence of these two giants. Nothing was said about the impending debate between them at the 2 p.m. plenary session, but Reichenbach related with much feeling the incomprehending hostility which he had encountered from the German academic establishment in philosophy, because of his championship of the relevance of the natural sciences to certain areas of philosophic concern. It was clear to me that even the passage of time and his evident success had not erased the scars that had been left in him by these traumatic experiences.

It was then a major trauma for me, when a colleague of mine at Lehigh University suddenly confronted me with Reichenbach's obituary in the *New York Times*. But we can all be grateful that the labors of Maria Reichenbach, Wesley Salmon, Robert Cohen and others are continuing to make his monumental philosophic legacy accessible to the intellectual world.

31. AT THE CHAPEL, 1953: ABRAHAM KAPLAN

We are all of us here as students of Hans Reichenbach – whether we only read his books, or sat in his classes, served as his colleagues, shared his friendship, or received his love. Whoever knew him, in whatever capacity, learned from him. He was a man of ideas, to whom ideas were all important, and he made their importance felt by all around him.

And this he could do because he was occupied with questions fundamental to philosophy: the nature of our knowledge and of the physical world. These questions he attacked with the full force of his intellectual powers, treating them with all the subtlety their complexity demanded. But he remained humanly anxious to make them intelligible to others. Among specialists he was himself an expert; but like few experts, he reached beyond specialization. We have all marvelled at his skill in giving profound ideas a habitation in simple speech and reducing remote scientific abstractions to the warmth and intimacy of shared experience.

For what *he* knew, he wanted others to know; the truth he believed in he wanted others to believe. For him, philosophical truth was not a prerogative of the specialist, but the basis of a shared understanding of a common world. In an age of science, he wanted men to look to the logic of science for a guide to belief and action. Of that logic he knew much, and discovered much, and so he had much to teach, and in his effort he gave his energies unsparingly.

To the success of that effort we who have known him can attest. In so doing we only give voice to the unspoken testimony of the many philosophical inquirers throughout the world who have felt the impact of his ideas.

To us, his friends, his teaching did not lie merely in the content of his work. That content he taught knowingly; he taught more unknowingly, simply by being the man be was. To us he was above all a lover of wisdom, in the only sense in which he felt it was in man's power to be wise: to understand with clarity and affirm with vigor the beliefs for which experience provides evidence.

For this wisdom, and for the hard work which its acquisition entails, he had unlimited passion and energy. His zest for the life of ideas was a vital part of his personality. It was expressed in an amazing productivity. It was communicated even in casual discussion. It became a part of all whom he could reach.

This enthusiasm of his was that of the genuine scientist. It was never enough for him simply to *have* an idea. With the true scientific temper, he had to work it out in its often bewildering detail, push it to its remote logical consequences, give it concrete application to the materials of experience.

In this endeavour he often faced resistance and criticism. But his integrity was unshakable. He never abandoned an idea merely because others did not feel its force, never accepted a belief because it was popular.

The great achievement of his life's work, which he left unfinished at the peak of his creativity, has not now been brought to a close, nor even his own share in it. His ideas have become a permanent part of contemporary philosophy

But it is not only his specific ideas which his students will continue to develop. His life and his teaching transcended those ideas. He did not wish an intellectual loyalty bound to his person, but rather to the philosophical ideals to which he was himself bound. This loyalty his death cannot affect, save to deepen it.

The loss is immeasurable – to his university, his profession and, more than we can say, to those who loved him. But, overwhelmed as we are by that loss, it is nevertheless fitting that we fix our thoughts also on what has not and cannot be lost: what he did and what he was. It is this which will remain for all of us the enduring presence of Hans Reichenbach.

32. HANS REICHENBACH, A MEMOIR: WESLEY C. SALMON

It is difficult, in writing a memoir of a man who has influenced one profoundly, to avoid writing an autobiography. I shall try to resist that temptation, for Hans Reichenbach is the subject of our attention. To whatever extent I find it necessary to write about my own experiences, the intent is to provide information about him. Modern physics has shown that the basic way of learning about an object is to perform scattering experiments – observing the ways in which other objects bounce off of the object in question. So it is in interpersonal relations; much can be learned about a person by seeing how others who come close to him are influenced.

Like many a philosopher, I found my way into the field via theology. During adolescence I was deeply involved in church and religion, but began shedding supernatural beliefs at a fairly rapid rate after entering college. Nevertheless, I persisted for a while in professional training for the ministry. Even though I retained none of the traditional dogmas, I still felt that the church could be an effective instrument for social and political reform. After a few months as a week-end pastor of a small church in a small town, I concluded that I did not have the psychological make-up to be a reformer, and soon gave up all thoughts of remaining in the ministry. In the course of theological studies at the University of Chicago, I had elected courses in metaphysics and philosophy of religion under Charles Hartshorne, and as a result of his enthusiastic instruction, I became imbued with Whitehead's philosophy. Here, in sharp contrast with most other courses in the theological curriculum, was an intellectual stimulus. Following the path of least resistance, I transferred from the Divinity School to the Philosophy Department and worked for a master's degree with a thesis on Whitehead. To my great regret I never had a course with Carnap, who was clearly the most gifted and productive member of the department. I suspect that my advisors did not wish even to acknowledge his existence.

Having been informed by the graduate advisor at Chicago that I had little talent for philosophy, I nevertheless resolved to pursue work toward the Ph.D. in that field. Chicago was out of the question, so I decided, on purely geographical grounds, to continue my education in Southern California. Since I was working my way, and tuition at the University of Southern California was beyond my means, I settled upon UCLA by a process of elimination. I knew nothing of its Philosophy Department, and when I got a brief run-down on it from teachers at Chicago, Reichenbach was mentioned in passing as 'a positivist' (a characterization which was, of course, mistaken). I had learned enough about philosophy at Chicago to know that positivism was superficial and evil. So I went to UCLA in 1947 a convinced Whiteheadian metaphysician; studying with Reichenbach was the farthest thing from my mind.

I am elaborating some of these details to illustrate concretely an aspect of philosophy in America in the late forties and early fifties. Such philosophers as Bergmann, Carnap, Feigl, Hempel, and Reichenbach (in alphabetical order) had come to America as refugees. All had been associated with either the Vienna Circle of logical positivists or the Berlin Group of logical empiricists. They shared a deep interest in philosophy of science and a common opposition to traditional metaphysics.

These logical positivists/empiricists were, for the most part, intellectually isolated, and they were regarded with suspicion and scorn. This kind of hostility was understandable, however, since they proclaimed themselves the opponents of traditional metaphysical and theological systems, and they espoused shocking (i.e., non-cognitivist) views of ethics. They were obviously brilliant minds, and they wielded such fearsome weapons as Occam's razor (the verifiability theory of meaning) and symbolic logic. Although *Principia Mathematica* had been around for several decades, symbolic logic was not a part of the standard philosophy curriculum, and it was widely regarded with distrust by philosophers of a more traditional bent. In the days before Quine's *Methods of Logic*, and Copi's *Introduction to Logic* even truth tables were considered pretty esoteric stuff.

Logical positivists/empiricists did not always make the moves that would reassure their colleagues. Ayer's *Language, Truth and Logic*, which was for a number of years the principal introduction in English to this philosophical approach, was a manifesto, as were other shorter pieces by various spokesmen for the group. So also was Reichenbach's Presidential Address to the Pacific Division of the American Philosophical Association in 1947 – consider the title, 'Rationalism and Empiricism: An Inquiry into the Roots of Philosophical

Error'. *The Rise of Scientific Philosophy* was no more conciliatory in tone. He expressed the view that philosophical truth demands logical analysis, while philosophical error demands psychoanalysis, thus making it quite clear that he considered differing philosophical approaches to be emotionally generated masses of confusion. In print and in person Reichenbach did not conceal his contempt for what he regarded as philosophical clap-trap.

The predictable reaction was to return contempt for contempt; moral condemnation was often substituted for philosophical argument. The phenomenon is not new. Hume, who had relegated most of traditional philosophy and theology to the category of 'sophistry and illusion', was the object of widespread abuse by his contemporaries. Similar attitudes had, I suppose, 'protected' me from Carnap at Chicago; later, I experienced them in Pullman, Washington. After receiving my degree and teaching for the subsequent year at UCLA, I went to Washington State College for my first position 'away from home'. I found that I had already been judged morally defective by some of my future colleagues (in other departments) before ever arriving on the scene, simply on the basis of my relationship to Reichenbach. I do not mean to say that American philosophers universally failed to appreciate logical positivism/empiricism. Certainly there were a number of people like Charles Morris, Ernest Nagel, and W. V. Quine, who understood the approach and made significant contributions, but they were in a small minority, and often had limited influence within their own institutions.

My academic career, up to my initial contact with Reichenbach, was a model of what rational decision-making should not be. I do not mean to discount the value of wrestling with Whitehead's metaphysical system; perhaps it is necessary to struggle through a commitment to a system like Kant's or Hegel's or Whitehead's if one is to appreciate the philosophical contributions of the logical positivists/empiricists. Whitehead, like Kant, was a deep and subtle philosopher who was thoroughly knowledgeable mathematically and scientifically.

During my first semester at UCLA I continued fairly intensive study of Whitehead's monumental *Process and Reality*. At the same time, I was taking Reichenbach's course, Philosophy of Nature, based upon his book, *Atom and Cosmos*. This course was not what would now be considered a standard introduction to philosophy of science. It provided, instead, a sort of scientific *Weltanschauung* – including a non-technical presentation of the basic ideas of relativity and quantum theory – along with an introduction to a few of the most basic issues in the philosophy of physics. As nearly as I can recall, it included the distinction between the context of discovery and the context of justification, the verifiability theory of meaning, the distinction between descriptive and inductive simplicity, the theory of equivalent descriptions, and the concept of the coordinative definition. This course was exciting and captivating; I must have realized before long that I had found what I wanted and needed – a philosophical approach and orientation that I could wholeheartedly embrace.

There was, to be sure, a period of intellectual turmoil in which I had come to grips with the opposition between Whitehead and Reichenbach; to the best of my recollection my intellectual crisis came during Christmas vacation that first year, when the Pacific Division meetings were held in Los Angeles and I heard Reichenbach's Presidential Address. I asked myself, as well as several fellow graduate students and teachers sympathetic to Whitehead, how one could hope to establish the truth or falsity of Whitehead's metaphysical claims. I still believe that this was the correct question to ask. All attempts to find a satisfactory answer failed. I abandoned Whiteheadian metaphysics to become a logical empiricist, a philosophical approach I have not subsequently been tempted to abandon.

It was sheer luck that led me to study with Reichenbach, but a better stroke of luck could not have befallen me. I am quite certain that, for intellectual or psychological reasons, I could never have remained satisfied with a philosophical approach radically different from logical empiricism or logical positivism, but I have often wondered whether a different stroke of luck at Chicago – getting into one of Carnap's courses by accident – might have made me into a highly confirmed Carnapian. As things turned out, on the major point of disagreement between Carnap and Reichenbach (induction and probability), it seemed to me that Carnap's view was untenable for clearly specifiable reasons. I wonder whether, under different circumstances, I might have drawn the opposite conclusion.

The net result was that Reichenbach became my philosophical hero; I was his disciple. At my final doctoral oral one member of my committee asked me what shortcomings I found in Reichenbach's philosophy. I had difficulty coming up with any kind of answer; I said something, but I do not remember what it was. I do not remember watching Reichenbach's reaction, but I do not suppose he was altogether displeased to observe my difficulty.

Throughout the time that I was a graduate student, Reichenbach was extremely kind, considerate, and understanding. For example, one of my preliminary examinations was in logic, and everyone knew that he would compose the questions. Before the examination he asked what issues interested me. I answered that the topic I found most interesting in logic was the problem of consistency. When the day came to write the three-hour examination, I saw that it consisted of one question: Discuss consistency. Another example relates to my lack of background in mathematics and physics; at that time I had taken one elementary course in physics as a freshman, and no college-level mathematics at all. He helped me to find topics in 'scientific philosophy', including a dissertation topic, which did not require more technical background than I could pick up in his courses.

When I received my doctorate, the Reichenbachs gave a party for me. Maria served an elegant dinner to a sizable number of guests, and each place card at the table contained a rhyme composed by Hans. He was a man of enormous charm and wit.

Reflecting upon my graduate student days, I must acknowledge two main factors. First, studying with Reichenbach was studying Reichenbach. In the various courses and seminars I had with him, we studied his books and papers almost exclusively. I can recall only a couple of exceptions. Shortly after Russell's *Human Knowledge* was published, we devoted a good deal of attention to that book, but we concentrated heavily upon those sections in which Russell discussed probability and induction, carefully trying to straighten out Russell's errors in his critique of Reichenbach's views. Another book we studied was von Wright's on *The Logical Problem of Induction*, mostly for the historical material leading up to the pragmatic approach of Peirce and Reichenbach. I am quite certain I was never required to read anything by Carnap or Hempel or Popper on probability, induction, or confirmation. Reichenbach told us what was wrong with their views; it remained for me to study all such material on my own after receiving the doctorate.

Reichenbach had the conviction that he was right. I recall one particular meeting of his seminar in which one of the students commented, in a rather light vein, on the fact that the rejection of 'the quest for certainty' played such a large role in Reichenbach's philosophy, and yet he seemed so certain that he was right. "Yes, yes," he said in a quietly jovial way, "we have meta-level certainty about object-level probability."

Second, however, I must add that studying Reichenbach, even if it was done in a rather parochial way, was a deeply rewarding experience. His philosophy was rich, comprehensive, and profound. His arguments were clear and incisive. He had what I still consider quite an incredible ability to cut through irrelevancies and unimportant details in order to go to the very root of any issue. His philosophy was not complete, not perfect, not faultless. Much work has been devoted to refining, extending, and correcting it, and much remains to be done. It has, nevertheless, been an intellectual gold mine.

My only real regret is in not having the technical knowledge and the intellectual power to appreciate it more fully and contribute to it more constructively. But, when you stand on the shoulders of giants you are lucky if you can see the ground. Reichenbach showed clearly the philosophical ground on which he stood, and it was solid.

After leaving UCLA in 1951, I taught at Washington State for two years. During that time I sent him papers that I wrote, and he sent me comments and criticisms. We also had a few opportunities for conversation. One of the topics that most interested me at the time was the use of Bayes's theorem as a schema for the confirmation of scientific hypotheses. Reichenbach had proposed this approach in his *Wahrscheinlichkeitslehre* (1935) and reiterated it in the English translation, *The Theory of Probability* (1949), but his presentation had been puzzling and confusing to many readers. He gave me suggestions which made it possible, I believe, to present the matter with much greater clarity. It is worth noting that all of this preceded the 1954 publication of L. J. Savage's landmark work in the 'Bayesian' approach, *The Foundations of Statistics*. And if Reichenbach was right, there is no need to resort to 'personal (i.e., subjective) probabilities'.

In the spring of 1953, Reichenbach died; I was shocked, dismayed, and grief-stricken. As it happened, I had been planning to spend the next academic year, as well as both adjacent summers, in Los Angeles. I offered his widow, Maria, as much help as I could in going through his papers and arranging for publication of various works. The main item was his unfinished manuscript, *The Direction of Time*, which was published posthumously (1956). Even though he did not live to complete it, the book contains rich resources which, in my opinion, are still not adequately appreciated.

There is no way in which I can adequately express my gratitude to him. I owe my philosophical career to him. He was a friend to me, but I was too much in awe to be a friend to him. The training and inspiration he provided were not perfect, but they were excellent, and I doubt that I could have found anything better with any other teacher. The intellectual legacy he left still pays rich dividends. Not having been a perfect student (or any approximation thereto), I can hardly complain of lack of perfection in my teacher.

If you have a hero, there is a temptation – perhaps bordering upon compulsion – to defend him against attacks, especially ill-founded and gratuitous ones. This impulse should be resisted in most cases, for it is probably better to let the insult be forgotten than to dignify it with a response. I shall, nevertheless, allow sentiment to override judgment. Shortly after his death, Reichenbach was subject to just such an attack by the novelist, Arthur

Koestler. In *The Invisible Writing*, the second volume of his autobiography *Arrow in the Blue*, Koestler wrote:

> In 1952, I met in Princeton an old friend, the late Hans Reichenbach, a leading mathematical logician and Professor of Philosophy at the University of California. I had not seen him for nearly twenty years. He had aged and become partly deaf; instead of a modern hearing aid, he used an old-fashioned ear-trumpet. He asked me what I had been interested in lately, and I told him that I had become interested in Rhine's work on extra-sensory perception. He said that it was all hokum, and I said that I did not think so – at least the statistical evaluation of the experiments seemed to show relevant results (meaning that they seemed to confirm the existence of telepathy and kindred phenomena). Reichenbach smiled and asked: "Who has checked the statistics?" I said: "R. A. Fisher in person." (Fisher is one of the leading contemporary experts in probability calculus.) Reichenbach adjusted his trumpet. "Who did you say?" I yelled into the trumpet: "Fisher. *The* Fisher." At that moment an extraordinary change took place in Reichenbach's face. He went pale, dropped his trumpet and said: "If that is true, it is terrible, terrible. It would mean that I would have to scrap everything and start from the beginning." In other words, if extra-sensory perception exists, the whole edifice of materialist philosophy crumbles. And for a professional philosopher that means the crumbling of his life's work (p. 294).

That this tasteless caricature is without any foundation in fact would not constitute reason for bringing it up again, after more than twenty years. Fictionalizing is the business of a novelist, and one could hardly expect him to desist even when dealing with such supposedly factual material as his own autobiography. But the story has a sequel.

Before I knew of this passage in Koestler's book, I had more than once heard the assertion by proponents of parapsychology that R. A. Fisher had pronounced Rhine's statistics significant. This seemed, in the minds of many, to place the seal of scientific acceptability on Rhine's theories. While I was teaching at Brown University, Fisher lectured there, and I had the opportunity to ask him about this claim. He confirmed the fact that Rhine had sent him the data, asking whether they were significant, and he further acknowledged that he had replied that the statistics were, indeed, significant. "But," he added, "*all* statistics signify *something*; Rhine did not ask *what* they signify, and I didn't tell him." (This is, if not absolutely verbatim, certainly a close paraphrase of Fisher's remark.)

The Hans Reichenbach I knew at UCLA from 1947 to 1953 was hard-of-hearing, but he used the most technically advanced hearing-aid available then. As a child, Reichenbach had dreamed of becoming an engineer; when he embarked upon his higher education, he went into engineering in the first instance; in the interim between conferral of his degrees (1915–16) and his university appointment in Berlin (1926) he worked in the radio industry and taught technical subjects. He had a life-long interest in and admiration for technological developments – and certainly no reluctance to use them. He carried the unit which contained the receiver and amplifier in the breast pocket of his coat. It was plain to see that he used a hearing aid, and it was obvious when he adjusted the volume. He was sometimes accused of turning the volume up when he spoke and turning it off when others spoke to him. The opposite was, of course, the case. Because of the location of the receiver on his chest, his own voice would have come in at an uncomfortably high volume had he not turned it down when he spoke.

To caricature Reichenbach as a fuddy-duddy old professor with an old-fashioned ear-trumpet is a gross but superficial distortion. To characterize him as a materialist involves a fundamental philosophical blunder. From his earliest student days until the time of his death, Reichenbach rejected all traditional systems of metaphysics, and he recognized materialism as one of the oldest and most persistent. Koestler refers to the 'edifice' of materialist philosophy; Reichenbach often reiterated his view that the traditional system-(edifice)-building approach of the metaphysicians had to be replaced with the approach of logical analysis. It is inconceivable that Reichenbach could have reacted with horror at the prospect of the collapse of *his* metaphysical edifice.

I do not recall ever hearing Reichenbach discuss extrasensory perception, telepathy, or any other parapsychological phenomenon. He had, however, an open mind with regard to the results of modern science. He took considerable enjoyment from the instances in which traditional philosophy was undermined by findings of twentieth-century science. The examples to which he directed most of his attention were, to be sure, taken from physics – especially relativity theory and quantum mechanics. But in no manner was he committed to the view that the whole of science is physics, or that physics is the only genuine science. In fact, he had a serious interest in psychoanalytic theory.

This was brought vividly to my attention during my last year as a graduate student when, because of stresses I was undergoing, I began psychoanalysis. After I had been in treatment for only a short time, I received an offer of a position at the University of Texas. At that time the employment situation was dismal, and I was fortunate to have any job prospects. After careful consideration, however, I decided to decline the offer because there was no analyst within 200 miles of Austin. It was a difficult decision, and I dreaded telling Reichenbach, for fear he would regard me as foolish and ungrateful in declining a position that had been offered completely as a result of his influence and reputation. Nevertheless, I felt I simply could not accept it.

I do not recall the reasons now, but I went to his home rather than his office to discuss it with him. He was genuinely sympathetic; we talked at some length, discussing my personal problems and various philosophical problems associated with psychoanalytic theory. He voiced the opinion, which is quite sound I believe, that anyone who wanted to do fundamental and extensive work in philosophical foundations of psychoanalytic theory would have to undergo analysis. I was strongly inclined to work in this area, and for some time studied psychoanalytic texts, but in the end found the problems of probability and induction more engrossing. I knew that I had Reichenbach's enthusiastic support had I chosen to follow through on the project.

From Reichenbach I learned the crucial role of Bayes's theorem in the structure of scientific reasoning – in the confirmation of scientific hypotheses. To apply Bayes's theorem we must have at least rough estimates of the antecedent (prior) probabilities. As already mentioned, I don't think I ever heard Reichenbach say anything about parapsychology, but I can well imagine the assignment of prior probabilities he would have made. Reichenbach was open-minded about scientific results, but he was logically critical as well. There could be no serious doubt that much of physics was good science, it remained an open question whether psychoanalysis could measure up in whole or in part, and I suspect it would have taken stronger evidence than was available to convince him that parapsychologists were close to any well-founded results. I fear that R. A. Fisher's little joke had unfortunate consequences with respect to the assignment of prior (and consequently, posterior) probabilities to parapsychological theories by quite a large number of people.

As I said, Reichenbach was, and is, my hero. I have made an attempt to defend him against one malicious attack, not because he needs defense, but because the occasion exhibits the sort of intellectual mischief against which he strove throughout his life. This is not to say that parapsychology is *a priori* absurd or false; it is rather that the acceptance or rejection of such theories must be based upon scientific evidence, objectively evaluated, not on the basis of wishful thinking. Whatever the final outcome, it offers no threat to Reichenbach's empirical epistemology which, by refusing to make transcendental assumptions about the nature of 'ultimate reality', can accommodate whatever conceptions of reality the experiential evidence demands. The philosopher who explicitly formulated and articulated the *principle of anomaly* in quantum mechanics could hardly be devastated by the results of experimental evidence in any area of honest scientific investigation.

University of Arizona

33. MEMORIES OF MY HUSBAND HANS REICHENBACH: MARIA REICHENBACH

I met Hans Reichenbach in 1934 in Istanbul, Turkey, where he was professor of philosophy at the university. When Hitler came to power in 1933, all persons deemed Jewish by the Nazis, whether full, half, or a quarter Jewish, lost their positions at the universities. Hans was half-Jewish and left the University of Berlin where he had been professor of philosophy of physics. At the hands of Mustafa Kemal Pasha, or Ataturk, the Father of the Turks, as he used to be called, Turkey experienced quite an intellectual revolution. He saw his great opportunity to build up the university of Istanbul with the help of about thirty-five refugee professors in all major fields: the physical sciences, the life sciences, the social sciences, the humanities, law, and medicine. Several of these scholars were 'Aryans' and came of their own free will, for political and moral reasons.

This foreign faculty was outstanding. Its members were given five year contracts and ample financial support to build up whatever institutes were needed. They were permitted to bring their own assistants and each was given an interpreter for every lecture, since hardly any professor spoke Turkish. Hans who was very gifted for languages lectured either in German, French, or English, depending on the language his interpreter spoke best. But he soon learned enough Turkish to get around, to do the required bargaining when shopping, and to use it later in his analysis of conversational language [see 1947c].

The first year every family of this new faculty invited every other family. It helped to establish some stability in this uprooted community. There were three major locations where they lived: either in Pera, the European part of Istanbul, or in Bebek, on the Bosphorus in the direction of the Black Sea, or in Kadiköy, on the Asiatic side. Commuting was done mainly by little steamers that criss-crossed the Bosphorus and the Marmora Sea and landed at the Galata Bridge, at the entrance to the Golden Horn. From there it was a short walk to the university that lay in the Turkish part of the city. Istanbul was beautiful, wherever one lived, and everybody boasted of the view he had from his windows. Hans lived with his wife Elisabeth and his children Hans and Jutta [Elisabeth] in Kadiköy, where I visited them often and went for walks with them.

After Mustafa Kemal Pasha's death the political climate in Turkey changed. It became more nationalistic, and most of the German and Austrian professors tried to avoid signing a new five year contract by securing positions elsewhere, mainly in the United States. Hans contacted his friends in America who were most generous in their assistance. He received an offer from the University of California at Los Angeles in January 1938 which he accepted. He left Istanbul at the end of the academic year of 1938 and, after a sojourn in Switzerland, where he did a great deal of hiking and climbing, arrived in Los Angeles in the fall with a serious coronary thrombosis. For six weeks his colleagues in the philosophy department at the University of California at Los Angeles gave the lectures he had been scheduled to teach. He recovered and began his assignment.

We met again in 1939 in Santa Monica and saw each other frequently. He taught me to drive a car and was a patient teacher. Although I had a Ph.D. from Germany I decided to go back to the university for one year in order to acquire my teaching credentials. I had married in Istanbul and had a young son who was by then old enough to go to nursery school. I took all the required education courses, United States Government and some psychology classes; for all my electives I chose Hans' classes and seminars.

In 1940 Bertrand Russell was visiting Flint Professor at UCLA and shared Hans' office. They became good friends and Hans took excellent photographs of Russell. He also tried to use photography for the purpose of convincing Russell to give up his two-dimensional, phenomenalist description of the world and of converting him to a realist and physicalist description. Hans took stereo-photographs of various things and then let Russell look at them through a viewer, waiting expectantly for the 'aha moment', when the two pictures would merge into a three-dimensional one and Russell's facial expression would indicate this fact. In spite of these attempts, I think Hans was not very successful in changing Russell's mind.

In 1946 Hans asked me to marry him. He said: "If you marry me, I will write a best seller for you." We both obtained divorces and married one day after mine. My reason was not the promise of the best seller; yet he wrote one – *The Rise of Scientific Philosophy* [1951a] – which grew out of his Presidential Address at the meeting of the Pacific Division of the American Philosophical Association in Los Angeles, 1947 [1948a]. It became a *Book Find Club* selection, in the meantime has been translated into ten other languages, and is still widely read.

In the spring of 1946 the UCLA Committee on Election of Faculty Research Lecturer made the following report to the Academic Senate: "The Committee wishes to report that, after a careful canvass of the membership of the Academic Senate, Southern Section, for the person best qualified by proved ability in research, its unanimous choice for Faculty Research Lecturer for 1945–1946 is Professor Hans Reichenbach of the Department of Philosophy." Hans delivered the lecture under the title 'Philosophy and Physics" [1948b] on March 25, 1946.

During the academic year 1946/47 Hans had a sabbatical leave. We lived on the ranch of friends in the San Fernando Valley. Hans worked on several articles and his book *Elements of Symbolic Logic* [1947c]. Mrs. Frances Deri, the psychoanalyst, had a weekend house on the same ranch. We became good friends with her and shared her friendship with her psychoanalytic colleagues, many of whom came out to visit her. Eventually Hans became an honorary member of the Psychoanalytic Institute and was invited on several occasions to lecture on the philosophy of psychology and related topics. Hans found his association with this group very stimulating.

For the spring semester 1947 Hans was invited by Professor Irwin Edman as a visiting professor at Columbia University; he also received invitations from the City College of New York and The New School for Social Research. He gave one course at each school and lectured on probability and epistemology.

In New York the only abode we were able to find was a tiny two-bedroom apartment in which no drawer or closet had been emptied. We rented it from a Greek couple. The lady's name was Xantippe. We contacted her and asked her to give us a little bit more room, but to no avail. We managed somehow and enjoyed our stay. Hans relished the little special food stores in our neighborhood in upper Manhattan. He was a cheese lover and connoisseur, and made many gourmet discoveries. We explored New York and its museums, had amiable social contacts with members of the philosophy faculties of the various schools, among them Ernest Nagel, Daniel Bronstein, Y. H. Krikorian, Philip Wiener, and Sidney Hook. We also renewed old friendships with fellow refugees from Germany. We visited Hans' brother Hermann who was professor of physics and music in Hartford, Connecticut. He had recently become the father of a son. I can still see Hans walking up and down the street with the baby in his arms, smiling benignly upon the boy. After the end of the semester, we were at last offered a large apartment on Morningside Drive, near Columbia University. We rented it and used it to give a big party for everybody we knew in New York. Hans was always a charming host and knew how to make young and old feel comfortable in his presence.

We decided to drive back to Los Angeles and finally found a car, which was difficult so soon after the war. We took our time and went via the Niagara Falls and Canada, with stops in Ann Arbor, Ames, Iowa, and Denver, to visit colleagues and friends, and went sight-seeing in Bryce Canyon, Zion Park, Grand Canyon, and Boulder Dam on the way. It was a leisurely seventeen days trip back to the ranch near San Fernando. Very soon after our return Hans entered into negotiations with the owner of a great deal of land in the Pacific Palisades. This suburb is located in the most western part of Los Angeles, about three hundred feet above the Pacific Ocean. We chose a lot with a magnificent view of the Santa Monica mountains and the ocean, decided on a plan, and began to build a house. We drove the roundtrip of fifty miles between San Fernando and Pacific Palisades every day in order to supervise the building crews who were mostly very inept. They were absolutely dumbfounded when, on the occasion of staking out the foundations, Hans figured out the length of the hypotenuse in a right triangle by means of his slide rule, i.e., without measuring its length. We noticed soon that blueprints did not mean very much to the carpenters, that right angles were a matter of luck, and that we had to remeasure all door openings in comparison to the specifications in order to see whether Hans' large desk and other pieces of furniture would go through. Hans was enormously practical. He made a drawing of every wall of every room in the house and measured the intervals between all wooden studs and the widths of all studs. We have often profited from his records when we wanted to drive a nail into a wall for hanging pictures or building shelves. The house was more or less ready by Christmas, 1947, and we moved in. As President of the American Philosophical Association that year Hans invited about thirty people for one of the evenings. We had no electricity yet, but had laid some cords to a neighbor's house and lit some candles which served us well when somebody stepped on the cords and interrupted the circuit. It was a very hilarious party.

In 1948 I was offered a position in the German Department at UCLA which I accepted with alacrity. But it lasted only one year, for the administration decided by then to enforce the so-called nepotism rule and I had to leave. In the fall of 1949 I started teaching philosophy at Los Angeles City College and Hans really launched me into this new career. Even though he himself never used any notes in the classes he taught and always spoke freely, fluently and, as it were, ready for print, he helped me in the beginning to get my lectures down on paper in outline, inventing most ingenious examples, suggesting some humorous stories, and investing the whole enterprise with so much drama and style that I felt very secure and enthusiastic and soon was able to sail under my own power.

Many people marveled at Hans's gift of speech. Carnap once told me a little anecdote corroborating the way Hans spoke publicly. They were returning together by train to Berlin from a congress in Königsberg [see 1930f], in animated conversation. Shortly before arrival Hans asked for a ten minute pause; after that he continued their talk. With rare awe Carnap said that Hans went directly from the train to the radio tower in Berlin to deliver, without a manuscript, one of his weekly lectures in which, according to Carnap, he explained modern physics and its philosophical problems in an exceptionally clear manner to a lay public. These lectures formed the basis for his later book *Atom and Cosmos* [1932h]. Hans was also a good model as a teacher. He never talked down to anybody and took all questions seriously. This trait enabled him to communicate easily with children and adults and answer their questions on the level of understanding of the questioner, without compromising the truth.

After we moved into our new house, Hans started to write *The Rise of Scientific Philosophy*. Every day he would read to me what he had written, ask for suggestions, discuss them with me and sometimes make changes. It was very gratifying for me to participate in his creativity in this manner. He set an approximate daily amount of work for himself; he regarded eight handwritten pages as a good average output. During the times of writing a book or an article, he would close the door to his study and not undertake any other major project. But when he had finished a work, there came a rest period and he would work with his hands, either build or repair something in the house, or make a fence for the garden, or devote his time to photography in which he was a master. He had provided for a darkroom in our house where he would develop, print, and enlarge his own photographs. Hans was very fond of music. His favorite composers were Bach and the romantics. He especially liked Mendelssohn's violin concerto in E, op. 64. He built a hi-fi set and constructed the cabinet of the same wood as the panelling of the wall against which it stood. He had bought an electric saw and used it for many purposes. He let my son help him with all this handiwork and this had a lasting influence on Frank.

Hans also loved to hike and we would take long walks in the Santa Monica mountains. We went camping quite often, on the shores of Lake Tahoe or in Sequoia National Park. It was luxury camping. Hans bought two very sturdy tents, one for us and one for my son. We took army cots along, and Hans made little shelves to keep things orderly in the tent. He would cook on the Coleman stove and make especially good hot chocolate at night. He encouraged me in everything I wanted to do; he was so positive and accepting, and awakened much in me that lay dormant. When I mentioned that I would like to paint, he bought me an easel, oil paint, canvas, a palette, and brushes and took all of it along on the camping trip. For company he sketched in pencil or charcoal. When we came home, he would frame my pictures and hang them up. I think, he was fooled about their value because he was colorblind. This fact made me an accomplice in a little deception. Every time he had to take a driver's test I had to stand next to him and whisper which square was green, or red, or yellow. He always passed his test this way. I did my whispering in German and he did his driving by spatial distinctions.

We attended a number of meetings of the various divisions of the American Philosophical Association and I often admired his ability to produce order out of discussions containing a welter of contradictory statements. In December, 1951 we attended the forty-eighth annual meeting of the Eastern Division at Bryn Mawr College where Hans read an invited paper [1952a]. It was an exciting occasion to listen to Hans, C. I. Lewis, and Nelson Goodman speak on the same subject, the question of the certainty of phenomenal reports. After being stranded in New York on New Year's Eve because of a violent snowstorm, we flew the next day of Spokane, Washington. Hans had been invited by Wesley Salmon to give an address and participate in a seminar at the State College in Pullman, Washington, where Salmon was teaching. He had been a student of Hans' and had taken his Ph.D. with him at UCLA.

Whenever we were in the East, we spent some time as guests of Hans' old friends, Dr. And Mrs. Paul Oppenheim, in their beautifully appointed house in Princeton. They knew all the famous members of the Institute for Advanced Study, the University, and many a prominent visitor that came to Princeton. The Oppenheims gave memorable luncheons and parties for us to which they would invite interesting people.

I should like to take this opportunity to rectify an account given by Arthur Koestler of a meeting between him and Hans in John von Neumann's house in Princeton. [See Salmon's account above – R.S.C.]

As soon as I learned about this passage I cabled Macmillan and wrote to the British and German publishers of Koestler's book requesting to eliminate it in forthcoming and later editions. I recall clearly that there was a discussion between Hans and Koestler, and that Hans did not show any reactions such as described by Koestler. At no time did Hans own or use an ear-trumpet and, therefore, could not have dropped it. At no time did he consider any statistical results concerning extrasensory perception conclusive enough to overthrow the basic elements of his own philosophical conception. At no time would he have shown any emotion of the sort indicated by Koestler, for his lifelong experience as a university teacher in half a dozen countries in the world had given him enough training to control himself and to organize his thoughts. At no time did he make or would he have made a remark as attributed to him by Koestler. It would need more than a conversation with

Koestler and his citing of somebody else's conclusions to have upset Hans' evaluation of the philosophy he considered his life's work. Even granting a writer poetic license, I believe it would have been better if Koestler had reported the event more truthfully. The only fact that has a kernel of truth in it is that Hans was somewhat hard of hearing. He wore an ordinary hearing aid.

Every time we were in Princeton we visited Einstein. I have mentioned some memories of these visits in another book [1965a]. It would sometimes happen that Einstein and Hans discussed a scientific topic during the day and that Einstein, after pondering it for a while, would call up around nine o'clock in the evening and continue the discussion for another hour over the telephone.

Hans and Carnap saw each other occasionally and Hans visited him also in Santa Fe, New Mexico, where Carnap spent several summers. Hans was very fond of Carnap and called him charming. ("*Carnap ist reizend.*") I met Carnap when he became Hans' successor at UCLA in 1954, after Hans' death. Hans had often told me about him, and Carnap was so understanding and compassionate the very first time I went to his house in Santa Monica that I had the feeling I was seeing an old friend. This close friendship lasted until Carnap's death in 1970. He was most generous in his advice and in writing forewords to posthumous works of Hans' that I published.

In the spring of 1952 Hans received an invitation from the Institut Henri Poincaré of the Sorbonne to give some lectures on quantum physics [1952d]. I had to stay behind a few more days because of my teaching commitments. When I joined Hans in Paris, he had already delivered his talks. The first one he had read from the French manuscript he had prepared, but after that he had been persuaded by Professor Jean-Louis Destouches to speak in French without a manuscript, which he did with great success. We had planned to spend our vacations in Europe after the lectures at the Sorbonne and for that purpose had bought a small Renault. When Hans picked me up at the airport in Orly, I was amazed how well he knew his way around in Paris, and how carefully he drove, stopping politely for pedestrians crossing the street, never losing his calm in spite of the furious honking going on behind us because for the Parisians such behavior was unheard of.

That summer we visited his old teacher Erich Regener in Southern Germany to whom Hans had dedicated a paper on the occasion of Regener's seventieth birthday in 1951 [1951g]. Later we drove to Stuttgart, where Hans had started his academic career and where he had been invited to give a talk, and on to Tübingen, where he participated in a colloquium with Alfred Landé, an old friend he met there by chance. We motored along the Neckar and stopped at Götz von Berlichingen's castle of Jagst where Hans looked with amusement at Götz' armor, remarking that it would fit him, since that great knight evidently had not been a giant. We were appalled at the ruined cities, for instance, Heilbronn and Darmstadt; it was the first time we were in Europe after the war. We went on to Göttingen, where Hans had been a student and which was unscathed by the war; he recognized stores and cafés in their old locations. We visited Otto Hahn who showed us his Institute of Physics, newly equipped with money from the Marshal Plan. We then went on to see Hans' sister Wendeli who lived in the Harz with her daughter, son-in-law and three grandchildren. It was an emotional encounter, for they had not seen each other for a very long time. The little car performed well. It even took us up the Cross-Glockner, the highest mountain in Austria, where we arrived in thick fog, but the next morning had breakfast outside in the sun, with a glorious view of the glacier. Hans was a great lover of nature and tried to share his feelings with me by taking me to some of the most beautiful spots on the continent.

We went back to Paris in August in order to attend a Logic Congress at which Hans gave a paper and participated in the discussion [1954b and c]. At the end of our sojourn in Paris we sold our car and flew to London in order to see Bertrand Russell. We visited him in his house in Sussex which was populated by children and grandchildren. The English were still suffering from the effects of World War II; food was rationed. But somehow the Russells had been able to get a large ham which Russell carved elegantly. After lunch we sat by the fireplace in Russell's study, the two philosophers talking and drawing contentedly on their pipes. From London we flew home to California.

I saw Russell for the last time in 1962 in his house in London. He was eighty-nine years old, as spry, lively, and alert as ever. He poured me a drink, half a glass of good old Scotch. He delighted in showing me his collection of caricatures of himself. We talked mostly politics.

We spent the Easter vacations of 1953 in the desert, near Los Angeles, swimming, walking, and horseback riding. Hans allowed Frank, who was only fourteen years old, to drive the car in unpopulated areas. Both were proud of his performance. We were back in the Pacific Palisades on Easter Sunday and attended a lively party in the nearby house of a colleague at UCLA. Tuesday, April 7, Hans gave an invited lecture at the California Institute of Technology. I always tried to minimize unnecessary exertions for Hans. Since the paper was scheduled for eight o'clock in the evening and the roundtrip to Pasadena about fifty miles, I had suggested to have a graduate student chauffeur him and Hans concurred. The Institute sent a student to drive him. Hans gave a very successful lecture. He seemed to be at a peak of productivity and enjoyment of life.

The following Thursday, April 9, I came home from my college unexpectedly early, shortly before noon. Hans had been working on his manuscript for *The Direction of Time* [1956b], but when he heard me come into the house, he got up from his desk and we went out on the patio together, to relax and talk a little before lunch. Suddenly Hans' head fell backward and he gasped for air. I knew immediately that he had a heart attack. Very gently I managed to bring him into the house and made him lie down on the bed. My telephone calls brought the doctor and an ambulance to our place in fifteen minutes. The hospital was alerted, and Hans carried out on a stretcher to the car; oxygen was applied the whole way while I was sitting next to him listening to the screaming siren. The cardiologist was waiting for us at the hospital; he did his best to save Hans' life. Hans died in my arms about two hours later. His last words to me were: "You have really done all you could. I'll be well again – but it will take a long time."

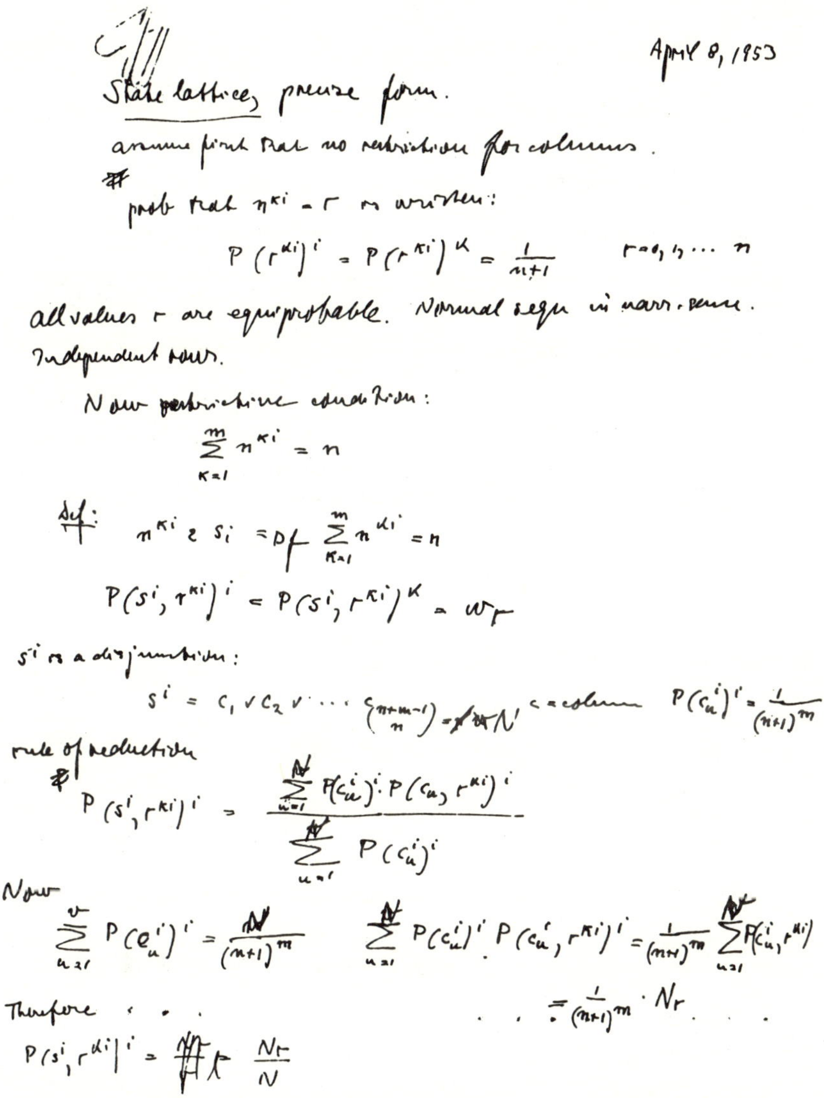

April 8, 1953

State lattice, precise form.

assume first that no restriction for columns.

prob that $n^{\kappa i} = r$ is written:

$$P(r^{\kappa i})^i = P(r^{\kappa i})^\kappa = \frac{1}{n+1} \qquad r = 0, 1, \dots n$$

all values r are equiprobable. Normal sequ. in narr. sense.
Independent rows.

Now restrictive condition:

$$\sum_{\kappa=1}^{m} n^{\kappa i} = n$$

Def: $n^{\kappa i} \in s_i =_{Df} \sum_{\kappa=1}^{m} n^{\kappa i} = n$

$$P(s^i, r^{\kappa i})^i = P(s^i, r^{\kappa i})^\kappa = w_r$$

s^i is a disjunction:

$$s^i = c_1 \vee c_2 \vee \dots c_{\binom{n+m-1}{n}} = N \qquad c = \text{column} \qquad P(c_u^i)^i = \frac{1}{(n+1)^m}$$

rule of reduction

$$P(s^i, r^{\kappa i})^i = \frac{\sum_{u=1}^{N} P(c_u^i)^i \cdot P(c_u, r^{\kappa i})^i}{\sum_{u=1}^{N} P(c_u^i)^i}$$

Now

$$\sum_{u=1}^{N} P(c_u^i)^i = \frac{N}{(n+1)^m} \qquad \sum_{u=1}^{N} P(c_u^i)^i \cdot P(c_u^i, r^{\kappa i})^i = \frac{1}{(n+1)^m} \sum_{u=1}^{N} P(c_u^i, r^{\kappa i}) = \frac{1}{(n+1)^m} \cdot N_r$$

Therefore

$$P(s^i, r^{\kappa i})^i = \frac{N_r}{N}$$

Facsimile of a handwritten page of Hans Reichenbach.

PART I

EARLY WRITINGS ON SOCIAL PROBLEMS

STUDENT YEARS: INTRODUCTORY NOTE TO PART I

[Prepared by Maria Reichenbach]

The overriding idea for bringing out the present volume in the *Vienna Circle Collection* was the intention of publishing a selection of those popular and scientific articles by Hans Reichenbach that so far have not been available in an English translation or that at this time are out of print in English.

Part I was written when Hans was a student at the universities of Munich and Berlin. These earliest pedagogical and political publications show the versatility and thoroughness with which he treated every topic and are witness to his profound concern for reform and reorganization of German education at the schools as well as at the universities. He stood for the freedom of the learner to learn and the teacher to teach, unencumbered by religious or political censorship. The articles are a mixture of idealistic aspirations and practical suggestions for social reform by combatting elitism, whether in the form of an aristocracy of birth or money. It is not surprising that his acuity and his equalitarian principles, expressed before and during the first World War, made him a natural leader of the Socialist Student Party in Berlin after the German Revolution in 1918.

Moreover, the future philosopher can already be discerned in these early writings which contain many seminal ideas elaborated in later academic works, such as his rejection of transcendental metaphysics and his theory of ethics which attempts to clarify the distinction between facts and values and opts for an explication of value judgments in terms of volitional decisions.

Since the first of the included articles deal with the German youth movement and various German student organizations that are not too familiar to the English speaking reader, some explanatory remarks about them seem to be in order. For that purpose I shall trace in broad outlines the German youth movement, its influence upon and its contrast to some of the student organizations, and Hans' role in them.

The *Wandervogel*, a group that originated the German youth movement, was founded in 1904.[1] The name 'Wandervogel' means 'bird of passage' and was applied to the whole movement as well as to each member. The *Wandervogel* with which Hans sympathized was a 'back to nature' movement,[2] originally consisting of high school students; it was essentially anti-authoritarian, aestheticist, and anti-intellectual. It repudiated sophisticated civilization (Hans was a teetotaler for some time) and parental guidance, however not in the form of attack but of withdrawal and escape to nature, as an alternative to the activities which the older generation had devised for its children. The youngsters hiked on foot, often for many days, throughout the German lands, camping out in the open.

In an article entitled 'Militarism and Youth'[3] Hans contrasts the *Wandervogel* with the national youth organizations (boy scouts and various paramilitary youth leagues) linked together 1911 in the *Jung-Deutschland-Bund* under the chairmanship of Fieldmarshal von der Goltz. First, he quotes Hans Blüher, a prominent leader in the *Wandervogel*:

> The *Wandervogel* is to the core of romantic blood, at once the noblest and finest and wildest that youth could ever have created without bothering one bit about the generation of its fathers. Youth had to create it because it was oppressed by a consistently fallacious upbringing. This development is a natural process, a regeneration, a great purification of the spirit; and in this manner the *Wandervogel* must be understood in its innermost meaning if one does not wish to remain alien to it. But it happened like all such great transformations below the threshold of consciousness. There is no design in it, no system, and what the individual pondered and planned has never been of any major influence upon actual events. The *Wandervogel* has never had a uniform tendency, a goal, an ideal unless it is romanticism itself. It has never been anything but the revolt of youth against the violation of its spirit (*Gemüt*).[4]

Hans calls the *Wandervogel* a healthy reaction against the school which suppresses the originality of the children, an instinctive, independent attempt to rebuild what the school has destroyed. The school had denied any proper value to youth and had tried to lock the pupils into a leveling system that seemed to the authorities to be the best preparation for future adulthood. But the members of the *Wandervogel* created their own world, outside, in the woods, around the camp fire where they could indulge in all the fantasies and joys of youth.

As the founders grew older and left high school, the *Wandervogel* expanded on the one hand into the *Freideutsche Jugend*, on the other hand – as the university branch – into the *Akademische Freischar*.

In another article[5] Hans gives us, in vignette form, a picture of the mentality of the paramilitary type youth organizations that existed contemporary with the *Wandervogel*:

> Meticulously ordered, the divisions make their way. Through the lonesome wood they march, past steep slopes with magnificent views of wide valleys. "Take cover", in sharp tones commands the group leader, a youth with a beautifully embroidered emblem on his sleeve. (The dress of the *Wandervogel* was more like the hippy style.) The boys converse softly so that the enemy will not hear them. At this moment the sun glides

behind a delicate layer of clouds and its rays throw a magic sheen over the woody slopes and the glittering ribbon of water in the valley. But not one of the boys stopped for a moment to look about him; they were all stomping along in their divisions, in neatly ordered rows, for obedience is the first duty of the *Wehrkraftsjunge*. After arrival at the entrenchment, some patrol scouts are dispatched to observe the movements of the approaching enemy. Some time later the enemy arrives; with hurrahs the camp is stormed. However, no fight takes place; the leader awards victory to the party that arrives in the most regimented order. Now that all divisions are together once more, they march in great columns back home again, drums and fifes at the fore.

Once I was marching along a country road with a happy-go-lucky crowd of *Wandervogel*. We had our guitars and sang to their strains all sorts of songs of the open road, of exhuberant wanderlust. I would like to say that the guitar, so harmlessly melancholy and merry, in a way represents the *Wandervogel*. But the national youth movement also has its symbol: the fife and drum have become their symbolic manifestation.[5]

Hans notes regretfully that the *Wandervogel* who initially had shunned politics eventually joined the *Jung-Deutschland-Bund*[6] in a complete failure to recognize the opposing views that existed between its adherents and those of the German national organizations. He comments that the *Wandervogel* was the reaction of youth to the prevailing school system. The paramilitary movement was the political reaction. No wonder it had the support of school and state.

But it was getting even worse. In a letter to the Editor[7] Hans responds to the antisemitic infiltration of the *Wandervogel*. Some leaders had published a pamphlet 'German or National?' ('*Deutsch oder National?*'). Hans calls it an outrage, a violation of the noblest principles of the *Wandervogel*, that such a tract could originate from its midst, one that attacks the Jews with insane racial hatred and demands for 'purity's sake' the separation of the Jews from the 'Arians'. "It is a disgrace to think that race has anything to do with human values. If such literary productions claim the name 'German' for themselves, one feels ashamed of being a German." (The date: November, 1913.)

Another branch of the German youth movement was represented by the *Freie Schulgemeinde Wickersdorf*, founded in 1907 by Gustav Wyneken.[8] The *Freie Schulgemeinde* was a boarding school in the style of the 'progressive' schools of today (Summerhill, etc.). It worked very well with a selected group of students. Wyneken was neither anti-intellectual nor romantic, but since his philosophy was anti-authoritarian he got into contact with the *Freideutsche Jugend* with whom he seemed at times to make common cause. This apparent common radicalism exploded at the first great convention which the *Freideutsche Jugend* held in 1913 on the Hohe Meissner in central Germany. It became clear that the romanticism and the almost hostile attitude of the core groups of the *Freideutsche Jugend* to rationality was not congenial to the Wyneken followers. After the convention conservative newspapers and speakers attacked the *Freideutsche Jugend* in a sort of witch-hunt, using largely material taken from the Wyneken publications. Thereupon the leaders of the *Freideutsche Jugend*, unwilling to engage in a political struggle for which they felt mentally and emotionally ill prepared, drew a definite line between their own movement and that of Wyneken's. The *Freie Studentenschaft*, founded in 1900, of which Hans was a member before World War I, offered somewhat ambivalent help to the *Freideutsche Jugend* which was also ambivalently received. Hans and his student friends were critical of the rejection of Wyneken and regarded the aversion of the *Freideutsche Jugend* to political argument as cowardly. The *Freideutsche Jugend*, on the other hand, welcomed the assistance of the more experienced *Freistudenten*, because they did not know how to meet the political attacks, but were suspicious of attempts to be drawn more deeply into politics than they liked. One year later, right after the outbreak of World War I, Hans, who earlier had tried to realize some of Wyneken's ideals in the *Freie Studentenschaft* and who had collaborated with him in some publications, chided Wyneken in an extensive exchange of letters for his extremely nationalistic stance. This correspondence (as yet unpublished) was circulated secretly among a small number of those adherents of the *Freie Schulgemeinde* who had not been infected by the hurrah-patriotism to which many other German intellectuals fell victims at the beginning of the war. It was impossible to have it printed at that time.

The *Freie Studentenschaft* originated at various universities as an organization of those students who did not belong to the *Korporationen*, [The German *Korporationen* have some similarities to the American fraternities.] almost all of which at the time of Hans' student days were very conservative and reactionary, antisemitic and socially exclusive; they submitted to the commands of the authorities, conscious that their members belonged to the social stratum that enabled them to enter the upper ranks of the state administration. All *Korporationen* had their colors (caps and ribbons) which they wore at academic festivals and almost all of them, except the Catholic ones, engaged in duelling. The students who belonged to the *Korporationen* looked down on the *Nicht-Inkorporierten* or *Finken*, and it was the original purpose of the *Freistudentenschaft* to fight this monopoly of prestige which was largely supported by the academic authorities.

Although non-political, the *Freistudentenschaft* had a definite ideology in which many strands of social development came together. More and more students hailed from the bourgeois middle class (not the petit-bourgeois or the working class; even in 1928 only 2% of the students came from the proletarian classes).[9] Due to the social changes stemming from the founding of the German Reich and the high industrialization of Germany there was a trend toward equalitarian reforms and democratization, most clearly expressed in the ideology of the *Freie Studentenschaft*. This organization tried to formulate an ideal common to both, the university and the community. It envisaged the university, in Hans' mind, as a community of students and teachers, opposed to all restrictions of academic freedom, insisting on equal rights for all students. This ideology was not shared by all *Nicht-Inkorporierten* and thus the question arose: How can an organization that wants to represent all the *Nicht-Inkorporierten* have an ideology which only some of them share? Felix Behrend, who wrote the first program of the *Freie Studentenschaft*, did not recognize the dilemma or did not want to see it. Hans criticizes him in Article No. 3 of the present volume. Walter A. Berendson, somewhat younger, leaned toward the position that the *Freie Studentenschaft* should not commit itself to an ideology. But some students around Hans, among them Carl Landauer (who wrote an essay in the same issue in which No. 3 appeared)[10] maintained that their organization should abandon the claim to represent all *Nicht-Inkorporierten* and declare itself an academic party with an ideology.

Although remnants of the style of the *Freideutsche Jugend* can be found in the articles which Hans wrote for the *Freie Studentenschaft*, the latter was very different from the former, sharing neither its anti-intellectualism nor its aversion to politics. Article No. 3 summarizes the major points in the program of the *Freistudenten* and clearly outlines the aims for university reform represented by Hans until 1918. The article stresses the ideal of the autonomy of the individual "who determines his own values freely and independently of others and who, as a member of society, demands this autonomy for all members and of all members." It is neither an accident that this declaration resembles the formula of the *Freideutsche Jugend* as pronounced the same year at the convention on the *Hohe Meissner*, attended by Hans, which spoke of "a life based on one's own goals and responsibility, guided by an inner sincerity"[11] nor that his formulation has an unmistakable Kantian ring to it. Hans had steeped himself in Kant's philosophy and his early scientific writings, including his dissertation finished in 1914, reflect Kant's influence of which he freed himself quite soon.

Among the aims of the *Freie Studentenschaft* was the fight against the *Lex Arons*,[12] which forbade Social Democrats to hold academic positions; a striving for greater pluralism with respect to the choices of academic courses and majors; juster disciplinary procedures on campus; freedom of assembly and opportunity; non-discrimination with regard to political views, race, creed, and nationality; the rejection of the university as a mere training ground for future civil servants; fostering of research in the sciences and the humanities for their own sake. The organization carried on many discussion circles and lectures for students by students.

All of these demands came under the heading of 'democratization' and mirrored the social and political developments under the reign of William II before World War I when in 1912 the Social Democrats received the largest number of votes in the election for the German Parliament (*Reichstag*).[13] On the practical side Hans aimed at the fulfillment of many student needs: formation of student governments that would concern themselves with scholarships, health and accident insurance, less expensive access to learning institutions and books, decent pay for working students, creation of housing agencies, sport, and student counselling. These articles often read as if they were written today.

A year later, in article No.5 of the present volume, Hans' goals were expressed in more apolitical and less mundane manner, partly reminiscent of the emotional and romantic ideals of the youth movement which he had favored: the university should not be interested in improving teaching methods – that might at best be a positive by-product. The ultimate goal was to awaken a hunger for knowledge and to make the students true servants of culture, thus giving fulfillment to both students and teachers by means of a genuine, direct, existential experience of the values inherent in the sciences which do not consist primarily in their applicability to practical purposes or in their precise results.

In Hans' pre-World War I articles we can see complex reflections not only of social and political influences but also of his character traits and talents: enthusiasm for purifying experiences in nature, contempt for the philistine, a strong defense of the autonomy of the individual paired with a deep feeling of social responsibility, impressive seriousness of purpose and discipline of mind, great poise and self-confidence, willingness to lead, courage to stand up for his convictions, the idealism, optimism and serenity of a halcyon nature (his own words to me), imaginativeness, independence of mind, and a great gift of expressing himself clearly in speaking and writing. I can attest to the fact that these characteristics stayed with him throughout his life.

World War I intervened and Hans went to the front. When he returned he came to the University of Berlin where he studied under Einstein. There he formulated the 'Platform' of the *Socialist Student Party Berlin* (SSPB; Article No.1 in the present volume), 1918. The word 'Party' was chosen deliberately for the following reasons: to stress unambiguously that the students were a group with the aim to reorganize the university along definite lines, to overcome the aversion of the students against the word 'party', and to express the politically aggressive character of the organization *vis à vis* the apolitical and reactionary students. Another incentive was his wish to prepare the way of realizing the aspiration of the SSPB to take the initiative for a future all-encompassing German Socialist Student Association. The concept of a party was not new since Hans had already spoken in 1913 on the *Hohe Meissner* of the *Freie Studentenschaft* as 'our party'. The Berlin movement included students of the university, the Institute of Technology, and the School for Business Administration.[14]

The SSPB originated from a group of socialist students of all persuasions (Hans was a non-Marxist socialist) who had come together secretly during the war and some of whom had participated in the November revolution. At the founding of the party Hans became its chairman. This fact almost cost him later a professorship at the University of Berlin. The negotiations lasted from 1925 to 1926 and it took Planck's and Einstein's interventions to produce a firm contract. This case is a good example of the reactionary mentality that prevailed even in the Weimar Republic at the German universities.

A difficult problem for the SSPB was its relation to leftist political parties which divided into a number of splinter parties after the revolution in 1918. Since these differences began to split the socialist student organizations, Hans, who was always reasonable and not fanatic, declared that the SSPB would restrict itself to university issues and participate politically only when all leftist parties were of one mind. In spite of this decision, the greater part of the Communist students left and founded an organization of their own which became a rival to the socialist organizations.[15]

In 'Student and Socialism',[16] 1919, a commentary to the 'Platform' he had written for the Socialist Student Party, Hans indicated what he saw as the major reasons for the necessity that the students should adopt socialism. He mentioned their personal interest, their feeling of justice, and their responsibility toward academic work. The first reason ought to be interpreted as a recognition that many students had personally suffered in World War I (Hans himself came back from the Russian front with a severe illness), branding all past wars as insane, and claiming that only socialism would be able to fashion an order that avoided the exploitive wars of capitalism. The second reason meant that admission to the institutions of higher learning should be based solely on intellectual competence, not on membership in a privileged class. Closely connected with this conviction was Hans' view that the education of intellectuals toward socialist and democratic attitudes was possible only in a common endeavor with working class youth. This was no formal matter for him, because he also demanded material assistance for all those who otherwise would not be able to attend the university. The last reason reiterated, in a more up-to-date manner, some critical ideas stemming from the time of the *Freie Studentenschaft*: in capitalism the Establishment had made the university into the instrument of a class society for training its civil servants, and its supporters and commentators in the churches.[17]

Hans saw the task of the socialist student movement as one of assistance to the Social Democrats who through their victory in 1918 had changed from being in the opposition to being in the position of power. According to him government needed the expertise of the leading intellectuals. He saw it as a problem that the end of a feudalistic aristocracy of birth and wealth, the carrier of an age-old cultural heritage, did not guarantee the continuity and further development of that culture by a relatively ignorant proletariat. The socialist students would be the most adequate catalysts because they were young and could assume the leadership of the proletarian youth who hungered for education. The new culture of the socialist society would, therefore, spring from this union of academic and working class youth.[18]

This vision of a socialist society, free from class antagonism, based on youth, and led by middle class students who were converted to socialism while retaining their social and cultural values may sound too idealistic and utopian, and hindsight tells us that it certainly did not work in Germany.

The 'Platform' spells out the concrete demands. What was new compared to the demands of the *Freie Studentenschaft* was a plan to structure education in such a way that the university would be linked to a state supported unified school system that would enable every competent student to go on to institutions of higher learning. The program was moderate and aimed at reform rather than revolution.

On the practical side, the SSPB had an office coordinating the various student agencies, such as the agency for political work at the university, scientific courses, lectures, party work, the youth movement, foreign propaganda, Berlin propaganda, publications on bulletin boards, and a book exchange. All of these activities were still largely modelled after the academic and social offices of the *Freie Studentenschaft*. The SSPB's political education at the university consisted in academic courses (lecture and discussion), weekly evening discussions by the group, and literary evenings. All these activities, however, were, in contrast to the principles of neutrality and pluralism of the *Freie Studentenschaft*, devoted exclusively to socialist theory and practice, led by leftist-pacifist or socialist intellectuals or party theoreticians. The SSPB saw its task more in enlightening students about socialism and in educating the proletarian youth than in political activism.[19] One must keep in mind that the whole movement was supported by only a relatively small number of students. Their adversaries were to a large extent the mass of students returning to the university after coming back from the war.

By April 1919 there existed about twenty socialist student organizations. For all of Germany the number of students who were members was between 600 and 800.[20] This figure must be understood in relation to the fact that up to 1923 all of the political student organizations represented at most 5% of all students at the universities. With the gradual domestication of the German November revolution, the leftist student organizations became more and more isolated. Already on the First General Student Congress of German Universities (*Erster Allgemeiner Studententag Deutscher Hochschulen*) July 17–19, 1919, in Würzburg only one representative of the SSPB participated.[21] He proposed a moderate program stemming from the tradition of the *Freistudenten*: student input for hiring and evaluating professors; abolishment of rank distinctions in the faculty and its rejuvenation and enlargement, linking of the university to a Unified School System (*Einheitsschule*); changes in admission requirements to the university; and consultation of the students concerning projected university reform. The response was totally negative. Politization of the student body was rejected as unacademic. The call for solidarity of scholars and workers found no echo at the universities. Thus the year of the revolution saw at once the rise of radical socialist student groups as well as the end of their effectiveness. The way was opened for the National Socialists' 'renovation of the university', their 'total university reform' which they, too, called a 'revolution of the university', in perversion of the leftist-liberal-socialist tradition.[22] The character of this 'revolution' is well known. Among many other victims of the Nazis, Hans lost his position at the University of Berlin in 1933.

NOTES

1 See Hans Wolf, 'Der Alt-Wandervogel als Traditionsbund der Jugendbewegung', *Jahrbuch des Archivs der deutschen Jugendbewegung*, Vol. 5 (Burg Ludwigstein, 1973), p. 27.

[2] In much of the following I am very much indebted to Professor Carl Landauer, Berkeley, who was a friend of my husband's, and studied with him in Munich. He sent me a letter explaining the various youth and student organizations. I made extensive use of this information.

[3] Hans Reichenbach, 'Militarismus und Jugend', *Die Tat*, Vol. 5, No. 12 (March 1914), pp. 1234–1238.

[4] Hans Blüher, *Wandervogel, Geschichte einer Jugendbewegung* (Berlin, 1912), p. 76. The translation of this quotation is by Hans Reichenbach.

[5] Hans Reichenbach, 'Die Militarisierung der Deutschen Jugend', *Die Freie Schulgemeinde*, iii (July 1913), p. 110. The translation of this quotation is by Hans Reichenbach.

[6] 'Militarismus und Jugend', *op. cit.*, p. 1235.

[7] Hans Reichenbach, 'Der Wandervogel und die Juden', *Berliner Börsen – Courier*, No. 539 (Nov. 16, 1913).

[8] I am also using some valuable information which I received through correspondence with my brother-in-law, Bernhard Reichenbach, London.

[9] Dr. Ulrich Linse, Munich, was kind enough to send me the page proofs of his essay 'Hochschulrevolution: zur Ideologie und Praxis sozialistischer Studentengruppen während der deutschen Revolutionszeit 1918/19', in *Archiv für Sozialgeschichte*, Vol. 14 (1974), pp. 1–114. I am relying heavily on it in what follows. I am citing and paraphrasing in English some of his findings without always using quotation marks or giving the quotation in German. But I always indicate the page in Linse's work where I found the information. I shall also cite primary sources according to Linse's essay because they are not available to me. I shall indicate this by [L.]. A great amount of the information in his book is familiar to me from conversations with my late husband. I am glad to have my memory refreshed and to have gained some additional information from Linse's excellent and very well researched essay.

[10] Carl Landauer, 'Die Verwirklichung der freistudentischen Idee im Rahmen der Gegenwartskultur', *Freistudententum* (Munich, 1913), pp. 41–53.

[11] Linse, *op. cit.*, p. 15.

[12] *Ibid.*, p. 17.

[13] *Ibid.*, p. 19.

[14] *Ibid.*, pp. 48, 49. See also 'Erläuterungen zum Programm [der SSPB]', in *Staatsarchiv Munich, Staatsanwaltschaft Munich* I, no. 2973, Vol. 2 [L.].

[15] *Ibid.*, pp. 46, 47.

[16] Hans Reichenbach, 'Student und Sozialismus', in *Der Aufbau*, Flugblätter der Jugend, No. 5 (Berlin, 1919). A letter of the SSPB to the Group of Socialist Academicians in Munich (*Gruppe sozialistischer Akademiker München*) mentions a propaganda article by Reichenbach: "In answer to your news that you are planning a publicity release for your university program we should like to inform you that our chairman, Dr. Hans Reichenbach, the well-known former leader of the *Freideutschen*, has written such an article some months ago, but was unfortunately not yet able to publish it.", Staatsarchiv Munich, *Staatsanwaltschaft Munich* I, No. 2973, Vol. 2 [L.]. This letter refers either to the above article or to a much longer work 'Die Sozialisierung der Hochschule' (Berlin, Dec. 1918) which I found in my husband's papers. I do not know whether it was ever published. [Article No. 7 in this volume – Ed.]

[17] Linse, *op. cit.*, pp. 49, 50.

[18] Hans Reichenbach, 'Student und Sozialismus'.

[19] Linse, *op. cit.*, p. 55.
[20] Jürgen Schwarz, *Studenten in der Weimarer Republic* (Berlin, 1971), p. 160 [L.].
[21] 'Tagungsbericht des Ersten Allgemeinen Studententages Deutscher Hochschulen in Würzburg vom 17. bis 19. July 1919', Göttingen, 1920, p. 65 ff.
[22] Linse, *op. cit.*, p. 113.

1. THE STUDENT

[1912f]

To speak of 'the student' is, from the outset, an undertaking doomed to failure, for the student body is not at all uniform. Rather, it is a great multitude composed of a variety of individuals, each one of whom possesses an entirely separate character and his own special leanings; a single type of student is not to be extricated from the mass. Yet it is a singularly agitated mass, full of deep stirrings and aspirations; it is, if I may so put it, vital through and through. Movements arise which aim to fire the student with enthusiasm for art, literature, and a broad cultural education; others wish to train him to be a citizen with a concern for politics; others, again, present to him an untroubled life of pleasure as the highest ideal; and still others wish to steel his character by means of systematic physical training. It is a play of waves, in constant motion, and there are as many waves as there are students. Yet there are greater waves and lesser waves, and some combine to form larger complexes. Parties are formed among the student body. Among others, we find the fraternal organizations [eine korpsstudentische und eine burschenschaftliche Partei] ; German nationalist, liberal, and socialist parties; Catholic, Protestant, Jewish, and free-thinking parties; organizations dedicated to strengthening the body and the will by means of physical exercises and abstinence from alcohol; parties aiming at the social activation of students; and parties promoting a more thorough professional education or a better liberal general education. The student finds himself right in the middle of this stormy battle of opinions, with no guidance at hand and no help in steering his course.

It is, I believe, precisely this aspect of student life – *that the student stands without a guiding hand in the midst of the currents* – that constitutes its *preeminent* educational value. The student is thereby forced to think for himself and to decide for himself what he will and what he will not do; he develops a critical judgment and a self-confident will. For this reason I reject all efforts to take fledgling students immediately under the wing of yet another external authority, as well-intended as these efforts sometimes are. To be sure, it is a rigorous regimen that the student has to follow, and in some cases the sudden deprivation of external support has doubtless led to a loss of internal stability. Yet these occasional unfortunate cases should prompt us rather to accustom people to independence in their early youth than to continue mothering eighteen-year-olds. For life in the outside world demands proud, self-governing people, in the best sense of the words.

People are forever praising the ancient tradition of academic freedom. And rightly so – but the foremost, *the true academic freedom is the freedom that makes one's own will the supreme authority*. I am a student. I want to hear for myself all the points of view which the people around me are defending so fiercely. And I want to decide for myself which view I will adopt. If I come to a conclusion that contradicts ancient traditions or even dogmas, I want the right to advocate my view, though I must defy the whole world. This is what I call freedom of investigation and teaching: no prescriptions are made regarding the results of my thinking.

The student's most precious possession is his freedom.

If we did wish, after all, to speak about the student as a single type, we could perhaps regard freedom of education as that which all students have in common, that which gives the student his special cast. For that is the difference between the student and all other young people in the nation, whose intellectual education is severely limited by economic restraints. But this characterization will not do, either, for there are still far too many opponents of freedom within the student body. At the present time, then, we can only speak of 'students': 'the student' is still a dream of the future. Yet we can come closer to the goal – if we fight for it.

2. THE STUDENT BODY AND CATHOLICISM

[1912c]

It is a noteworthy fact that the Catholic Church exercises a persistent influence not only upon the general population but also upon the leadership of academic circles. The Church maintains this influence with the help of the powerful edifice of her philosophy, which she calls the only true philosophy, and she expresses this influence through the large number of Catholic student fraternities, as well as through the more loosely organized groups of Catholic students, such as the Free Association of Free Catholic Students and the various Catholic student organizations that perform charitable work. All the young students who are banded together in these groups appeal, in scientific disputes, to the Catholic philosophy. It is the scientific shield with which the educated Catholic, who acknowledges the validity of scientific thinking in other spheres, protects himself whenever he is attacked on religious grounds.

This mode of defense in itself clearly constitutes a sweeping admission. For if religious dogmas can be scientifically defended, they can also be scientifically attacked and refuted and do not lie outside the confines of rational knowledge. This rationalistic bent is peculiar to Catholic thinking, distinguishing it sharply from other schools of religious thought that, as their 'defense', are given to fleeing into that realm in which reason allegedly has nothing to say. The Catholic claims it is possible to recognize rationally that the systematic teachings of the Catholic Church are objectively correct – this proposition is itself a dogma.

At this point, the following mode of proof is employed. The Catholic possesses two forms of knowledge. Certainly, people can, and should, have scientific knowledge – that he will admit. But more important and profound than this is intuitive knowledge, or knowledge by revelation. This kind of knowing is immediate and 'supernatural' and possesses the advantage of offering unconditionally valid truths. Only a few people are party to such revelations, which are presented in the 'inspired' Bible and in dogmas. The Church is the defender of these truths.

Now, since these dogmas are absolutely true, they have an easy position *vis-à-vis* science. They cannot come into contradiction with science, for science, too, aims at the 'one, absolute' truth. Some dogmas lie completely outside the realm of scientific proof; others, such as the existence of God and the immortality of the soul, are allegedly capable of direct scientific proof. And for the Catholic, the whole question is herewith settled.

But not for other people. For they allow themselves to ask why so-called revelation offers true knowledge. The Church has no answer to this question, or at least none with a logical foundation. She demands of reason instead that it suddenly withdraw, recognize its own 'inadequacy', and acknowledge revelation as the source of true knowledge. Submission to the authority of the Church is requested – certainly not a logical, but an arbitrary demand. At this vital juncture, upon which the entire problem, admittedly subject to reason in every other respect, depends, logic is suddenly rejected.

With that, the whole edifice collapses – for anyone who wishes his knowledge to be coherent. The assertion that the dogmas contain the truth is totally unproven and vague in the eyes of a thinking person, who must therefore reject them as completely unscientific. For science desires coherent knowledge; what is known through so-called supernatural means is not worthy of bearing its name.

And that is not all. A theory that is directly opposed to scientific knowledge creates a manifest danger for science. For if the dogmas are not proven to be true, the possibility exists that scientific research will come into conflict with them at some point, and the Church demands that science give way in such cases. The following quote is from the *Dictionary of the Catholic Church* [Wetzer and Welte, *Kirchenlexikon*, 2nd ed., Hergenröther and Kaulen (eds.), Vol. 3, Freiburg, 1898], which can be taken as absolutely authoritative.

> Insofar as theology treats of truths derived from divine revelation by means of supernatural belief, it is independent of all other sciences and absolutely supersedes them . . . In these truths, revealed by God and infallibly propagated by the Church, theology possesses the supreme and certain criterion against which she must test the correctness of human research. That which contradicts divine truth cannot be true, even within the sphere of terrestial science.

So long as the institution of the Church still possesses secular power, the possibility exists that she will forceably compel the scientific scholar to renounce what he knows to be true. Cases of this kind have occurred again and again in history. Galileo, the genius who founded modern mathematical physics, was thrown into prison when he was an old man of seventy, stricken with gout, and was forced to renounce the Copernican theory. Another profound thinker, Giordano Bruno, had to burn at the stake as atonement for his enthusiastic espousal of the theory of the unity and infinity of the universe. Thousands of free-thinking men were slaughtered in the prisons of the Inquisition for their honesty. Science has been most severely hampered in its course by the unfortunate theory that the dogmas of the Church contain the truth.

I hear the Catholic theologians interrupting me with cries of, "But that was in the Middle Ages!" For they leave the mediaeval Church in the lurch and admit that these trials were a mistake. That is, they involved a misinterpretation of the dogmas – the dogmas themselves remaining, of course, true. The passage in the Bible that says the Earth stands in the center of the universe has been misunderstood. The idea really contained in the Bible is different. Very well, but where in the text do we find this genuine idea? Afterwards, when science has brought about the acknowledgment of its truths in spite of all dogmas, it is easy enough to say that the 'interpretation' of the dogmas was incorrect. This recourse is always available, for Catholic theologians are so cunning that they never fail to find an escape route through the back door. Nonetheless, science continues to be hindered if it is always required to break through barricades before it can attain to new truths. Unlimited dominance of Catholicism would be the death of science.

Freedom from every extra-scientific authority is thus a condition for the existence of science. And there are other aspects to this matter. A person who wishes to do scientific work must himself be likewise free from the authority of dogma. A Catholic scholar cannot be regarded as a scientist in the true sense of the word, even if he happens accidentally to achieve something in a remote specialized subject. For it may come to pass some day that his science must give way to dogma. It is even more important for the student to demand freedom from the authority of the Church. For he is still in a state of becoming; he has yet to penetrate into science and to look for a philosophy, and it is doubly significant for him if a foreign authority dictates his views on the most important questions of life from the first or presents to him a one-sided view of the world through the dark glasses of Catholicism. For the Church follows this latter practice, also. The Index forbids to every Catholic the reading of a whole row of dissenting works. To be sure, it is possible to obtain a dispensation from this prescription, but, again, the individual is not permitted to make this decision, but rather an official of the Church. A student needs freedom to pursue his education in every possible direction – this is the demand that must be put more urgently than any other to an institution of higher learning and to its student body.

Among students, this idea has been most emphatically and persistently espoused by the Free Students [*Freie Studentenschaft*]. From the beginning, this organization has supported a certain form of student education, demanding that every student become familiar with every point of view, that he develop his own point of view in the battle of minds, independent of every alien authority. It has expressed this ideal by the nature and form of its educational programmes: it has given representatives of every intellectual orientation an opportunity to present their views in lectures and has required of each student in its various chapters that he choose his own direction on the basis of his own rational knowledge. This thought was very clearly expressed in the programmatic declaration presented by Dr. Felix Behrend in the charter of the German Free Students. As he put it,

> If the acknowledged truth must remain locked within the bosom of the thinker because the laws of the state forbid freedom of speech, if the Church can dare to hand over to the *auto-da-fé* those theories that contradict revelations, how, then, can knowledge blossom?

It is all the more peculiar that, despite this openly expressed viewpoint, Catholic students have been able to penetrate the ranks of the Free Students. The blame for this must fall upon the awkward form of its organization, which grants indiscriminately to every student *not* belonging to a fraternity the right to vote in the 'general assembly', the supreme decision-making body in every Free Student group. This regulation was instituted with good intentions, but unfortunately it was not clearly thought out, and its undesirable consequences are now gradually becoming apparent. Some of the groups have not yet come to realize that the movement urgently needs to limit the right to vote and to take a clear position with a firmly delineated programme; within some individual organizations, a struggle is already beginning for dominance among the Free Students. In the Munich organization at the present moment a large group of Catholic non-fraternity students are attempting to prevent the movement from taking this decisive step. Over against them stand opponents ready to do battle; the issue will be decided soon.

May the Free Students succeed in carrying out their cultural mission despite internal and external enemies!

3. THE FREE STUDENT IDEA: ITS UNIFIED CONTENTS

[1913c]

That a cultural movement requires fifteen years from the time of its origin to develop and clarify its philosophical foundations does not tell in any way against the value of that movement. On the contrary, that action resting upon certain definite values precedes a clear recognition of these values, that an ideal clearly and distinctly enters the consciousness only long after the person acting under the influence of the emotion to which it gives rise has begun to attempt to realize by means of his actions his vaguely divined rather than clearly acknowledged end, is a frequently observed phenomenon. It is not clear, rational knowledge that gives an act its moral value, but rather the motive for it, independent of the intellectual processes which accompany the action. Yet that does not mean that a clear formulation of one's own volitional decision is unnecessary. On the contrary, a person undergoing progressive development will endeavour to determine clearly and explicitly the precise nature of the goal of all his laborious efforts, which so often are repaid only with ingratitude. He will make this endeavour because he wishes his action to be consistent, because he wants all his activities to conform to his one great ideal, because nothing but the clear recognition of his own ultimate end gives him the proper strength to fight for it.

It is just such a clear definition of purpose that the organization of Free Students (*Freie Studentenschaft*) has lacked hitherto. It has undertaken much and accomplished much, it has established chapters and arranged for lectures and created welfare agencies and worked for the creation of student committees – but to what end it has done all this, why 700 German students who hold Free Student office have dedicated to it their often hard-won spare time and sacrificed to it their energy, has never been properly articulated by the organization. Individual leaders have arisen and spoken of ideals, tossing valuable ideas into the chaos of Free Student ideology, but what they have never offered is a unified compendium of all these ideas, what they have never discussed is the single idea that is the basis for all these ideals. Behrend offered a good deal in his *Ideenkreis* [see Note 1], yet this powerful little book fails to formulate clearly the ideal as an ideal; it suffers from the unfortunate notion that this ideal is not a strictly delineated subjective goal but an 'objective' interest of a large number of people – viz., students who do not belong to a fraternity – who cannot do otherwise than joyfully embrace this 'objective' institution, once they have discovered it, as their main purpose in life. This conception represented for a time the official justification for Free Student organizations, yet it was therefore attacked all the more vigorously from all sides and abandoned all the more quickly by almost all groups.

The fault in the system could no longer be hidden. There is, for once and for all, no such thing as an objective interest; interest always consists in a subject's taking a position with regard to an object. There is no universally binding rule determining how a subject will decide. Only the individual himself is able to say what he considers to be his interest. This depends upon the nature of his evaluations, upon the stance he takes respecting values in general, and nobody can expect to refute a person's values by means of reason. Evaluation has nothing at all to do with logic. Should it turn out that certain interests are common to a larger number of people, it would simply mean that they are the subjective interests of this group of people – that is, of those people who embrace them – but never in any way will they become objective interests, interests that every other person similarly situated must acknowledge. Referring to them as 'public interests' does not remove the logical fallacy contained in this theory. From the mere fact that a person is not a member of a fraternity there follows nothing concerning the makeup of his interests. Instead, it must be clearly and distinctly laid down that, no matter what interests the Free Students represent, they are invariably the interests of a particular group of people; only the free volitional decision of the individual can determine membership in this group.

It is the task of this article to work out the contents of the interests represented by the Free Students, to put them in the form of a set of ideals. It is in the nature of the case that nothing original will be said here, for it is not my aim to force new goals upon the Free Students, but only to present the same goals they have always fought for in a clear and unified form. And yet I do have one further aim: to lay bare the spirit that has lived in their work, to set forth the unified structure of ideas that, even when not clearly perceived, has been the true foundation of Free Student accomplishments.

The desired end of the Free Students can be summarized as follows:

> *The supreme moral ideal is exemplified in the person who determines his own values freely and independently of others and who, as a member of society, demands this autonomy for all members and of all members.*

This ideal is purely formal, for it says nothing as to the direction the individual should follow in choosing for himself. No contents ought to be stipulated, for the very reason that it is intended as an ideal. Only the form of an ideal may be put forward categorically: sketching in the contents is the personal duty of each individual. The fascination of the human character lies precisely in its complexity; it is the very variety of special interests and personal viewpoints that gives life its zest. The individual may give his life whatever form he finds to be of value and may set for himself particular goals, as, for instance, to follow the profession of an artist or a mathematician, but to demand that others pursue the very same goals is to overrate one's own particular gifts to the exclusion of others, is to be both petty and pedantic. Only one universal demand can be made: the formal ideal. We require the autonomous creation of the ideal; that is, we require that each person, of his own free will, set the goal to which he will aspire and follow none but a suitable course of action. The individual may do whatever he considers to be right. Indeed, he ought to do it; in general, we consider as immoral nothing but an inconsistency between goal and action. To force a person to commit an act that he himself does not consider right is to compel him to be immoral. That is why we reject every authoritarian morality that wants to replace the autonomy of the individual with principles of action set forth by some external authority or other. That is the essence of our morality, that is the fundamental idea underlying our moral sensibility, and only those who hold this view from the depth of conviction may count themselves among our ranks.

If, in the formulation of our ideals, we put forth a second point of view, concerning society, that is not to be regarded as contradicting the principle of autonomy just presented. It is incorrect to speak of a contradiction between individualism and socialism, and it is also incorrect to view the ideal that has just been sketched out as a synthesis of the two, as a sort of compromise joining two mutually antagonistic positions. When we demand the autonomy of the individual and require at the same time that the individual grant to everyone else the same right to self-determination, we are really presenting one and the same thought from two different aspects. The second is an extension that is necessary to complete the ideal, an addition that transforms what is desired for the individual into a universal law. The recognition of the principle of autonomy as a universal precept – that is, one applying to everybody – does not entail a constriction of this principle; it merely gives the principle its proper content. What we demand, then, is moral behavior as a right and a duty for everyone; each person is to decide himself what his course of action shall be, and the sole criterion for morality is the correspondence between goal and action.

The task of the Free Students is this: to educate students to the acceptance of this ethical ideal.

The principle guiding Free Student activity may be summarized in these few words. It is apparent that the two basic directions followed by the Free Students in their work can be derived from this principle.

First of all, the genuinely educational work must be accomplished. Possibilities for education must be set up for the student; through lectures, chapter meetings, tours, and the like, the student must be offered opportunities to form his own judgments, to establish his own values. This is the Free Students' task at the university with respect to *education*.

Second, the external conditions within academic life must be structured in such a way that the student can educate himself according to the ideal of autonomy as a universal precept. Freedom with respect to teaching must prevail at the university, and all students, equally, must be granted the right to live as they will and the opportunity to have an influence upon the structure of the university through student committees: these provisions must be guaranteed by the constitution of the university, and most especially in the disciplinary code. To the extent they are able, welfare agencies must combat the limitations on possibilities for education under which students of limited means, who must earn their bread, suffer as a result of social inequalities. This is the Free Students' task at the university with regard to *politics*.

The educational work of the Free Students ought to culminate in the successful raising of free, self-determining people within the student body. The word 'raise' is singularly inappropriate here, for such people cannot be forceably 'raised'. They must, rather, raise themselves on their own; the educational organization can do nothing towards this end except present the individual with the opportunities needed for such an education. Yet in order that really every student may attain freedom of judgment, the institution must not present a one-sided selection in creating its educational possibilities, must not set up a biased course of education. Neutrality must be the ruling principle, for it is only upon this basis that our educational activity can bring about the moral ideal. For the Free Students, neutrality is not a merely incidental side-effect of our educational efforts, not a compromise with other, similarly directed efforts, but a conscious expression of our educational ideal. We *want* to be neutral. The concept of self-education, presented so clearly and explicitly by Behrend in his programmatic writings, requires as a necessary condition the neutrality of the educational organization. We call for a neutral course of instructions; that is, the course must be so organized that no points of view are excluded. This principle has but one opponent: the notion of breeding to a single point of view. And we do battle only with this one opponent, through our words, our writings, and our practical work. The significance for the university of the application of our basic principle and the consequent requirements for the appointment of the faculty will be presented later in this article. It will be demonstrated that our principle means nothing more nor less than academic freedom and that, as a state institution, the university must be so organized that every scientific viewpoint is accorded its due rights. Our own chapters must be similarly organized. Every point of view may be presented within them; in lectures and discussions, proper representation must also be given to the opposition to a neutral education, that is, to the opposition to our highest principle. To be sure, the advocates of this view take a peculiar position, for they oppose the Free Student educational organization, the very institution that offers them the opportunity to communicate their views, and therefore they cannot become voting members of the Free Student organization. However, as auxiliary members they take part in the proceedings of the various chapters.

Our educational work is carried out through the organization of mass lectures, discussion evenings for smaller groups, tours of every kind, student trips, and athletic activities. As the theme of this article prevents my treating questions of individual methods, I shall not say very much here concerning the detailed execution of such projects. But there are some important points to be emphasized in this regard. The work of the chapters, which represents the core of all educational activity, is unfortunately still badly neglected; it is often forgotten through preoccupation with insignificant trivialities. Yet this is precisely the area in which the Free Students can still gain important achievements; indeed, they must if they are to continue to justify their existence in the academic world. It is the vocation of the Free Students to offer an important extension of the educational work of the university. Today, the university is becoming increasingly a school for specialists in each subject. Its accomplishments respecting the technical training of the student are in part excellent, yet it does virtually nothing towards furnishing him with a liberal education, towards helping him to a coherent mastery of our cultural heritage and to the development of his moral character. It is in just this sphere that the most fertile ground for the work of student organizations is to be found: it is in just this sphere that working together with one's peers is the most beneficial mode of education, for it teaches the individual to form his own independent judgments. Here, the grave problems of taking a position respecting the fundamental values of life in general become for the first time truly alive, for they demand a solution in a battle with similarly armed opponents. The arrangement of lectures open to large numbers of people has above all the advantage of allowing the student to become acquainted with prominent public leaders. For instance, training in political economy can be valuably complemented by offering the student, in addition to instruction by his university teachers, an occasional opportunity to hear a practical politician discuss issues of economic life. And the best introduction to general philosophical problems has always been discussion with fellow students not excessively advanced beyond oneself. In this era of sports, it is probably unnecessary to say much regarding the value of cultivating physical exercise in the athletic chapters – an area that the university has up to now virtually ignored.

All our work in education culminates in the single notion of self-education. This idea must be expressed through every one of our activities; it is the one genuinely new concept that the Free Students have brought into academic life. It is also the decisive issue that separates us from the old-style fraternities [*Korporationen*]. For their aim is to educate by means of authority. Older students are expected to train the younger to be worthy members of the community by means of compulsion: initiation rites, drinking ceremonies, fines. One of our most important tasks is working to effect the disappearance of this form of education from academic life. To be sure, we have a different attitude towards certain newly founded fraternities – not properly so called – for they reject the principle of education by authority, just as we do. We have in mind most particularly the *Freischaren* (Irregulars), the temperance unions, and the academic organizations. We must not forget that the Free Students can never provide more than a portion of the necessary educational efforts and that there are, in addition to their ethical ideal, other ideals respecting aesthetics, science, and religion. The creation of organizations for the promotion of these ideals among the student body is doubtless worthwhile. Aesthetic education has a particular need of close relations between sympathetically minded persons, as is most apparent in the creation of new and valuable forms of social intercourse. These questions lie beyond the proper task of the Free Students. Thus the organizations that attempt to find solutions to them need not be opposed to the Free Students. On the contrary, they may be convinced that the moral ideal of the Free Students is right and may see them as furnishing an important complement to their own activity. A coordination of efforts would be advantageous to both parties.

Educational activity at the university is one of two fundamental directions taken by Free Student work. This activity is aimed directly at educational objectives; we set our sights forthrightly at educating students to our ethical ideal. The second fundamental direction must follow a parallel path; our political activity at the university must create within the academic world the external circumstances necessary to enable the student to embrace our ideal through self-education. This second approach contributes only indirectly to the realization of our ideal; what it brings about is not an end in itself but simply a necessary foundation upon which the ideal is to arise. This distinction is not to be taken as casting aspersions on political activity. Quite the contrary; anyone who is aware of the tremendous influence of the milieu, the environment, upon the development of the individual character, anyone who has learned to conceive of the evolution of a culture not as the work of a few prominent persons but as a systematic and progressive development of the various strata of society will know how to appreciate properly the value of the political activities carried out by the Free Students at the university. These activities are characteristic of the organization, and distinguish it from the modern corporations, which carry out related activities. The Free Student movement is a movement of the people: it has a sociological orientation and is not content with working towards its goals through a series of individuals. Nonetheless, the goals of the movement's political work are the same as the goals of its educational work in the university. In both instances, the ultimate aim is the emergence of human beings who are self-determining and who grant to all others the right to be the same. The final end is not the creation of an artificial structure, a student society, but people who live and act in accordance with our moral idea.

The Free Student movement ought immediately to turn its attention to certain circumstances of modern life within the state that interfere with student life and stand in the way of the development of the student towards the moral ideal. For instance, economic inequalities have become today a partial determinant of student life. Students entering the university today no longer come almost entirely from a privileged class of society, as was the case in the past. Consequently, there are now a large number of students who cannot count upon adequate financial support from their parents and must provide for themselves. They do so largely by giving private lessons. However, compensation for intellectual work has become totally inadequate as a result of the excessive number of teachers. The Free Students must create better living conditions for underpaid students by handling in their employment bureaus only tutoring positions that offer adequate reimbursement, for the student can only carry out scholarly work, can only develop towards our moral ideal, when he is free from economic pressures and anxieties.

Although the afore-mentioned social activities of the Free Students pertain to the promotion of our ideal solely in relation to the student, that is not true of our work in social welfare in the broader sense, which we will now touch upon briefly. We are cooperating in relief work in the juvenile courts and in the provision of courses for workers; these are the only areas in which our work is not limited to the student body and related concerns. To be sure, this work, too, promotes the realization of our moral ideal – as becomes immediately apparent to anyone who considers that the education of workers consists in the dissemination of independent learning among the lower classes of society and that our work with the juvenile court consists in protecting minors against mistreatment and against inaccurate assessments of their misdemeanours. Whereas otherwise our work is confined by our very nature as an academic organization to a single portion of the population, the student body, these activities extend to other classes of society. This does not involve us in any inconsistency, for the limits of a sphere of activity are in the end set only by the effectiveness of the means, and if our means turn out to extend beyond the academic circle, precisely because we are students and command the necessary scientific knowledge, we should gladly grasp the opportunity to work for our ideal in the new area as well. On the other hand, the same reason doubtless accounts for the fact that the workers' courses and juvenile court work are generally only loosely associated with the Free Students and have in many instances dissolved the tie altogether.

The other, more important aspect of our political activity concerns the constitution of the university. The Free Students have the highly significant task of bringing under the sway of our ideology the institution that has the most decisive influence upon the intellectual and moral development of the student.

Thus our foremost demand respecting the university is the freedom of research and teaching from influence by outside authorities. We insist that every student and every teacher have the right to declare his own personal convictions, both within and without the academic community; in particular, every teacher must be permitted in his lectures to present as true such ideas as he recognizes to be true upon the basis of his personal powers of judgment. We oppose bitterly all ideologies that wish to bind the scientific investigator in advance to establishing through his thinking certain predetermined results or to exclude an investigator from academic teaching because of the contents of his intellectual outlook. Hence we are opposed to political clericalism at the university, which forbids to the scholar the free acknowledgment and dissemination of scientific truths through the imprisonment of his conscience, and we are opposed to that nationalistic tendency that repeatedly tries and frequently succeeds in excluding scientific investigators from teaching posts because of their political convictions. We demand that intellectual ability be the sole consideration in the selection of university teachers.

In defiance of all nationalistic and Ultramontanist movements, we hold freedom of inquiry to be the necessary foundation of every truly scientific approach. Behrend totally mistakes the essence of all science when he says, in his essay entitled 'The Student and the Student Body',[1] that Catholic scientific theory is on a par with ours. To be sure, it is true that every science makes presuppositions, namely those which are necessary to make it a science at all (for instance, the principle of causality). But the Catholic theory of knowledge does something quite different. At the very beginning of a science it makes presuppositions about the results of research; it prescribes for the investigator what the contents of the results of his work must be and forces the scholar to subject himself to Catholic dogma by using all the powers of the Church, especially the strict sanctions of the conscience. That is the decisive, irreconcilable difference between ourselves and the advocates of academic clericalism. We do not in any way hold against a Catholic philosopher his presentation of certain views as scientific which we believe to be easily refuted[2], but when he commits himself from the first to certain findings in his investigations, when he binds himself forever to avoiding all other results – and this is frequently done by oath – he forfeits, in our eyes, the right to use the term 'science' for his mode of thinking.

We demand that the university adopt a form of organization that permits every teacher to develop freely his intellectual view in accordance with his scientific knowledge. We also demand that the university not appoint to teaching posts men who in any way are committed to obtaining certain specified results from their research.

We have culled our concept of a free university from our moral ideal; it forms the basis necessary for the growth of autonomous human beings. Autonomy and freedom of the individual have dominion only where every intellectual standpoint is permitted representation; the student may construct freely his own intellectual convictions, uninfluenced by any biased selection in the means used to educate him, only where every intellectual standpoint is permitted expression from the lectern.

Certain elements in the Free Student movement are quite wrong-headed in that they continually assert that, as a student organization, we have no concern with the occupation of academic posts, holding this to be a function of the state in which we ought not to interfere. This attitude can only be regarded as reflecting a most naive view of university politics. For if we advocate certain ideals among the student body, if we fight for equality for all convictions and the self-education of the individual within the student body, then we must also advocate freedom of knowledge within the constitution of the university. If we do not, we may as well be drawing water in a sieve; all our laborious efforts among the student body will be undone if the teaching practices of the university, which is far more powerful then we, work in the opposite direction, and all our educational activities will be meaningless if they are not accompanied by a parallel battle for freedom in teaching. Moreover, the usual retort that academic freedom is guaranteed to the university by the constitution of the state does not prove that we, too, ought not to do the utmost to press this demand. Even if freedom of research and teaching actually did prevail in our universities, we would have to write our demand for these freedoms into our platform because it belongs there logically. The only difference would ensue in our practical operations, where the battle for freedom of teaching would be superfluous, the goal having been realized already. But we must be quite clear in our minds about this matter: good as the guarantee of freedom of teaching may look on paper, independence of research does not, in fact, exist today. Science is not free, at least not when taken as a whole. One entire faculty, the Catholic Theological Faculty, is opposed in principle to freedom of investigation. The teachers in this faculty have made their life's work not the acknowledgment of truth, but the proof and proclamation of the dogmas set forth by the Church. Take careful note: it is not to show whether they are true or not, but to prove *that* they are true. The results of Catholic research are laid down in advance; the only problem is to establish the method of proof. It was not long ago that the Church issued to Catholic scholars a demand that they once again bind themselves, by means of a special oath, to submission to the teachings of the Church and to opposition to certain recent theories, a request to which the majority of Catholic theologians acceded. And it was even more recently than an agreement between the German empire and the papacy was made public according to which one chair in the philosophy faculty at a German university must be occupied by a teacher of Ultramontane persuasion. The influence of nationalism likewise holds sway in the selection of teachers for our universities. The era of the *lex Arons* is not yet forgotten. [The *lex Arons* was a Prussian law excluding Social Democrats from university teaching positions. – E.H.S.] Consider further that there is scarcely a single socialist among the full professors of political economy, a fact that can scarcely be attributed to a chronic intellectual inferiority on the part of these gentlemen. The tendency to view universities as institutes for the training of civil servants – the state being considered, in this context, as identical with whatever party happens to be in power at the time – still plays a decisive role in our universities. All this we must bear in mind in including a demand for intellectual freedom in our platform.

While the demand that has just been explicated concerns the teaching staff of the university, the ones that follow concern the position of the student body in the structure of the university. This demand is expressed in the form of a student code of rights.

It must be said straight off that the existing student code still falls far short of what we require on the basis of the Free Student ideal. Here as elsewhere, our demands naturally culminate in the one single point: the student code must be so structured that the student may be educated to be a self-determining, virtuous human being, and that this possibility is apportioned equally among all students. From this there follows a series of particular demands that will be indicated only briefly here, as a wealth of Free Student literature concerning them has already appeared. Indeed, we already possess a closely detailed plan for a new student code.

First, the proceedings laid down in the existing disciplinary code are totally inadequate. None of our universities has at present truly unambiguous stipulations concerning penal proceedings against students; everywhere we find only a few very general stipulations, saying in some cases nothing as to the manner of interrogation of the accused or the written execution of the judgment along with the presentation of evidence. In the case of minor punishments, the penal powers lie solely in the hands of the chancellor; the appeals procedure is far from perfect; and, most dangerous of all, there is not the slightest guarantee that, when the judgment is handed down, the accused receives all his rights, that the punishment is suited to the misdemeanor, or that considerations other than academic order and custom did not lead to the sentence. We demand, therefore, a fundamental reconstruction of penal procedures.

We demand, further, a change in the penalties; detention in the lock-up, where it still exists, ought to be eliminated. We view it as an expression of the ancient view that sets students aside as a special caste lying outside the population as a whole.

Third, we direct our demands towards a reform of the students' right of organization and of assembly. The code prevailing up to now places the student on a lower rung than any other German citizen of the same age. Although he is more highly educated, the student does not possess the right held by every other young man over eighteen years of age to establish independent organizations or to arrange meetings. He is required to get the approval of the academic authorities. This instance makes quite clear how all our efforts to reform disciplinary legislation follow necessarily from our fundamental idea. For the student is truly autonomous only when he may openly declare his opinion in groups and at meetings as well as in private, when he may fight for his ideals free of all regimentation. It is the same with all the other points discussed. We need a guarantee that all students will be indifferently judged with equal justice, and we must demand assurances that the judgment does not infringe upon the individual's right to intellectual conviction and that students are not reprimanded by the authorities because of their intellectual viewpoints or their political activities.

We must further concern ourselves in our critique with the regulations governing the admission of students to the university. If we wish to realize an ideal among the student body, we have a vital interest in the material of which that body is composed. If we wish to ensure that the very selection procedure does not work against our ideas, we must demand that intellectual studies are equally available to every person intellectually capable of undertaking them. Considerations of political persuasion, religious affiliation, race, or nationality ought not to determine admissions to academic study; to permit a decisive role to elements other than intellectual ability contradicts our notion of a university. On the same grounds we demand a maximum degree of separation between the system of qualification and the university itself, which exists not for the purpose of training civil servants but solely in order to cultivate and propagate learning.

Finally, I come to a discussion of the new body that the Free Students would like to see introduced into the constitution of the university: a general student committee. The wealth of literature concerning this matter that already exists relieves me of the need for a detailed investigation of the problems involved. In particular, I do not wish to examine the question of whether a bicameral or a unicameral system is preferable. Here I wish only to shed some light upon the principal elements of the problem to demonstrate how the student committee could serve as a valuable means towards the realization of our fundamental idea. By student committee, I mean a public, legitimately constituted organ representative of the whole of the student body: a student parliament in relation to which every student possesses the same rights. My presentation is quite independent of the purely technical question concerning the form in which this committee can best be realized.

The significance of such a representative body for the realization of the political goals within the university that were set forth earlier is immediately apparent. The reform of the disciplinary code, the influence upon the teaching staff of the university, the improvement of the economic situation of students – the translation of all this into fact can be carried out quite differently if it is backed by the acknowledged representative student organ rather than by one small group. The committee will possess a disproportionately greater position of power than the Free Students. Its considerable economic resources, to come from fees levied upon all students, will enable the committee to operate on a large scale. It will be able to run an employment bureau, using paid clerical personnel, and negotiate for concessions in a manner very different from what the Free Students have found possible. The execution of a reform of the student code will succeed much more quickly if the student parliament takes up the battle than if it is backed by a mere student organization. Consider, for instance, the relatively great success that has been attained in this matter by the Stuttgart student committee, which, according to our notions, is still quite imperfectly organized. The influence of the committee on teaching will likewise be extensive. Call to mind, for a moment, the fact that up to the present time no thought has been given to the wishes of the student body in setting up the schedule for lectures and seminars. Recall, too, that the important question of how the material is to be handled – whether it is best presented in seminars or in lectures, how it is best divided, how it can best stimulate the student to productive work – has hitherto been decided entirely by the teachers: no one has ever interrogated the students as to the effectiveness of teaching. The general inquiry instituted by the law students represents the first attempt to tackle this situation. Just think how differently this commission could operate if it were a publicly recognized organ representing the entire student body instead of the bold undertaking of a few brave students!

Yet there is a second point that is decisive in the endeavor to establish general student committees: the immediate effect that the parliament will have upon the student body.

In Free Student circles, the unification of the student body has often been set as the highest goal, the creation of a general student committee being regarded as the most important step towards its realization. But unification in and of itself means nothing. Unification by itself is just a word, and can be coupled with any content whatsoever and, therefore, with none at all. People can unite only in support of specific viewpoints or in order to pursue specific ends. Only by setting up goals can one unite the student body – unite them, that is, in a common recognition and pursuit of these ideals. Behrend and his group dimly sensed this fact, but they then fell into that curious error that led to the unhappy marriage of two utterly different ideas, the notion of representation and the idea of education. For they believed that these goals, around which the student body were to be united, could be objectively established as necessarily to be recognized by every student, and they consequently believed that the unification of the student body would be achieved once the public representative student organ was forced to acknowledge and to fight for these ideals. It was hoped that with the help of official authority the Free Student program – essentially the one I set forth below, though it be in another guise – could be made the program of the student committee. This approach was thought to be justified because the ideals had been found to have an objective quality. The fundamental error contained in this idea is readily apparent: we simply cannot assume that every single student is ready to underwrite our desired goal. The same idea has surfaced at the present time in a somewhat different form[3]. The fiction of an objective interest has been abandoned and, with it, the demand that our program be introduced by force; instead, the majority within the committee are to determine its program. This is, likewise, a fundamentally mistaken idea. A parliament cannot have any program at all. It can only issue resolutions concerning each matter as it arises, each resolution according with this program or the other, but it cannot bind itself permanently to one definite direction. The essential point about any program is precisely that it is the program of a party: i.e., the members of the party have committed themselves permanently to the program and will be ousted from the party if they offend against it. Every program is a program for all time. Hence an institution whose members have been selected not because of their recognition of a program but as representatives of a numerous and varied mass of people cannot have a program. The national German parliament does not have a program, either. Rather, the parties represented there have programs and each of them attempts to make every resolution passed by the parliament accord with their subjective program.

Thus the general student committee cannot have a program. It may do, instead, everything it wants to do and is able to do; it may act in whatever direction the majority determines. It is the task of the Free Students to direct their efforts within the committee towards making each resolution accord with the Free Student program. But they can do this only if they present themselves in the parliament as a determined student party.

We wish to unite the student body. We want to win the largest possible group of students over to our party. We want to disseminate our ideals among the student body, to make *our* desired ends the goal around which the student body unite in common efforts. By means of the struggle in the student parliament we want to penetrate the masses, to make our ideology the common possession of as many people as possible. We want to gain supporters not by means of external force but through the clarification and dissemination of our ideals and positive accomplishments for the collective student body. In particular, we believe that the keen struggle for election each semester will bring vitality to the student body and that the great significance of all parliamentary resolutions will induce every student to vote and therefore to deal critically with the problems of student life; we believe, too, that with the heightening of interest in student problems in general a large flock of new supporters will join our group, more than any other. For we hold our goals to be those most valuable for students – otherwise, we certainly would not be fighting for them.

The efforts of the Free Students are permeated by a remarkable harmony.

As I have shown, our educational activities at the university are aimed directly at educating the student body to our ideal. Our political activities, on the other hand, fulfill this task indirectly, by creating the external conditions needed for the realization of our ideal.

It is a fact that the work we do to further the realization of our political goals for the university also operates directly to educate the student concerning our ideal. This would not necessarily have to be the case. It is conceivable that efforts aimed at shaping external conditions might be worthless or even harmful to the student. It is conceivable that the work in welfare and employment agencies, the struggle for reform of the student code, and the concern with questions about university teaching might in itself hold no educational value for the student whatsoever, that it might even do him injury by distracting him from his studies. These effects are thinkable, but the reality is quite the opposite. These very activities hold great value for the student. In discussing the student committee, I showed how this institution has not only an indirect but also a direct educational effect upon the student body, how it engages their interest in totally new problems for the first time, how it broadens their frame of reference, how it enables them to make independent critical decisions. It is the same with all our other political goals at the university. Whoever at some time or other has worked as a Free Student, has grappled as a committee member with the great variety of problems attendant upon student life, or has worked in employment agencies or sat on commissions knows well the spiritual value he has derived from such work, even though it is often directed at purely external concerns; how this work has made him, for the first time free, far-sighted, and self-confident; how this work has educated him. I call this a rare harmony, for it need not have worked out this way. But it has.

It is that harmony which is revealed to the man who faces life freely and bravely and fights for his ideals. This is the man who, rather than regarding mankind as a heap of worthless noodle-heads and human existence as perpetual suffering, expresses an unconquerable affirmation of life by constantly sacrificing himself for his ideals in order that his fellow men may partake of them. Every form of social action rests ultimately upon a profound optimism.

This we must bear in mind as we freely and boldly proclaim our ideals as self-chosen and fight for them in the happy confidence of victory. For they are too good to perish.

NOTES

[1] *Akademische Rundschau*, Verlag K.F. Koehler, January, 1913. How different Behrend's earlier views appear to have been; he wrote in *Der Freistudentische Ideenkreis* ([*Free Student Ideology*] Munich, 1907 and 1912), "If the acknowledged truth must remain locked within the bosom of the thinker because the laws of the state forbid freedom of speech, if the Church can dare to hand over to the *auto-da-fé* theories that contradict Revelation, how, then, can knowledge blossom?" See also my 'Studentenschaft und Katholizismus' [1912c], [translated in this volume as 'The Student Body and Catholicism'].

[2] For instance, I believe that since the time of Kant there has been no justification for supposing that the existence of God can be derived from the law of causality. However, no one should be denied the right to the view that it can be so derived; the view only becomes unscientific at the moment at which the Church uses its powers of jurisdiction to declare it a dogma, to doubt which is forbidden on pain of the harshest punishments of the Church.

[3] Cf. Paul Roth, 'Die einheitliche Idee des Freistudententums' ['The Unitary Idea of the Free Student Movement'], *Dresdener Studentische Blätter*, No. 8, 1912.

4. WHY DO WE ADVOCATE PHYSICAL CULTURE?

[1913g]

If we were to list the elements that characterize the life of the general population in our time, we would undoubtedly have to place at the top of the list a strong interest in the cultivation of the body, an effort to improve physical health. There is no question but that the numerous athletic movements, such as hiking, gymnastics, fencing, swimming, ball games, skiing, boating, have won over a large portion of the population. The power of the old fogey who previously enjoyed great respect because of the great circumference of his beer belly and his immense physical sloth has gone into a decline and is receding daily. Slender, sinuous bodies that have willingly developed a capacity for strenuous tasks are becoming increasingly the ideal of our people. As a result, numerous societies and unions have sprung up with the intent of bringing about physical development, especially in the young, in a great variety of ways.

What has brought about this transformation, this rejection of the hothouse cultivation of the pure intellect? We could investigate the extent to which it arose from a purely rational recognition of the fact that a healthy body is far more useful than a sickly one and, further, the extent to which the influence of particular men or the example of other nations, especially England, contributed to this movement, but, in the end, the reason is probably to be found in the fact that a wholesome natural instinct cannot be indefinitely suppressed, in the universal law of nature that what is healthy wins out in the struggle for existence. A more detailed study of the gradual development of the notion of keeping fit would certainly be most interesting and would provide a valuable task for a sociologist. Yet for the individual it is more important to find an answer to the question, why this development? Why do I myself pursue physical culture?

One answer has already been indicated. If it is a natural instinct to be physically strong and powerful, if purely physical muscular activity gives psychological satisfaction, then I have a right to secure this pleasure for myself, while at the same time the goal of a harmonious education demands the liberation of this urge, so long, at least, as it does not come into direct conflict with other overriding desires. To comprehend this delight in physical activity, we have only to look at the gymnast as he deftly swings up unto the horizontal bar and observe the unmistakable expression of pleasure that spreads over his face. I need make only incidental mention of the mountain climber's yodel as he finally masters the top of the mountain. That sensual pleasure undoubtedly gives indication of a value that we have quite reason enough to enjoy.

But that is not the sole value that physical activity secures for us. I am setting aside altogether the purely physiological value that requires care of the body simply as a means necessary for the maintenance of an able mind; this has been stressed quite often enough by doctors and scarcely remains a matter of dispute. I prefer to speak of the purely spiritual value of physical exertion and of the influence of the cultivation of the body upon the development of the character.

Today's sport-loving youth displays a character quite different from that of their physically indolent contemporaries. They are firmer, more determined, more clear-headed, more cheerful and more modest, and, above all, more active than their comrades. They are more vital – and there are profound reasons for this.

Let us take a look, for a start, at the sport that is most important – because cheapest and most popular – hiking. I must state immediately that, in speaking of hikers, I do not have in mind those who, after taking a walk in top hat and high collar, end up enjoying nature from behind a mug in a beer garden, nor do I mean those who, because they are more affluent, visit the renowned beauties of nature by railway and sit on the hotel veranda behind their bottle of wine in order to have 'been everywhere'. The snob is, unfortunately, a byproduct of every movement, once it becomes fashionable, yet that does not make him a part of the movement. When I speak of hikers, I mean those who set out on their way for the pure joy of walking and looking about, who refrain from excessive luxury in order to come really close to nature. The preparations in themselves are most edifying. These involve providing for the whole day, for there is nothing to buy on a lonely mountain top or in a remote valley. The quantity of the provisions must be carefully considered, but the quality of the nourishment must also be intelligently selected, for a good hiker has such respect for his own body that he knows the great extent to which enjoyment of observation and clarity of thinking depend upon his purely physiological condition. He learns prudence and self-discipline. These characteristics are generally required of the hiker; if he wanders heedlessly out into the day, he will quickly find himself in the most unpleasant situations. I once encountered on the heath, at the height of the summer heat, two hikers sitting by the wayside, parched with thirst. It had not occurred to them that they would not encounter any more water along their way. Cooking out in the open also has the value of teaching the hiker a proper regard for the aids furnished by nature, such as water, windbreaks, shade, and various topographical features. In this way he enters into a far more intimate relation with nature. And of course there is great educational value in the preliminary planning of the hike and in learning to find one's way. The hiker learns to take an overview of the landscape and get a unified sense of it, he comes to understand the appearance of the countryside from its characteristic features. He learns to think in terms of causes and to observe – the latter being an activity that constitutes the necessary foundation of a sense of direction. And, finally, the aesthetic value: what a lavish profusion of beauty nature reveals to us! It is like a healing spring from which the artistic sensibilities, weary of the pressures of the big city, can drink again and again, coming away each time refreshed and restored to youth; it is like a clear fountain that washes away from us all the dirt and filth of the city that remains clinging to us.

The objection will be raised here that all these values are promoted by hiking in particular and are not a result following from the pursuit of sports in general. This is true to a certain extent – which is why I place hiking and skiing, which is closely related, above all other sports – yet other physical activities offer their own advantages. Think of the presence of mind cultivated by any sort of ball game. The opponent must be prepared to return the throw with lightning speed, and must actually do so as well. Or think of the courage and determination required of the gymnast when he performs a dangerous exercise. Purely gymnastic exercises are primarily important as a complement to hiking, for they develop all the muscles in the body, and an elastic body is a necessary condition for the performance of every other activity; above all, it is a condition required for an elastic spirit. The close relation between body and soul is also demonstrated by the success of rhythmical gymnastics, as taught by Dalcroze. The rhythmical movement of the body in time to music, the sudden change in rhythm when the time is altered, and the great concentration it demands, have a liberating effect upon the spirits. They remove the physical inhibitions that have accumulated alongside feelings of listlessness, producing an equable feeling of well-being. Water sports inculcate presence of mind and a recognition of one's own limitations, but also fearlessness and a justified self-confidence.

A sense of self: that is, in general, the most valuable fruit of physical work. We see our own achievement before us, in concrete form, and recognize that we need not underrate ourselves, for there we have a measure of our own worth. Yet this measure serves, too, as a preventive means against any leaning in the opposite direction, for we learn the limitations of our ability. What a difference there is between this wholesome sense of self and that of the young literary gentleman or aesthete, who considers himself an unacknowledged genius because he has no concrete standard by which to measure his own work!

The benefits that accrue to the individual as a result of physical activity would, if enjoyed by an entire people, have a social value of the first order. Every friend of physical culture is aware of this, and thus it is that a distinct inclination to spread the enthusiasm among the masses usually goes along with the enjoyment of sports. For healthy social instincts are usually closely bound up with the character of the man who cultivates both body and soul. Indeed, the camaraderie presupposed by most varieties of sport, especially games requiring large teams, itself constitutes an education in mutual respect and social organization. Hence the leaders of the movements furthering physical culture have written upon their standards, 'The Health of the People'. They differ only in their conception of how to bring it about. For instance, one large organization today places beside this goal another, namely that of increasing our military power. Certainly, it is true that an increase in our military strength through cultivation of wholesome bodies among our youth has great value for us, but it is never more than a byproduct, never the purpose of the movement, which is rather the maintenance of health in itself. Quite the contrary: all leaders of the movement ought to support the maximum development of nations beyond the condition of mutual hostility, for war, with its mass killing of precisely the healthiest and strongest elements in the population, destroys their work. They should endeavour to maintain today's youth free of the one-sided social conceit that makes of a man's own people a host of gods vastly superior to all others. It is an error to permit social sensibilities to stop at national boundaries. That in the present state of international relations an army is necessary and that an increase in military ability cannot but be useful will be clear enough to our youth.

One last point. Anyone with a genuine concern for the people's health must also not overlook the negative means required to this end. What can be the use of gathering together working class children to play games and do sports in the daytime if at night they sleep together with six other people in a narrow room in a dank corner of the big city, if they are undernourished, if they remain under the brutalizing influence of wretched family relations that have reduced an entire people to alcoholism and sexual suffering? And what is the use of rearing children to be free, self-regulating individuals if in school they must bow under the sway of a traditional system that suppresses independence or if, because of their parents' position, they are introduced into a social class in which the highest ideal is the faddish person who imitates the example of everyone else? I use the term *negative means* for the breaking of this cycle, which compels our people to remain within the traditional system and which consists in the sharp separation between classes, in the school system that suppresses the individual worth of the child, in the unjust distribution of property, in sexual dishonesty, and in the tyranny of alcohol.

The man who considers these opponents to be his own opponents and who on this basis takes up the cause of the physical development of our people deserves to be called a fighter for the people's health. Only upon a completely altered basis can a superior people develop, a people with greater health, strength, and vitality than all but a few individuals have attained up to the present.

5. THE MEANING OF UNIVERSITY REFORM

[1914c]

University reform must begin with a critique of science [*Wissenschaft*, or organized knowledge]. Not a logical critique, such as is offered in epistemology, but an evaluation; that is to say, the ultimate meaning, the ultimate significance of science for humanity must be set forth. For one thing is clear: service to science remains the essential task of the university, and any move towards reform that does not keep such service squarely at the center of its field of vision is doomed to failure from the start.

A wealth of scientific work is currently carried out in our universities. The successes in which almost all individual sciences have taken part in the past century have outstripped by far the boldest expectation expressed by reflective philosophers of the nineteenth century.

It is unnecessary to illuminate with examples the wealth of new discoveries that have been made since that time. The active detailed work that occupies so much of the life of the investigator today, constantly leading him to valuable results while at the same time presenting him with new problems; the extensive production of technical literature; the numerous conferences concerning each speciality – all bear ample witness to the fruitfulness of the scientific life in our time. Yet for all that, it still seems as though something is lacking, as though there is a gaping void that has yet to be filled. We behold and admire the numerous scientific results, yet they do not move us spiritually, we do not *experience* them. It is as though they were infinitely removed from that within us which is human. Thus we separate technical professional work from human concerns, forgetting that we are therewith sounding the death knell of science.

For ultimately, it is human experience that matters. The affirmation of values must stem from our deepest immediate sensibilities, if it is to be a genuine recognition of values, not a tacit subordination to traditional laws. From this stems the ultimate viability of all values: that they are cherished in the pure and genuine feelings of a human being, that they come to life within his soul. Every science that is not, in this sense, an ultimate spiritual need of man has lost its significance and its vocation. The true value of science does not lie in the success of its practical application in technology or in the admissibility and precision of its particular results, but in the perception, by a living mind of the broad regularities that permeate all existence, a mind that sees all details solely as an expression of universal law and that recognizes and experiences the unity of which all reality is a part.

It is from this vantage point that university reform must be viewed. The issue does not turn on improvements of methods or trifling reforms of teaching, as is thought in some circles today. The usefulness of good teaching techniques is not to be denied, but it simply does not get to the root of the problem. For the question whether science is really experienced as a unity, whether it means the ultimate fulfillment of the lives of its servants and students, must be the focal point of all considerations. Today, we find just the reverse. Extensive groups of students view scientific study as a necessary evil to which they must submit in order to attain a respected position, yet others, too, who undertake scholarly work with real interest, view it more as a sacrifice of their own personal desires than as a fulfillment of their deepest impulses. As a result, today's student seeks the main focus of his life – perhaps not with respect to time, but certainly with respect to interest – outside the university, finding it in participation in the life of student politics or in the outside world of the general public. He came to this through a healthy reaction, for the present scientific pursuits of our universities are really not capable of offering the student a deep spiritual life. In disputes with other student movements, in struggles to implement political demands within the university of a new order of economic life without, he has found a sphere of truly spiritual work into which he can throw himself whole-heartedly and which holds meaning for him as a human being. These experiences have given rise to those movements that have disseminated social concerns among the student body, recalling the student to a sense of his membership in the greater academic and in the general community and to the ultimate offer of his life to its service. This was quite correct and all to the good, for it signified, in the face of the trifling shop-talk and the philistinism of scientific nit-pickers, a return to higher aims. It was the call to the student to place his life once again in the service of *ideas* and to fashion in his own person an expression of his new attitude. For that is the real point: not to pronounce in resounding tones one's ideals, but to display by one's own existence the living reality of such a crusade. The socially concerned student leader came into being as a type, signifying, certainly, victory over the primitive form of societal egoism that made of the old 'scientist' in truth a representative of the class-conscious bourgeoisie.

Yet this student type represents only a preliminary stage, for with all his spiritual vitality and his gallant identification of person and object, he still lacks that basic attitude towards learning that views it no longer as the means, but as the end, that ceases to see science as an instrument of social progress, considering, rather, that social progress will only come about through total dedication to science. That all civilization must ultimately bring about the realization of the supreme spiritual values, which are its sole measure, is a principle that cannot find expression as long as the student body seeks to be regenerated by means of ideas other than the experiencing of knowledge. As the institution charged with the transmittal and further development of the treasury of scientific knowledge, the university must be the focal point of all reflections concerning reforms of student life, for it alone has control of the means that give this life its fulfillment and its value. The real tragedy of all student reform movements is that they carry within them, by their very nature and as a matter of principle, the seeds of their own destruction, for they are *student* movements, which is to say, movements arising outside the university, which the universities therefore cannot absorb. All that can be done is to place the demand for the *capture of the university* at the center of the program and, by gradual steps, to realize this takeover, taking as a starting point the struggle for principles of justice.

It must be a genuine *takeover*; that is, the new university must be the spiritual home of the student body. The opposition between professors and students must not be allowed to continue; it must give way to the relation of friendship between a leader and his young comrades. Above all, the great gaps between those who teach and those who learn must be bridged by bringing younger teachers and even advanced students into a closer relation with the mass of students. Groups must come into being whose life purpose is joint work on science. This penetration cannot take the form of mere discussion. The whole student life must, rather, be attuned to a new key, must express at every moment the dignity of pure service to the mind, the spirit. Each person must feel himself to be a member of a great community of culture that is charged with the mammoth task of allowing science to become the vital element within it. For there is no more unsullied place in which science can find its ultimate fulfillment than here, where it has become the whole life to a host of those dedicated to its service. A new spirit, a breath of fresher and nobler air, must waft through these halls. The university must truly become the temple of science and the home of a higher and nobler humanity.

6. PLATFORM OF THE SOCIALIST STUDENTS' PARTY

[1918b]

The Socialist Students' Party is a coalition of students and faculty members who believe in socialism.

1. First and foremost, the Party asks the cooperation of every citizen in bringing about socialism within the state. The execution of this project is left, however, to the various socialist parties that are open to the general public. The Party is, furthermore, not committed to any one particular method of promotion of its goals and leaves it up to its members to join one or another of the socialist parties.

2. The Party sees as its special task the conversion of the intellectual class to socialism. To this end it disseminates socialist ideas among students. The Party is well aware that its members can only train their fellow students to think in socialistic and democratic terms by working together with members of the proletarian youth movement. It is equally conscious of its obligation to begin immediately acquainting the youth of the proletariat with the intellectual values of scientific work.

3. As the party of the student community, the Party considers its task to be the reformation of the university in accordance with the socialist platform. It therefore demands an end to the influence presently exerted by class differences over the composition and operation of the university. The following are guidelines for the carrying out of this goal.

Guidelines for the Socialist Reformation of the University.

I. ECONOMIC DEMANDS

Abolition of all fees for lectures, tuition, registration, and examinations. Cost of printing dissertations to be carried by the state. On the other hand, well-to-do students to pay higher fees.

State support of students to the extent that they lack private means. This could take the form, e.g., of dormitories, the residents of which would form a self-governing community. Extension of state support to students who have attained student status through other than the normal educational channels. And extension thereof to needy foreigners on the basis of reciprocal arrangements with other states.

Just compensation for lecturers and assistants, and an increase in the number of such positions.

II. LEGAL DEMANDS

(a) *Basic Freedoms*

Autonomy of the university in matters concerning scholarship.

Freedom of expression for lecturers and students in both the spoken and written word, and extending to pronouncements from the lectern.

Freedom of assembly and freedom to organize for lecturers and students, without obligation to register or give notification.

Selection of lecturers and students without regard to their membership in any class, party, church, sect, race, sex, or state.

(b) *Admission*

Formulation of the conception of enrollment so that it signifies participation in the intellectual and spiritual life of the university.

Linkage of the university to a basic uniform system of public schooling in such a way that all intellectually qualified persons are given equally the opportunity to study.

Connection of the university with the system of mass education. Creation of legal procedures making it possible for any capable person to take part in university studies even without following normal educational channels.

The right of lecturers to admit to their own lectures and classes any person appearing capable of participation; the duty of lecturers to admit any scholastically qualified person.

Selection of lecturers by other professionals in their subject throughout the country, without participation or confirmation from the governmental ministry. Creation of legal forms that will eliminate the predominance of particular groups.

An increase in the number of examining professors.

(c) *Administration*

Establishment of student committees, to be put together on the principle of proportional representation and elected by secret ballot, the franchise being extended equally to all. Tasks of the various committees: self-government of the student body with respect to economic matters and disciplinary powers, and provisions for representation and mediation *vis-à-vis* the university.

Establishment of means through which the student body, acting in an advisory capacity, can influence the curriculum; e.g., through faculty assemblies or special commissions.

Granting the use of temporarily unoccupied lecture-halls to student organizations, for the purpose of holding meetings, without special permission from the chancellor, the sole requirement being notification of the department of buildings. Also, granting of the use of lecture-halls to scholars outside the university, for the purpose of giving scholarly lectures, upon petition of at least twenty students. These arrangements to be available on a term-long basis as well.

Bulletin boards to be made available to student organizations free of the chancellor's control.

Self-government for lecturers with respect to disciplinary powers and economic affairs.

Establishment of an appeal procedure for students and lecturers leading directly to a judgment by the regular courts of law, without previously going through the channels of the state ministry.

III. EDUCATIONAL DEMANDS

Fundamental reformation of instruction in accordance with pedagogical principles with the participation of the student body. Massive extension of the seminar system by drawing upon assistants.

Modification of examinations in such a way that passing them signifies proficiency in a subject rather than the mere possession of memorized information.

Abolition of groundless holdovers from the past, such as the preference accorded certain subjects.

Division of the university by subject systematically and in accordance with pedagogical principles. Abolition of the faculty of theology; the scientific content of the subjects in this field to be covered by the departments of history and philosophy.

Establishment of new chairs in hitherto supressed and neglected subjects, such as socialism, sociology, and education.

Fundamental reformation of the teaching of certain subjects that have ceased to be systematic and become merely descriptive or of merely historical interest; e.g., philosophy, jurisprudence, and philology.

University appointments for independent scholars or men from public walks of life whose work is pertinent.

Introduction of practical courses for specific professions, e.g., for lawyers and teachers.

Establishment of elementary lectures open to participants outside the normal channels of education.

Extension of scholarly institutes with the goal of letting students do a maximum amount of independent work.

Expansion of libraries; vast extension of reference libraries for all individual subjects.

7. SOCIALIZING THE UNIVERSITY

[1918a]

TABLE OF CONTENTS

FOREWORD

The Socialist Students' Party has given me the task of outlining in this initial essay their university program. This essay is the first in a series of pamphlets presenting to the public the demands and the plans of the Socialist Students.

The essay does not set out to present an exhaustive account of the wealth of individual demands, to establish their contents once and for all. It is our expectation that public discussion will generate new suggestions, and improve upon existing ones. This presentation is intended only to give the drift of the suggestions, to identify the spirit through which alone a new order in the university can be carried out. The concept of society and its legal organization appears to us to be vital to this presentation. For this reason we are furnishing the reader with the detailed set of principles underlying our system of demands; it was, after all, upon this basis that we were first able to give a systematic order to these demands.

Berlin-Lichterfelde-West/Ziethenstr. 2 Dr. Hans Reichenbach
December, 1918

I. SOCIETY AND COMMUNITY

The thinking subject's awareness of self constitutes the starting point of all laws governing the external world. Yet, in saying this, we must note that, among the objects of the external world, there exists one special class that can never be fully exhausted with the concept of an 'object', that can never be adequately characterized through observation of the object. Invariably, this remains a phenomenon that cannot be further analyzed, that can only simply be pointed out: that, in addition to our own 'I', our own self, there exist individuals that are centers of consciousness in the same sense but which are nonetheless eternally separated from consciousness of our own self. We may attempt to describe the process of knowing more precisely, to formulate its psychological aspect in laws. But in the end, we will inevitably be left with the phenomenon of our awareness of other selves, which can only be personally experienced and can only be justified by experience.

Given that we acknowledge this phenomenon, we may differentiate two forms of appearance that are in principle distinct. However foreign the individuals, the simple fact that another center of consciousness exists alongside my own self forms the basis of attention to the existence and functioning of an alien consciousness, following laws that are unknown, yet are known by force of experience to exist in a definite form. But this first form of side-by-side existence occasionally gives way to a second form, for there are moments when the contents of our consciousnesses are found to be identical, when, suddenly, strangeness is replaced with the knowledge that the same thought, the same law, is ordering our mutual sensations. When this occurs, the feeling of respect loses its cool aloofness and is transformed into a warm sense of community; the contiguity of persons side-by-side grows into a contiguity of minds, of spirits. Where it is the experience of a compelling idea that binds several selves together, the sharing of the experience grows into a consciously willed bond, into friendship.

We must clearly bear in mind that this duality cannot be overcome. Friendship rests upon the rare chance experience of mutually conceiving the same fundamental aims. It is meaningless to describe the totality of human beings as friends. The title is too stern, the duty that it proclaims too hard for that. For it signifies a bond in the name of the will, a union drawing its strength from common knowledge of an idea. Sympathy, concern, a kindly disposition are not sufficient bonds for this relation. Because of the heavy task allotted to friendship in human history, it should never be acknowledged without the sharpest scrutiny.

Community consists in that unification of human beings grounded in the experience of a common content of wills. Resting upon friendships between individuals, it is that bond in which spiritual [*geistige*] values can become directly creative and which, exercising an outward influence, is able to bring about the extension of spiritual values to other persons. Cultural development will always rest basically upon community, and all creative periods will find their support in communities.

Yet the community cannot embrace the totality of mankind, precisely because it is so selective. The community is invariably surrounded by a world that would, no doubt, like to penetrate and absorb it, but which instead spreads in endless ripples around the small group comprising the community. Alien though it is to the community, it would be a mistake to conceive of this surrounding environment as an enemy to the community. On the contrary, the efficacy of the community presupposes the existence of ordered relations in the surrounding environment and between it and the community. *Society* is the name for such an ordered association of people, mutual respect the basis upon which intercourse between the various willing subjects is possible. Opposed as the orientation of various wills can be to one another, a knowledge of the centeredness of another mind is sufficient to prevent our side-by-side existence from slipping over into unbounded hostility. Legal justice consists of the rules according to which co-existence is ordered. Consequently, legal precepts are laws based upon the foundation of respect for the person but applying also between persons of irreconcilably different characters.

Society and community are the two categories under which groupings of men must be viewed. That other centers of thought may arise outside the thinking consciousness is the phenomenon in which both these categories are grounded; while the one category sees the persons involved solely under the aspect of formal equality, the other constructs the relation upon the basis of the identity of the contents of their wills.

This division into two parts is not to be so conceived that any actual human organization can be definitively placed in one or the other of these categories. Rather, every organization may be viewed under either standpoint. Every human association seeks, after some fashion, its justification in a goal, feels itself to be a band of warriors battling for its realization and, as such, set aside from the great mass of humanity, to which its members are bound solely through legal forms and to whom they stand in the relation of a community to a society. This community itself forms, in turn, a small society within which still smaller circles may be marked out as communities. Insofar as this is the case, it requires a system of justice in accordance with which the subgroups may associate with one another. The clearest illustration of this duality is found in the party system in the state and the consequent endless subdivisions. A political party may be conceived of as a community with respect to the totality of the population, yet it separates into innumerable separate factions, and from these intermediate steps it is still a long way to that narrowest form of relation that we designate as a circle of friends, as a community in the strictest sense. Society and community, then, are correlates. Every society requires a separation into communities; every community must be subject to a society if its existence is to be sustained.

Admittedly, this duality becomes almost unrecognizable at the upper and lower boundaries. The state as such consists of such a large complex of persons that it can scarcely be regarded as a community and invariably appears only in the role of a society. It must be noted, however, that it once again assumes the character of a community just as soon as it becomes seriously opposed in its ideology to the international world; an example is the Russian socialist-proletarian state. On the other hand, small groups such as student associations with narrowly circumscribed educational aims and strong personal relations present themselves so clearly as communities that their character as a society is scarcely noticed. That they do, nonetheless, possess this aspect is to be recognized in the fact that even these associations set up regulations for themselves – set up, that is to say, a legal system that stipulates ordered relations for instances of disagreement. The less this legal system is used, of course, the more intimate and livelier the community. Theoretically, however, a body retains the characteristic of being a society until the number of persons is reduced to one – that is, it ceases to be a society just when it ceases to be a community.

We must also note that it is a mistake to link the concept of a society with the concept of a purpose, an end. Communities, too, realize ends. Ends are steps along the way towards the realization of an idea and may therefore constitute goals of the community, even if they are of sheerly utilitarian value. Indeed, we must describe every union formed around a purpose as falling under the category of community, for it endeavors to attain a definite material goal. In calling the state, that prime example of a society, a union dedicated to the preservation of economic existence, we are pointing specifically to its character as a community. Society is never more than a purely legal union, independent of a common content of will. The union may also serve a purpose, but this is not the ground of its legal validity, as may be seen from the fact that with the disappearance of respect, e.g., during a war, not even the utilitarian value is capable of maintaining legal relations.

The significance of society consists in its serving as the precondition for the existence and expansion of communities. It is, then, never to be regarded as an end in itself. For the meaning and purpose of human existence is always the realization of spiritual values. Which value system is to be preferred will be left open here, but that there is one superior system, and that man's supreme duty is to pursue it, will be taken for granted throughout these remarks. The reader is consequently asked always to bear in mind our basic tenets: that the building of communities working towards the perfection of values is the most important achievement and that the fulfillment of human tasks is possible only through this achievement. Nonetheless, if only the category of society comes under discussion here, the reason is that the realization of the precondition appears to be both more urgent and easier at the present time than the realization of the goal.

The materialist conception of history is the view that the evolution of spiritual and intellectual values is a direct function of economic conditions. Politicians who espouse its principles rest content with improving economic conditions, for they are convinced that a spiritual transformation then will follow and that battles for intellectual reform are therefore superfluous, serving merely to dissipate available forces. Their outlook begins with the fact that stratification of human beings according to their education and culture essentially coincides with stratification according to their standard of living; and they seek to employ for purposes of social reform the sociological law which gives expression to this fact. That the existing state of affairs is as described must be conceded. However, exception must be taken to the way in which it is interpretated in historical materialism, for this sociological law has validity only as a proposition about average conditions. The intellectual and spiritual variety within each social stratum is so great that a poverty of intelligence and cultivation is encountered as frequently among the rich as cleverness and creativity among the poor – not to mention moral qualities, which are surely not a function of material possessions. Thus the relation between economic well-being and culture, insofar as it exists at all, is comparable not to an unambiguous one-to-one functional relation in physics but to a law of probability which is compelled to reckon with a broad distribution of individual results. The economic leveling of human beings will not eliminate the differences in their worth. The problem of the order of values will remain, and the battle for culture will not end with socialization. On the contrary, the view taken here is that this battle can become a general confrontation only after socialization is completed. Up to now, barriers have been put in its path. Up to now, indeed, it has been impossible for certain logically coherent schools of thought to declare their opinions in public. Up to now, the *de facto* effective power has remained in the hands of a social stratum who can cite nothing other than their own ideology as a justification of their historical position and who assuredly do not offer the most fertile ground for the realization of intellectual and spiritual values. From this point of view, a justly organized society – which has never yet existed – may be called the precondition of culture. In what follows, we must look for the conditions that this society will have to fulfill if it is to become the precondition for the development of spiritual and intellectual culture, i.e., if the effects of the intellect are to be manifested in communities, if organizations are to be based upon mutual respect, if the just society is to arise among people who differ completely in material and intellectual respects.

Socialism has already undertaken to solve this problem, and it is possible to adopt its solution without accepting historical materialism, even for those who see the justly organized society as only a necessary and not a sufficient condition for civilized conditions. Back in the eighteenth and nineteenth centuries, advocates of liberal democracy believed they had found a solution. They called for abolition of political privileges tied to social position, which excluded certain classes of people from exercising political power in administration and in the government, and they believed that the ideal legal system would result from the introduction of formal legal equality, which found expression most particularly in the granting of the same active and passive rights of suffrage to all citizens. Socialists were distinguished by their belief that these formal legal regulations were inadequate; they held that the inequalities arising from the accumulation of the means of production in the hands of a few capitalists lead to their having effective control over the bodies and the time of the working class. Marx supplied the classic formulation: a worker works five or six hours a day for his own existence and two to four hours a day for the prosperity of his employer. Thus political equality is merely illusory; the reality is that the propertied classes have the privilege of being nourished without working with their own hands. Marx therefore demands the nationalization of the means of production. This idea has been further extended in modern socialism: only through economic equivalence – not equality, for work is to be paid for in accordance with its quality – will circumstances be created in which formal legal equality becomes actual equality. Socialism, then, may be described as an organization of society in which effective legal equality of all persons is brought about through the existence of economic equivalence alongside formal democratic equality.

It is not my intention here to go into the economic politics of this doctrine. I prefer to leave this debate to general political discussion, particularly the disagreement between the socialists and the advocates of agrarian reform, who wish to achieve the same goal in a simpler way, without first nationalizing capital. This pamphlet has a different object in view: to derive from a socialist foundation the consequences of socialist doctrines for one important component of social organization, the university. To this end, it is not necessary to specify the concept of socialism in one of its more detailed forms. The difference between these forms is in any case merely methodological. Throughout this piece, socialism is to be understood as the organization of society that brings about the effective equality of all persons in the sense defined above.

II. THE UNIVERSITY

The university is the most illustrious educational institution to be found in society. However, its significance extends far beyond the idea of society.

For knowledge, which it bears, is one of those spiritual goods that is a fulfillment of human existence rather than an expedient of the state. The existence of the ordered society is justified by its being a precondition of communities having an intellectual or spiritual basis. Knowledge is one form of the intellect around which a community is built, and the place where knowledge is forged and experienced is therefore divorced from all considerations of utility, bearing its value within itself. That the state is able to educate its officials at the university, that laws are discovered there which the state is able to exploit to its own technological ends, is incidental. The very existence of the university constitutes its meaning and value. It is one of those achievements that in no way serve as means, but are ends in themselves.

Not many people are granted the privilege of accomplishing productive scientific work. Those few, however, acquire through knowledge an intellectual structure of specific value. Accustomed as they are to approach everything given as though it were a problem, they never adopt a positive or negative attitude straightaway, but start out by criticizing, doubting, analyzing. They analyze every idea into its conceptual constituents, they distinguish between presupposition and consequence, between means and end. They search doggedly for the genuine nub of the problem, and, in critically testing its validity, they organize the whole problem according to its true significance. Yet, for them, criticism does not mean abstention from decision, nor does ordering mean abstention from ranking. Quite the contrary: their decisions will be sounder just because their critique is well-grounded and their judgment based upon analysis and ranking in accordance with value rather than upon emotion-tinged attitudes. It is also notable that thinkers of this calibre by no means become personally insecure through viewing all matters as though they were problems. Indeed, the ability to take the critical approach on any occasion renders them quite unafraid of any new ideas, and their consciousness of their capacity for conceptual ordering gives rise to a freedom and security of decisions never granted to fanatical believers. He who is wise need no fear even his own past errors, for he knows he has the power to correct every error uncovered with fresh knowledge; while the fanatic cannot but lose faith in his mental acuity through discovery of his own errors, such a discovery will merely serve the wise man as a stimulus to rethink the order previously assumed to be true. We admire the clarity of thought displayed by great scholars, their breadth of vision, the sure aim of their judgments, the precision and simplicity of their language, and we are aware that this surety is based not upon mere dull schematism, but upon consciousness of a capacity for a systematic approach to problems. This is the deepest spring of scientific productivity, for creativity is found only in the decisive mentality that is bent upon the uncovering of the connections between objects and the foundations of knowledge. The riches flowing from the resulting reordering of knowledge is living proof of the capacity of the critical approach. Not until we have seen the spiritual architectonic of such scholars and scientists with a real vocation can we grasp the idea that knowledge is truly a good, truly a form of human life. Just make a brief attempt to imagine the kind of community that would result if the university were populated by such a company of scholars! We see before us the circle of pupils, wide-eyed, with conviction written upon their faces, hanging on the words of their mentors. And then we begin to understand that the scientific community represents one possible form of fulfillment for human existence.

We have no intention to seek the laws governing such a community. To attempt to construct such a university by means of organization and the formation of parties, through propaganda and technology, would be a childish undertaking. We must clearly bear in mind that the laws of these islands of culture can only be gotten from experience, that their origin is connected with the appearance of leading figures who breathe life into the scientific community by means of their own lives and the example they set. Nor do we propose to offer a list of means whereby such communities could be created. Contemplating the model offered by Plato's academy, we are well aware that it was founded by Plato, not by academics. Clearly as the hope for such a university may hover before us, we have no desire to create it through social reforms of one sort or another. We wish, instead, to wait until it is called for by those who have the right to demand it.

If, nonetheless, we are to discuss university reform, then our observations must set out from quite a different starting point. For this is not a philosophical inquiry into the structure and meaning of science. The purpose of this pamphlet is simply to present the concrete measures that can be taken today in order to steer the university onto a new course. We are convinced that today's university is far from living up to its philosophical model, and we know that only a revolution will be able to create the mentality from which the new academy can grow. But we know, too, that this revolution must be spiritual, intellectual, and that, in this era of political revolutions, its day has not yet come. We cherish the idea of the university too highly to attempt to implement it directly through liberal discussion of education. Given the human material presently filling the universities and the intellectual level of the average academic, for whom comfort and a respected position are still the determining factors, we believe that the intellectual revolution cannot be carried out, and we consequently do not advocate forming a party representing the university in the philosophical sense. We would regard as possible no more than the formation of small, intimate groups, with a carefully selected membership and genuine leaders at their head, which would take on the task of founding an isolated, private university *community*. But we demand something different. We pose the following question: what conditions would at least make it possible, even if not inevitable, for a university of the proposed kind to come into being? Up to the present, formation of such an academy has actually been *impossible*. We therefore see the university as falling under the category of society. As it is in the hands of a large number and variety of people whose aspirations and attitudes range from one extreme to the other, we must inquire into its legal organization, as we do with every society. And as the members of the university are one part of that greater complex, the people, we must also investigate the position of the university in the political society surrounding it. Only if university reform is directed at these relations will circumstances arise making possible the construction of a university centered on ideas. On the other hand, such demands as we make present feasible goals, i.e., goals that can be realized even today, given the present state of public insight and the public will. Like every form of technology, politics is the art of the possible. Let us deduce from our theoretical model those presuppositions the realization of which, as social demands, is possible in the present phase of our society.

The reason the university must be viewed as a society lies in the vast expansion of academic education. The very number of students, now over ten thousand in large universities such as Berlin, is sufficient proof that university complexes can no longer be organized in accordance with the assumption of a common will. But even the number of scholarly viewpoints that must be taken seriously is so great that there can be no question of unity on the basis of a single doctrine that is to be accounted true. Were scientific truth so unambiguously apparent that none could fail to acknowledge it, the question of, e.g., which scientific opinions are admissible would be cleared up. In declaring such unity to be impossible, we are certainly not asserting that there is no such thing as unambiguous scientific truth. On the contrary, the existence of scientific truth is presupposed by every assertion that claims recognition, including the ideas expressed here. Nonetheless, the situation in scientific fields calls for such a plethora of detailed knowledge and such rigorous investigation before a final judgment is made that only a few minds possess the truth; who these few are cannot, unfortunately, be determined from without. But that is just the way it is: the *de facto* power in society lies in the hands of the ignorant, who have the job of deciding whom among the learned they will admit to the university. Formulate the requirements for entry as you will, they derive their force from the state executive who stands behind them. The problem can only be solved upon the basis of legal stipulations that, instead of restricting selection to those holding one particular scholastic view, admit as equals a whole series of conflicting opinions so long as they possess those formal qualities designated by the expression, 'scientifically qualified'. It is relatively easy to establish the presence of these qualities, and the verification can be carried out by state officials, the examining teachers. The concrete decision as to which particular view is correct is left to the individual, and hence to insight into the actual content of the discipline, and the construction of general scientific views will consequently come about through factual agreement between those able to make sound judgments.

What applies in principle to research and theories applies likewise to each single facet of the life of a university. The fact remains that the differences involved are so great that they can only be ordered into a legal organization, never eliminated. The attempts of the Catholic church to establish a single system of scientific truth dogmatically and to support it by means of the powers of the state must be passed over as having failed in the event. That the real truth is certainly not chosen in this process has been shown far too clearly. Let us, then, rest content with a system that allows the true theory at least the possibility of attaining recognition.

The problem of giving to the university the form of a legally organized society within the legal organization of the state must be solved in accordance with the answer to this question: what are the most favorable circumstances for the growth of a scientific university? In facint this problem, we may once again make use of the concept of effective legal equality, discovered and introduced by socialism. In addition to formal democratic equality for all members of the university and all scholastically qualified citizens, we must also demand economic equivalence, for differences in income and standard of living have a profound effect upon the scholarly achievement and outlook of the individual, for which effective allowances must be made. It is obvious, then, that the problem facing the university cannot be solved without a simultaneous solution of the problem facing the state; only a socialist state will be able to guarantee economic equivalence for all who are scientifically qualified. In demanding the socialization of the university, we are well aware that this demand is meaningful only within the framework of the more sweeping demand for socialization of the state, and we raise it here only as a partial demand, albeit for a clearly circumscribed area. That we do so at the very moment when the socialization of the state is getting under way bears witness to our consciousness of the connection between these demands.

Socialism is the overcoming of classes. By a class, we mean a stratum of the population whose function in society cannot be justified on the grounds of its distinguishing characteristics. For instance, the capitalist class occupies a leading economic position, although its distinguishing mark, ownership, fails to reveal any connection whatsoever with qualities of economic leadership. The socialist ideal is to eliminate all legal privileges based upon secondary characteristics, in order to allow for the ranking of people in accordance with their potential. It is vital to emphasize here that a class structure has a restrictive influence on the development of the university, as elsewhere. For up to the present, scholars and scientists have only been chosen from a narrow stratum of the population, the intellectual powers of the remaining strata of the people being thereby lost to science. Surely these powers are not negligible; insofar as it exists at all, the functional relation between ownership and intelligence is certainly not univocal, and given the broad distribution within the individual strata, we can expect to find numerous intelligent minds among the proletarians, indeed, probably more intelligent people in absolute numbers. Exclusion of the proletariat from the university, then, means a restriction of the possibility of scientific development which is, after all, always tied to people.

In opposition to our conception, a theory has been developed according to which the existence of an affluent social class is actually a necessary condition for intellectual culture. Were all goods equally distributed, the average amount received by each citizen would be so small, it is said, that it would no longer be possible to maintain any intellectual standard. In order to preserve the spirit, we are to tolerate, alongside the vast class of workers engaged solely in economic production, a small group of superior persons who, supported in luxury by the labor of others, like drones in a society of bees, give the state its really fruitful power. This view is correct insofar as it holds that the performer of intellectual work must be released from economic tasks, but it errs in assuming that this arrangement is to be carried out by segregating a single class. For a class takes in members on the basis of birth and possessions, not scholastic qualities; even if it takes in a relatively high percentage of intelligent people in the process, the absolute number of such people will still be greater among the working class. The objection that scientific development necessarily presupposes a certain level of education and aesthetic cultivation is contradicted by the fact that a large number of scientists do come from the ranks of the proletariat or the petit-bourgeoisie. The weaker claim, that this level is merely conducive to intellectual achievement, must be conceded but we draw quite different consequences. For such a level cannot be maintained in a class containing a large percentage of worldly, materialistic people in which, indeed, these very people, having the greatest capitalist capacity, exercise the greatest influence. The mediocrity of the propertied class, the banality of their mental outlook, which has reduced aesthetic and scientific forms of expression to nothing but relief from the strain of nervous work is the best proof that a creative spiritual level is *not* guaranteed to exist in such a class. On the contrary, the maintenance of a level fostering the creative spirit requires that members of the cultured class be selected solely according to their mental characteristics. Those who are capable of receiving and continuing cultural traditions must stream forth in a never-ending procession from the collective whole of the people. The educational system itself must bring forth the cultured stratum, whose membership must be drawn from every circle of the population. The great accidents of birth must no longer be permitted to carry out the selection of creative minds; rather, they must be counterbalanced by a system of rational selection.

The exclusive hold upon education that the propertied classes have maintained up to the present has not only restricted science and scholarship, but also carries a serious danger for them. Under the influence of the ideology of this class, a certain caste mentality has developed in academic circles, excluding from universities certain intellectual schools of thought that undoubtedly possess proper scientific qualifications. I am thinking, e.g., of Freudian psychoanalysis and socialistic theories of political economy. Their exclusion is not grounded upon scientific objections, but on objections of morality or of taste derived solely from the class attitudes of this narrow bourgeois stratum. This caste spirit has run so rampant that family connections with incumbents in university posts have played a role in the selection of occupants for vacant posts. Consequently, men lacking all scientific ability have frequently attained academic positions while those who are qualified have had to stand back. Similarly, socialists and intellectual radicals have been barred from becoming students, while the sons of well-to-do parents have often been assisted through their school-leaving examination and admitted to university study by this means, rather than because of intellectual suitability. As a result of this closed system, not only is the number of scientists and scholars chosen too small, but the selection process within the narrow class is also unjust. Just examine briefly the extent of the connection between the degeneration of certain subjects, such as philology and jurisprudence, and the caste mentality. At the very least, it is true that the scientific development of the university has been most severely hampered by this outlook and that the old legal order of the university represented the most unfavorable conditions imaginable for learning. If we ask after the most favorable conditions, our inquiry leads naturally to socialism. For if our response to the question, "How is the university to be justly constituted?" is "Admission must be granted to as many as possible of those who are intellectually qualified, and to none who are not", then we must add that socialism alone can guarantee such equality of opportunity, as it places economic equivalence alongside formal equality.

Socialization of the university is a necessary condition even for those of a purely scientific orientation who wish to realize the idea of a university. These thinkers, too, will have to regard the university under the category of society, and to modify it accordingly if one day they are to make a community of the university. That we look to the preachers of socialism is a matter of course, but we will also turn to those for whom socialism is not a religion and who regard the realization of intellectual and spiritual values as a task on its own, aside from economic development. We assume in them only that absolute sense of justice which tells us that everyone who is able should participate in the fruits of the mind, that science is not an amusement or a recreation for the aesthetically cultivated affluent classes, but a good belonging to humanity, the riches and ennobling influence of which is desired by everyone to whom nature has granted an inquisitive soul. We presuppose nothing more than this clear sense of justice in demanding the socialization of the university. Thus we seem called upon to unite all those who think justly about the goal of the justly ordered university. Let each continue to pursue his special plan for the intellectual development of the university, and for the community of scientists over and above this! For the time being, it seems impossible to attain a unity of greater proportions without producing mediocrity. But let every one of these thinkers recognize that the just university is a precondition of his particular university community, too, that the most urgent task is to give the university the character of a justly ordered society.

We direct this rallying cry to all thinking scholars, to all educated men of discernment, and to all proletarians who already have such a vivid sense of the spirit of science which has hitherto been withheld from them that they sense the significance of a pure higher scientific institution. The creation of a just order for the supporters of that most honorable and worthy of all goods, scientific knowledge, must be our goal. Let us therefore turn this idea into a solemn, unifying vow that no undertaking is more serious and urgent than the socialization of the university.

III. THE NEW ORDER

Socialism not only teaches us a new form of society, but also shows us the way to reach it. It starts from the idea that we cannot sit and wait until mankind as a whole undergoes a rational awakening and introduces socialism of its own free will, as the rational form of society; instead we must exploit the economic development of present-day society in such a way that socialism will be forcibly introduced rather sooner. Thus the social class that is oppressed and bled dry under the present system and that therefore calls for a socialist state not only – or not yet – on rational grounds, but also out of a healthy egoism, must be organized to do battle against the old order of society. This stratum is the proletariat, and its egotistical battle is the class struggle. The impetus to this battle, stemming, as it does, from a natural drive for self-preservation and the elementary need to live, will be so powerful that in a relatively short time it must lead to the overthrowing of the oppressor; in defiance of all deeply-rooted bourgeois conceptions of property, the resistors will then be relieved of their private possession of the means of production. By means of a power struggle, through the rise of the lowest class, that form of economy will be achieved which alone makes possible equal justice for all, which is to say, for the other classes as well. We are not taking the view that rule of the proletariat must be a dictatorship. Dictatorship is government by a minority; but in European industrial states, the proletariat represents such a vast majority that their rule can be realized constitutionally in a true democracy, in which full political equality is found.

The revolution in Germany has substantiated the teachings of socialism. The class struggle of the proletariat has led to shaking off of the yoke of the ruling class, has eliminated the anti-democratic privileges of the nobility and the capitalists, and is currently in the process of creating a legal structure for the democratic introduction of socialism. The question is whether the same method is applicable in bringing about socialization in the university.

Just as the means of production are in private hands, rights of ownership in education fall to the capitalists. Attending academic high schools and institutions of higher learning involves such heavy expenditures that the proletarian is excluded from them. His long working hours make it impossible for him to acquire any cultural goods during his few hours of leisure. Education and culture are a monopoly in the hands of the wealthy, and it might be thought that, like every other monopoly, they could be expropriated only by force. And yet there is a decisive difference. Education [*Bildung*] is a spiritual, an intellectual good, not a material object that can be grasped in the hand. It is not related to the bourgeoisie in the external way that holds between them and economic goods, which they own only by virtue of a legal fiction; rather it is a definite intellectual disposition or cast of mind, a definite structure of the brain. The bourgeoisie gained ownership of education not by theft or by a legal declaration, but by a gradual incorporation of maturing youth into the ranked order of human cultivation. Unjust as the restriction of higher education to bourgeois youth no doubt is, its success cannot be eliminated all at once. Education cannot be seized in the way in which vast property holdings are seized. It will not be possible to drive the bourgeoisie out of the universities and leave the new order to the occupation of the empty building by the proletariat. Education would be driven out along with the bourgeoisie, culture driven out along with their intellect. Riches that have accumulated for three thousand years would be destroyed along with the bearers. If socialism is not to begin by perpetrating the supreme act of economic squandering, the annihilation of culture, it must find some means other than expropriation for achieving its ends in the university.

We propose the institution of a legal order whereby proletarian youth will be fitted into the process of education, culture gradually passed on to the collective population by its bourgeois carriers. This process will take at least one generation, probably more, but we choose it because there is no other way. If we are accused of wishing to employ the spirit of the bourgeoisie for the education of the proletariat, we can only say: we choose it because it *is* creative spirit, creative intellect. If intellect were located among the Hottentots, we would seek it out even there. We would go wherever the spirit dwelled and learn from those who unjustly possess it until the day we knew it to be firmly in the possession of ourselves and our comrades. We would have no choice: Are our longing souls to starve, that we may enjoy the fanatic's righteous sense of triumph?

A certain favorable circumstance in prior sociological developments comes to our assistance here. Industrial development has increasingly deprived capitalists of peaceful enjoyment of their surplus value, forcing them to frenetically increase their capital, so that they have had no spare time for the pursuit of intellectual occupations and have abandoned education and culture to their children and their children's teachers. Gradually, capitalists and culture have become alienated; the administrators of education have become functionaries of the capitalists and have, generally speaking, been the architects of the ideology proclaimed and upheld by capitalism as the necessary form of society, yet in this process a part of the social stratum dispensing education have themselves become proletarians, have, that is, become workers who are paid no more than the minimum required for subsistence. This intellectual proletariat, which is gradually becoming conscious of its position, will join the industrial proletariat, wage class struggle to the very end along with them and then take on the role of the stratum destined for leadership in the cultural education of the proletariat. Among them are, too, all the proletarians who through unwavering diligence and admirable powers of self-restraint have managed to find the path to learning, even in the old society. They will also be joined by all those who, though born into the bourgeois class, have had their eyes opened by insights into science and into the deepest springs of culture, who thereby cease to belong to the bourgeoisie, recognize the senselessness of the old order, and strive for the overthrow of class domination and work for the institution of socialist society as the most vital goal, for the sake of truth and culture. Led by this company, the proletariat will be able to enter the world of culture and to take over the university. And if, in the process, the bourgeoisie take on the role of intermediary, their influence on social morality will be offset by the afore-mentioned leaders, their pedagogical activities confined to that area in which they are competent to fulfill them: to the transmission of the scientific and aesthetic fruits of culture.

With these thoughts in mind, we have worked out the following program for a new order. Legal forms are suggested which are tailored to the present-day university and presuppose the existing human resources but which, at the same time, will remove present restrictions and bring about the gradual suffusion of the university with gifted persons from every class. In addition to this change in admission requirements, we point out various possibilities for vitalizing the scientific spirit and reviving the scientific atmosphere, which is largely stagnant. At the same time, we will investigate the problem of developing without a break in continuity the new society of scholars out of the old caste of professors.

In accordance with the socialist idea that the just society cannot be constituted solely through formal laws but also requires auxiliary economic measures, we will place certain economic demands on the university, as well as others pertaining to legal structure. A third group of demands, those concerning pedagogy, will be added to these two. But in this investigation, as we have set out solely from the concept of society and have avoided any particular form of scientific bias, these demands will be exclusively of a socio-pedagogical nature; that is, they will represent only such instructional measures as can be applied in a society of diverse wills. The totality of our individual demands, which contains transitional measures along with the fundamental forms of the new order, is summarized in a table at the end of this work. [The table has been omitted, as its contents coincide with the outline presented in *Platform of the Socialist Students' Party*, in this volume. – Ed.]

We anticipate that the political reform which started with the revolution will give a new constitution to the university, too, and it is with this purpose in mind that we sketch our model. The constitution will be laid down by state authority. To be sure, that is the most the state can do; it cannot determine the intellectual form that institutions of higher learning will assume in the future. But in providing a legal structure for the universities, the state will at least create the conditions most likely to foster the flowering of the intellect. Restriction of state power within these boundaries is the greatest service that socialist democracy can perform. Imposition of this restriction would allow the political and economic development, the class struggle of the proletariat, to create the pre-conditions that will culminate in a cultural order born of the intellect, the spirit. This new order itself we shall then leave to the growing insight of reason, to its unlimited expansion and the persuasive powers of its heralds. We will, however, consider that society exemplary that does not mistake its legal precepts for divine laws and the directives of its leaders for ultimate commands of reason.

IV. ECONOMIC DEMANDS

Economic inequalities influence the student body in two respects. On the one hand, the cost of university study is such that only a certain percentage of those citizens who are qualified are able to attend a university; thus the student body does not have the proper composition from the standpoint of scientific and scholarly ability. On the other hand, the high expenses create inequalities within the student class itself which in practice result in unequal justice.

It is generally recognized that, unlike other citizens, the student cannot use the main part of his working hours for economically productive work, that is, work that is immediately reimbursed. It is quite unnecessary to offer any further proof that scientific study is too extensive an undertaking to be carried out in evening hours of leisure, not even in the socialist state where these leisure hours are increased. As a consequence, the student is dependent upon support by other people's money until he reaches the age of about twenty-four. The costs of study as such, the university fees, are small in comparison to living expenses. Parents have usually met the costs; the few, slender scholarships which are restricted by irrelevant stipulations such as the requirement that the student belong to a certain religious community or come from a certain part of the country, have not been sufficient to satisfy needs. This situation has made university education a preserve of the well-to-do middle classes. This is so clearly intolerable from the standpoints of justice and of science that we need not emphasize that fact once again. What does need to be pointed out is that existing economic inequalities result in gross injustices even within the class of students. For there have always been a large number of students who must earn their keep in part or altogether. Observe these young people, and how they fill their free time with badly paid private tutoring, how they allow their bodies to become run down through poor nourishment, how they are excluded from the aesthetic goods of life – theater, concerts, travel, cultivated living. It is clear at a glance that they are indeed deprived of certain possibilities of development that have served others as routes to freedom of criticism and maturity of judgment. Nowhere is the connection between material existence and intellectual development more evident than it is here. Some really good minds have run aground on this problem; their critical discernment has stultified into unstructured scepticism, the multiplicity of their interests has been paid for by the inadequacy of their skills. To the objection that the strong will persevere, we reply that the best are not always the robust, that no one has ever counted up the number of noble spirits who have been able to save their souls in the economic struggle only by abandoning the pursuit of learning.

It is as obvious that these injustices must be done away with in a era that proclaims a system of justice based on principle as it is difficult to indicate the way in which they are to be eliminated. Once the socialist state has been established, education of the youth of the nation must take place entirely at public expense, for herein lies the real injustice of past society, the root of all class differences: that the cost of raising, of educating children is charged to parents means that social position is hereditary and thus that education is tied to class structure. Even if income is uniformly proportional to economic production and social rank is identical with ranking according to value for the parents, the classification of the children might be quite different. In order for the correct structure to hold for the next generation and to continue to hold for the whole population, the education of children, indeed, their entire support, must be financed by the state in proportion to their intellectual abilities. The taxes required for this purpose cannot be greater than the amount that would otherwise be directly spent on raising the children. It is harder to decide what should be done during the transitional period before socialism has been carried out completely. We must simply insist on the principle that means *must* be found. Even for the modern state, which is still struggling for its constitution, there is no more serious task than to lead its youth into the universities; no era is as dependent as the present one upon leading minds. Of course, there have always been leaders, and especially political leaders, who have discovered and proclaimed original ideas without having had a university education, but that simply proves that the university does not yet include all those who spiritually belong to it.

Students without private means can be supported more economically through the establishment of dormitories. With these, another valuable institution can be created. The problem of education will undoubtedly find its intellectual and spiritual resolution through the form of communities – this much can be confidently predicted even today. For establishing dormitories will offer a direct means of forming student life into living communities, through athletics and dancing, through music and other forms of artistic expression, until they comprise a genuine cultural community. True to our principles, we will not give any further indication of the concrete solution to the problem of education; we consciously restricted our inquiry to the social and legal conditions necessary for its solution. We may, however, reasonably demand that the numerous attempts now being made to create a student community be granted a possibility of realization. It should also be noted that the dormitories to be set up are not to be under the surveillance or control by the state, but are to be self-governing communities, determining their own development.

The state's obligation to support students must also be extended to those who come to the university without having approached it through normal educational channels. It must also include foreigners of insufficient means; until the day when socialism holds sway throughout the world, reciprocal arrangements with other countries will probably be expedient. Later on we will give a justification for these demands.

If students are supported by the state, some supervision of their education will become necessary, for there will have to be some guarantee that the student fulfills the purpose for which he was released from the need to support himself. It will be the job of pedagogical reformers to work out how the controls can be combined with the maximum degree of freedom for personal development. We stipulate here only that the same controls must be extended to those students not enjoying state support; otherwise inequalities will reappear.

At present, the situation could be improved immediately by eliminating, at least for poor students, all forms of university fees, that is, those for matriculation, registration, attending lectures, and sitting for examinations. Moreover, the state should pay printing costs for dissertations – if these works, which serve the purpose of an examination, continue to be printed in the university of the future.

The other aspect of the economic question concerns the teachers. We have at present three classes of instructors: full [*ordentliche*] professors, associate [*ausserordentliche*] professors, and lecturers [*Privatdozenten*]. [For ease of reference, we have assigned these rough descriptions in English of the German academic titles, but they have no exact counterparts in the English-speaking academic world. An *ordentlicher Professor* is holder of a chair in a major subject, with full rights of a member of the academic corporation, and full membership in the administration of his faculty. (A German 'faculty' is the equivalent of a division, e.g., of the social sciences, or a school, e.g., of law.) He receives a regular salary and the benefits of a state employee. An *ausserordentlicher Professor* has restricted rights and benefits with respect to both teaching and administration somewhat like the British 'Reader'. A *Privatdozent* traditionally had no participatory rights, benefits or regular salary, and Reichenbach describes this position in detail below. – E.S.] Unless they happen to have special teaching assignments, the lecturers [*Pivatdozenten*] receive no income except the money paid by students attending their lectures. A German lecturer, a scholar and a teacher of youth, a worker fully engaged in intellectual pursuits, of the average age of twenty-seven to forty years – and he may well continue in the same position until he is seventy – receives a yearly income of a few hundred marks. The result of this situation is that only large-scale capitalists supported by investments are able to become such lecturers. Thus the selection of academic faculty is even more plutocratic than the selection of students. These circumstances are so absurd that we must look for their original purpose if we are to grasp at all how they could ever have arisen. Originally, only professors had university teaching assignments. Then the position of lecturer was invented in order to circumvent the restriction of scientific and scholarly circles to state appointees and to extend the possibility of teaching to every scholar outside those circles. The intention was throughly democratic; it was the freedom of teaching. But, like every other democratic measure that is not supported by economic compensation, it achieved its opposite and became plutocratic, with the result that lectureships were taken up, not by the intellectually and scientifically qualified, but by the economically independent. This result is thoroughly unjust, and the injustice is increased still further by the caste segregation arising from the plutocratization of the class of academic teachers. (For every professorship is filled from the ranks of the lecturers.) Conditions deteriorated to the point where not even all economically qualified applicants were accepted as lecturers. Those representing certain schools of thought were excluded – principally, of course, the intellectual radicals. The resulting lack of freedom in teaching is recognized today and in many quarters has prompted demands for free lectureships; that is, the granting, on principle, of permission to become a lecturer to every scholar who applies, without selections being made by the faculty. The ineffectual nature of this typical liberal demand is readily apparent. It would only clear up the symptoms of the caste system, not cure its causes, and would continue to maintain the influence of economic power over the university. Even quite aside from this, it is a fundamentally mistaken demand. Admission to lectureships must be made dependent upon possession of outstanding scientific qualifications, which is by no means guaranteed by the passing of an ordinary academic examination. Thus the decision to admit must be made by the faculty. The problem is to find a system that renders domination by individual groups impossible. Such an arrangement will be described below.

To get rid of the economic influences at their source, we must demand that all teachers be paid a full salary. They must also be included in retirement pension schemes, as long as old-age pensions are not universally mandated and civil servants are granted special privileges because of their low salaries. We know full well the dangers entailed by such measures. They will make the academic lectureship into a tenured profession, attracting many who are simply interested in a comfortable berth and have no true vocation. Nonetheless, this liability does not constitute a viable objection, for no one, surely, will maintain that a better selection results from the past system. The very object of the social organization of the university is the discovery of a selection system securing domination of the scientific, scholarly point of view. Why should we suppose this task to be incapable of solution? In any case, it is a necessary condition that pecuniary considerations be permitted *no* influence upon selection, and this can only be guaranteed by payment of lecturers. Nor must we lose sight of the underlying principle. Scientific work is productive work for the people and must be treated as the equivalent of economic work. The scholar is not to be left out in a state that offers payment for productive work only, for it is recognized that he cannot practice an economic occupation from which he lives, in addition to doing his scholarly work.

What applies to lecturers and students applies equally to assistants. While they are not yet actual university teachers, they are employed in teaching – far more in the modern university than they were in the past – and must therefore be paid. Moreover, the state has a special interest in supporting them, for today's assistants are tomorrow's lecturers. Pension arrangements will have to be omitted, however, as it is not yet certain that they will remain in the academic world.

The economic requirements we have outlined will mean a substantial increase in the state's university budget and their feasibility may therefore appear to some to be dubious. We will not at this time go into the exact way in which the state's income can most suitably be extended in order to cover these particular expenses. A special education tax has been suggested. It would also be possible to increase revenues by requiring substantially higher fees of well-to-do students, proportional perhaps to their monthly allowance or to the tax bracket of their supporter; for the current fees, we must recall, by no means cover expenditures, with the result that the state is largely supporting the education of the wealthy. But how the state orders its house in the future must be left to public politics. It is possible that in the foreseeable future all forms of indirect taxation will be replaced by direct taxation, so that one-sided measures of this kind become unnecessary. Let us emphasize just one point: increased expenditures for education are inevitable. The economic demands made here are fundamentally necessary. If the society of the future wishes to have a class of leaders consisting of free minds, socialist and democratic, which is to say just, it must come up with the subsidy necessary for their education. If it refuses to comply with this demand, it will cut itself off from the path to that just legal system which can only be shown by [such] leaders.

V. LEGAL DEMANDS

The legal principles of the university begin with a system of fundamental freedoms that form the presupposition of all positive legal forms.

First, we demand the scientific autonomy of the university, that is, independence of scientific research from surveillance by the state or any political party. This demand still requires emphasis today, for it has never been fully realized in the past. The Catholic church has, for instance, had an extensive influence over the assignment of teaching posts, and it even demanded that Catholic professors vow to uphold the canonical truths of the church. Moreover, it extracted an agreement from the state whereby a certain number of positions were reserved for instructors to whom they gave their mark of approval. Therefore, this must be made a fundamental principle: the university is completely free to make judgments concerning scientific truths and may even make decisions running counter to the advantage of the state or of political parties.

What is true for the university as a whole applies equally to teachers and students: for these, too, we demand freedom of speech and of the written word. Freedom of thought we need not demand, for it has always existed, but freedom of dissemination has been limited. The teacher must be free to set forth every opinion he holds to be true even from the lectern, that is to say, even in the course of his university teaching duties. We further demand for students and teachers freedom of assembly and freedom to form associations without giving any formal notification – the registration requirement being, after all, nothing but a covert way of requiring permission. This demand does not need much justification at a time like the present. We have already pointed out that we cannot count on a clear univocal selection of scientific truths and must therefore be content with formal legal stipulations. But no group today seriously contends that it would be purposeful, or even possible, for students – and thereby the whole population – to be under a guardian, to be protected from 'erroneous' teachings.

Finally, we demand that all academic rights be made independent of class, party, church, sect, race, sex, or citizenship. Only the last two require more detailed explanation. Equal rights for women are widely opposed, and alongside the crude objections of the 'women-belong-at-home' variety, some serious grounds of opposition have been offered. Recently, some young intellectual radicals, calling themselves the anti-feminists, have formed a movement for the exclusion of women from science and learning. Hans Blüher, its leader, makes a distinction in principle between intellect and Eros and calls intellect a secondary sexual characteristic of men, while reserving for women the realm of Eros. That Blüher's aesthetically and psychologically sophisticated ideas do considerable justice to the warmth of the male and the female Eros cannot be denied, yet we must equally dispute their claim to represent the nature and the laws of the intellect. For to make the intellect the mark of a biological class is to confuse philosophy and natural history. The fact of the existence of another individual, just like that of my existing 'I', can in principle only be grasped intellectually. Intellect is the precondition of society, not its function; and it is, equally, the precondition of every erotic relation (no matter how unspiritual). Qualitative differences between the sexes can pertain only to secondary functions of the intellect, such as productivity, imagination, activity, spontaneity, continuity and others. In this sense, certainly, there are essential differences between man and woman. If we scrutinize Blüher's representation of the male spirit more closely, we find that for him such secondary qualities are identical with the concept of the intellectual spirit. For him, the intellect is productivity, creative experience, imagination, artistic creation; a sum of characteristics that could, perhaps, be called masculine but which are found in only a few individuals, even among men; possibly only among men of letters, the literati, in the best sense of that word. That this conception of the intellect cannot be a basis for categorization, but simply sets off particular communities, is readily apparent. If the anti-feminists would restrict themselves to bringing together in an intimate community people of their persuasion, they might succeed in creating something of cultural value, but their conception can in no way bring about a fundamental division of society. We must add that there is by no means an invariable and univocal connection between the artistic, creative spirit and the male as a sexual being. Insofar as it exists at all, it is merely a probability relation which may fail to hold in numerous cases: a wide distribution of results in to be expected with such laws of psychological correlation. Thus, a law of this nature may in no way serve as a basis for a division of society into classes. The exclusion of women from certain social functions would mean repression of a class in just the same sense that exclusion of the proletariat has amounted to repression heretofore. The fundamental principle of every justly organized society is the distribution of functions according to the ability of the individual, independent of any other characteristics. Even if it were proven that women are not suited to govern, women as a class could not be excluded from the government on principle; the selection would, instead, have to be made for each person anew on the basis of that person's qualities. The alleged law of nature could only be expressed as the fact that relatively many men and relatively few women have been found to be suitable.

Thus equality of men and women must be a principle of the legal organization of the university, for science must never be the preserve of a class – and, we add, not the class of men. If women really are less suited for scientific work, this fact will show itself in the proportional number who are admitted as qualified. Under no circumstances must university selection be allowed to rest on a personal conception of the intellect, such as that of anti-feminism, which, e.g., gives no place to the mathematical sciences and formulates a completely one-sided view of philosophy as artistic experience. If anti-feminists find it impossible to carry out serious scientific work in collaboration with women, they should be free to form their own communities and male associations. But such personal sensibilities have nothing to do with the law. We know that the organization of the university must be conceived under the category of society. Consequently, equality of the sexes must be a fundamental demand.

Our other controversial demand is that of equal rights for foreigners. No one seriously doubts that science has an international value, yet the view has grown up that the university is essentially an institution of the state for the education of its citizens, who consequently enjoy a privileged position. This notion can be criticized from two viewpoints, for it is contradicted by the idea of science as much as by the idea of socialism.

If knowledge has truly an international value, this means that it is anchored in the nature of man as such, not in a particular kind of man. It is, after all, one of the fruits of culture, possession of which is not a means, but an end of human existence. The university is, accordingly, not essentially an educational institution, but an institution that directly realizes the fulfillment of the goals of culture. It is not a state organization, like a court or a parliament, which the state sets up for itself as an instrument for fulfilling its functions, as the citizens' tool for maintaining social order. It is not even true that the younger generation is brought up by the university; this task is left, instead, to the schools, while the university accepts a small group comprised of only the most gifted. The question of nationality plays no role for schools, for the number of aliens is negligible in comparison to the number of German children. It is nonetheless noteworthy that no one contemplates restricting the rights of foreigners in schools – a proof that no basic objections exist to equal treatment. The state erects the university for a select circle because it is able to achieve the ultimate realization of its intellectual, its spiritual goal only in its best citizens. If, then, it means to gain such a goal, it must permit the laws of this institution to be determined, not by what is useful for the state, but by the requirements of science and learning. To bloom fully, these must be upheld by persons not restricted in any way, and selected from every attainable group. Bear in mind the thoroughly international character of scientific research, how theories conceived in Zürich are experimentally confirmed in England and taught in Germany: think of the rich yield in which scientific investigators take part through contact with thinkers from other schools. To reduce a science to one nation today would mean to limit, to distort it. It is a solemn duty of the state to employ internal revenues for the creation of an institution that works for the fulfillment of the highest human values and gathers to itself the whole wealth of supra-national culture. Then, let the university throw open its gates to all those who seek knowledge, from whatever country; let it embrace all as citizens of equal rank and create in its organization a model of the justly ordered cultured society, that recognizes no differences between persons other than their achievements.

Development of the university as an international institution will go hand in hand with the development of socialist society. For national boundaries, which no longer section off even a race, but merely an historical complex, lose their significance at that moment in which the contradictions of economic class dominate the world, ushering in a fundamentally new organization of society. The new society can no longer rest content with populating its cultural institutions with the citizens of a single nation. Perhaps the hour has arrived already; perhaps the socialization of Europe has begun with the revolutionary ending of the World War. If so, it will no longer be necessary for us to derive demands from the concept of knowledge which society will in any case realize in the course of its legal development.

After this discussion of the basic freedoms forming the precondition for a just university system, let us proceed to the positive demands. Admission requirements are the first consideration.

Up to now, certain demands have been made of the students upon matriculation, quite aside from scholastic aptitude. In addition to signing the university register, they have been required to register for a certain number of course hours. Moreover, they have not been permitted to engage in a remunerative occupation. That these external stipulations have excluded from the university a number of academics who belong there from the intellectual standpoint has been shown. Many who do work for a living are far closer to the scientific life of the university than some of those who fulfill the formal requirements, yet never put a foot inside the lecture halls and institutes. And the large number of special students and auditors are in many instances more serious and more mature in their handling of scientific work than those who successfully completed their school-leaving examination. We must insist that the rights of the academic citizen be extended equally to all such people.

The problem of selecting those with scientific ability essentially concerns schools rather than universities [At that time, German children attended either a popular school (*Volksschule*), leading only to vocational schools, or an elementary school (*Grundschule*) leading to a high school (*Oberschule* or *Gymnasium*) of one kind or another. Thus they were totally segregated, essentially along class lines, throughout their education. Those who completed nine years of high school and passed the school-leaving examination (*Abitur*) had the right to enter the university. The popular institutes (*Volkshochschulen*) were public institutions for adult education, open and free to all, but offering neither credit nor degrees. – E.H.S.] We know very well that, quite aside from the social injustice involved in the distribution of education, selection by means of the school-leaving examination is totally inadequate for the university. On the one hand, this examination is too easy, for it opens the doors of the university to anyone of the most mediocre intelligence who possesses something of a memory and a bit of tenacity; on the other hand, it is too complicated, for it burdens those who are truly gifted and intelligent with a senseless load of facts and a vast hodgepodge of random information. Moreover, this examination, which is tailored to a particular form of schooling, offers the only route to the university; it is closed, for instance, to teachers in lower schools, who from childhood on have received an entirely different form of schooling, yet often display a strong impulse towards learning. Only a fundamental reorganization of the education system can resolve this problem; there should be a single school for everyone, and all those with the necessary gifts would have an equal opportunity to proceed to the university. A detailed expansion of this idea would exceed the bounds of this essay. Let us simply say that the problem of university admission in general can be solved only in conjunction with the problem of schools and that here, too, it is only the socialist state that can create the legal forms necessary for the emergence of the new university.

Until a unified school has been instituted, formal arrangements must be made for admitting to university study gifted people who have had a different form of preliminary education. Even in the socialist state there will be many people who in their youth are not among the most gifted, who are not really destined to be scientists or scholars, but who firmly wish to dedicate the whole of their leisure hours during their working life to partaking of intellectual and spiritual goods. Many of these people will already be qualified to participate in certain special courses of lectures. Even if the satisfaction of these needs must be left to the popular institutes, a temporary rescue operation, at least, may be effected by granting to the faculty the right to admit to their own lectures and classes anyone they consider qualified, without requiring matriculation or tuition fees. This freedom is really one that any scholarly teacher assumes as a matter of course; it must take the place of the former system of nepotism, whereby the sons, daughters, brothers, sisters, and wives of professors had freedom of entry to the lectures.

The problem of admissions is more easily resolved for students than for the teachers. We indicated earlier that the selection of faculty members has been plutocratic heretofore, and we therefore demanded remuneration for all faculty. We rejected the notion of unrestricted admission to the faculty, for it would not decrease economic injustices and runs counter to the demands of scholarship. Now the time has come to give positive details of how selection is to be carried out. That only the highest scientific qualifications warrant bestowal of a university faculty position goes without saying. Only the community of scholars can decide on these qualifications, for they alone possess the necessary scientific insight. Nor can the decision be made by means of an examination, for an examination is basically inadequate for a choice of such moment. That leaves only selection by the faculty assembly. This is the arrangement that existed in the past, and we all know what has come of it: intellectual inbreeding, exclusion of certain important schools of thought, an exclusive professorial stratum bearing all the marks of a caste. Consequently, many new suggestions have been made, for instance, that from the democratic standpoint, the student body itself should select the professors. However, democratic as the reorganization of the formal structure of the university ought to be, its structure with respect to science and scholarship is the very aspect that cannot be democratically settled. For scientific insight is so closely tied to certain select minds that there can be no question of voting and selection by the majority. Universal election of the faculty by the student body would without question lead to a lowering of the intellectual level of the university. That the faculty as a group are intellectually superior to the students as a group, and that the two cannot be granted equal influence upon scientific decisions, is a fact so firmly grounded in reason that it must be made a principle of scientific organization. This separation into two groups has nothing to do with class structure, for a class groups people together on the basis of material [*ungeistiger*] characteristics, such as property and heredity, directly contradicting classification in accordance with intellectual qualities. Moreover, it grants privileges to its members that have nothing to do with the principles of classification. It would, for instance, be quite wrong if the division of the university into two groups were accompanied by the granting of certain general prerogatives to the faculty exclusively, say, freedom of speech or certain civil rights. The faculty may have a superior position with respect to scholarly concerns only; that is, decisions should be their exclusive prerogative only in matters concerning the organization and dissemination of scientific knowledge. But within this sphere they alone are competent judges. Scientific qualifications can be judged neither by a popular political election nor by the body of pupils. The assembly of university teachers should represent the model of a corporation that grows only by taking peers, and that nurtures and sustains from generation to generation the good entrusted to it: knowledge.

We must not be discouraged by the bad consequences of the earlier selection. For the intellectual inbreeding of which we presently complain arose more from causes external to science than from the ill-will of scholars. The number of applicants for lectureships was limited by economic restrictions, because lecturers were not paid. Academic teachers as a class were a part of the strata of civil servants, with their bureaucratic morality and political bigotry. Social conceptions of standing had a greater influence upon occupancy of academic chairs than their occupants realized. And science and scholarship, too, have suffered far more under the class structure of the population than any of us suspect. The grand illusion that lent such spiritual power to the occupants, according to which the well-being of the people was supposed to be tied up with the unearned income of the employer, was handed over to the apologists for capitalism even by science, scientific scholarship, the guardian of truth. When the restrictions fall away, when the social structure of the people has no more classes, the assembly of scholars will be able to make judgments freely in accordance with scientific principles.

The solution for the immediate future seems to us to pose difficulties. We face the necessity of leaving in the hands of a caste of professors of bourgeois mentality their expansion to an independent collegium of scholars. To this end it would appear expedient to free this group from at least the worst of the narrowness that has hitherto restricted it, for in the past selections were made by each university, for itself, and admission was often refused because of personal animosity. If everyone in the country in a given field were permitted to decide on the occupants of chairs, the very fact of the greater number of participants would balance out against the influence exercised by personal animosity or petty family interests. We recommend annual election of a committee of professionals within each field, charged with the task of selecting new faculty members. A similar arrangement already exists in Italy. We must spell out further what we mean by 'professional'. It will probably be quite sufficient to include under this label every university teacher in the field and also every scholar in the field with an appointment at one of the academic research institutes, such as the Physical-Technical Federal Institute [*Reichsanstalt*] and the various research institutes attached to the Kaiser

Wilhelm Society. The resulting group should be quite large enough to have a refreshing effect upon the university. University teachers must also have full freedom to move about. Those qualified to join the faculty should be permitted to choose their university and also to change it. This will make possible the formation of working teams concentrating on particular lines of scientific research, as only intimate communities can. The new selection procedure will also yield the fruits of scientific individualization, as displayed by the Göttingen school of mathematicians and Hermann Cohen's Marburg school. At the same time, contact between the universities will become closer, more personal, and the dissemination of new thoughts easier. We need not fear an excessively hasty change or discontinuity in instruction, for the working groups will tie the faculty to one place; older faculty members in particular will not be anxious to leave the circle in which they have a firm foothold. On the other hand, the system of free movement contains tremendous impetus. Younger faculty who have recognized the narrowness of the older order of scholarship with its caste ethic will join forces and enter the universities that are mostly strongly imbued with the petit-bourgeois mentality and start an intellectual revolution. The enlivening of scientific instruction, the creation of new research methods will be the consequence of the predominance of the younger, more lively element over the conservative forces resulting from the freedom to change locations. A university led by new forces will thus gradually emerge, and we may assume that within a few decades we will witness the existence of a faculty selected purely from the standpoint of scholarship.

Up to now, professors have been granted life-long tenure, subject only to disciplinary grounds for dismissal. It has been urged against this that faculty members should also be subject to dismissal if they demonstrate a lack of academic ability. However, we should not fail to recognize that tenure of position guarantees great freedom to carry out scientific work. We would like to suggest that every faculty member be given an appointment of a few years' duration at first; only after this trial period, during which he is to enjoy all faculty rights and to receive a full salary, will he receive final confirmation as a permanent faculty member. This matter is, however, more of a pedagogical than of a legal nature, for it is related to the possibility of keeping a check on scientific achievement and should therefore be left to pedagogical reform.

Let us proceed from the question of admissions to the problem of university administration.

The university is to be autonomous in scientific matters; therefore the university corporation must independently administer all administrative regulations insofar as they affect the intellectual function of the university. The natural separation into the two groups, students (including assistants) and faculty members, should serve as the basis for administration. Within each group, all members must be treated as equals. Just as the whole is to be autonomous, each group shall take over administration of its own functions. To this end, each of the two shall elect a representative body; where their affairs overlap, the two bodies shall work cooperatively.

Taking self-government of the student body first, we demand the establishment of student councils as its instruments. A membership of between twenty-five and forty will be suitable, depending on the size of the student body. An executive committee must be selected for the conduct of business, and special tasks can be put in the hands of special committees, chosen from among the members. On the matter of student councils, the reader is referred to the Free Student literature, which has for years put special stress on this demand; see, e.g., the writings of Kühnert-Kranold. Student councils are to be chosen on the principle of proportional representation by equally weighted, universal, secret ballot. This method of selection is, in our eyes, absolutely vital. We consider it quite wrong to let the student body be represented by the Korporationen [fraternal organizations], or to have a bicameral assembly with a chamber for Korporation members and another for non-members. *Vis-a-vis* the university there stands only the students as a whole, not represented by the spokesmen of these numerous small social organizations, but only by a body in which every student has an equal say. On the other hand, the student body is too large to elect individual persons according to their personal abilities, as would occur with a system of direct voting. Large numbers of voters will be moved to agree on particular candidates only through concrete agreement on their approach, on their basic attitude towards society, education, and the university. That means setting up parties among the student body, replacing the old fraternal cliques by associations with concrete programs corresponding to national political parties. In demanding the introduction of proportional representation, we are asking nothing other than that the student body be politicized. To anyone who, like us, sees the social organization of the university as the immediate end, the only possible means will, in fact, appear to be the formation of student political parties, dividing essentially along the same lines as the general political parties. For the problems of social order are the same in the state as in every other association of people with varying aims and opinions. Wherever we look, we can distinguish between socialists, i.e., builders of society, liberals, i.e., advocates of formal democracy, who attract all kinds of intellectual reformers who otherwise have no voice in society; and conservatives, i.e., historically minded persons who wish to preserve traditional forms. These three fundamental forms must likewise find expression in the student body. We will find that there are socialist students, who wish to organize university society so as to foster *de facto* justice, and liberal students, including the innumerable reformers in the student body, who advocate formal legal equality. The conservatives will find representation in the *Korporationen*; romantics who are discontent with the present, they believe traditional customs to offer the best training for a better future. Social organization will begin when, at last, proportional represention forces a division of students in accordance with these concrete groupings. For then the student, in finding his personal position, will have to choose between genuine alternatives; his choice of intimate companions will no longer be determined by the accident of personal acquaintance. He will be forced to give thought on his own to the basic problems of society, which will lay the foundations for the emergence of a clear moral sense and independent insight. Let no one object than the necessity of making a decision stunts human development. Insight develops only through making clear decisions; why should we not be able to change our opinion when we experience its concrete influence in our own person? "What then is the worry – that politics destroys character? But that is not true. It is possible only so long as no men of character engage in politics" [Rudolf Leonhard, *Die Sezession der Universität*, Munich, 1916].

The breaking up into splinter groups that has been occurring among the political parties in the last few years must be regarded as a healthy development among the student body. We have witnessed sensitive and serious students cutting themselves off from the whole and forming small communities in the belief that they could revolutionize the whole student body through the gradual expansion of their circles. Referring back to the concepts discussed at the beginning of this essay, we maintain firmly that the ultimate realization of intellectual and spiritual values is possible only in a community, yet it is equally certain that the community can make its influence felt only in the justly organized society. Thus, when the day finally comes in which every social organizer is attracted to a party, to cooperative efforts in the construction of society, a substantial step forward will have been taken for the future of the university. It will in the same sense constitute an advance when the *Korporationen* get together to form a conservative party. For then they will be stripped of their romantic halo and appear as what they really are: the apostles of a crass form of education based upon indiscriminate acceptance of traditions. Let us just hope that these contrasts are clearly outlined, for face-to-face experience is an indispensable element of education.

The task of the student council shall be the governing of the student body in disciplinary and economic matters. The old academic disciplinary code – which set up a special system of academic justice – granted the rector extensive discretionary powers over student organizations and arrangements, threatened individual students with severe punishments, most notably expulsion from the university, and frequently worked as an instrument of repression of free activity among the student body, and this code, by its patriarchal nature, conflicts with the demand for a socialized university. As the law of the land guarantees freedom of association and assembly, just as it guarantees prosecution of common crimes, and as freedom of opinion is in principle granted to every citizen, and certainly to every student, the most important provisions of the former disciplinary code have been superseded by civil law, and it might almost appear advantageous to eliminate the entire university code along with it. Yet this is in principle impossible, for a self-enclosed body such as the university necessarily gives rise to relations and problems of organization that can only be handled by a formal code. The old system was wrong in content, and also in form insofar as it placed jurisdiction in the hands of a body free of democratic controls, the academic senate, but the concept of a university code cannot be neglected. The sole qualification is that this code is to contain no stipulations that are at variance with valid civil laws, either in a positive or a negative sense. The general section of the university code should outline the afore-mentioned basic freedoms and regulations concerning admissions and formal organization; the specifics, which the old disciplinary code attempted to give in detail, should be reduced to a sort of house code, giving directions as to the arrangements in university buildings and the duties of individuals as members of university society. Any sort of guardianship over the students is to be avoided, the regulations confining themselves to cases in which concrete collisions are possible. This internal portion of the code must, however, be left to the society itself and will therefore be handed over to the student council insofar as it affects the student body. In any situation involving faculty members as well, the faculty council should, according to our principle, also have a voice. Student councils will have the job of working out this house code for the student body and of finding ways of implementing it. In the event of disputes, the council will institute a court of arbitration whose decisions will be binding on all students. We must note, however, that the jurisdiction of the council is to be confined to matters concerning the house code, that it shall have no right to make judgments in matters concerning belief or opinion, including matters of honor. It shall, consequently, possess no penal powers. Above all, expulsion must be eliminated in principle, for it means excluding a person from education, a power which no body may exercise. Even in cases of common crime, civil punishment is quite a sufficient measure, and should not be strengthened through further restrictions of freedom, such as exclusion from the university. On the other hand, the council must have certain powers of execution if it is to implement its organizational resolutions. These will be upheld, we believe, through the fact that the student council in conjunction with the faculty council will be the representatives of the university house code, having recourse to regular courts of law if their regulations (insofar as they lie within the boundaries stipulated) are frequently or continuously violated.

Along with the elimination of the disciplinary code in its old form, we demand the elimination of university judges and state-appointed trustees.

The other task facing the student council is the management of economic affairs. Here very little preparatory work has been done, and it will be up to the students to seek the path to economic development. Establishment of cooperatives would be beneficial; living arrangements for students are in need of organization; dining halls and dormitories need to be set up; and employment bureaus must be organized, and will be needed until such time as students without private means come to be supported by the state. It is possible that the serious financial situation will, for the next few decades, prevent the state from actually giving full support to proletarian youth during their years of study, but efforts in this direction could be met by the creation of opportunities for students to do economically productive work two or three hours a day for good wages. This work might consist of, e.g., managing their own dormitories, which would offer students experience with domestic economy; technical work, such as construction and making installations and repairs; and business management, involved in the purchasing of groceries. For the academics' old belief that working for money was demeaning, a class prejudice having more to do with their close relations to the caste of officers and civil servants than with the nature of their own intellectual work, no longer has meaning in an era in which the sons of directors and councillors may become manual laborers and which sees nothing fundamentally degrading about working for a living. It would, of course, be regrettable if intellectual achievement were diminished because of economic functions. But if, in a truly socialistic state in which there are no more unjust gains, there should occur general difficulties with production such that adequate means were simply not available, students will be quite prepared to participate part-time in production and to reserve any budding resentment for the old feudal order that plunged us into economic catastrophe.

In addition to these detailed administrative tasks, the student council will have to perform representative functions. That the romantic vision of student fraternity officers in dress-uniform with sword and waving standard will be buried along with feudalism goes without saying. It is possible that the student body will find some new form of expression for their communal spirit, but until then civilian dress will serve as a living expression of the bond between their representatives and the workers.

If students are granted self-government in all respects and required to put their intellectual life into public practice through the formation of parties and participation in parliaments, public assembly halls must also be made available to them. No building seems so appropriate for this purpose as the university itself, and considering how little use has been made up to now of the many lecture halls, it seems reasonable to ask that these halls be given over to students for their meetings. The head of building administration can take charge of the distribution of meeting rooms, which may, of course, not be permitted to disrupt academic instruction. At present, students hold party meeting and scientific courses in the gloomy backrooms of restaurants with the landlord constantly urging them to drink, while pleasant, comfortable university rooms lie empty, kept carefully under lock and key by the building superintendent. A stop must be put to this situation, once and for all. The student body must finally come to realize that these rooms are *theirs*, that their intellectual endeavors and the activities of their political parties are in no way demeaning to the halls of science. On the contrary, such activities can only enrich the life of scholarship. For the students are the bearers in the university of the refreshing, lively spirit of youth it has so sorely wanted in the past, and the scientific courses and evening discussions that they have organized in the last few years signify an earnest endeavor to apply academic fields to the burning questions of social order and justice, of how to lead a meaningful life. Students will invariably look to figures in the public sector for leadership; indeed, they will generally go beyond faculty circles and give scholars and teachers an opportunity to disseminate their ideas. And this will serve to break the final fetters represented by the limitation in the number of university faculty members. Earlier on we rejected the notion of unrestricted entry into the ranks of the faculty and suggested instead that lecturers be paid and faculty be selected by the totality of professionals in each field. We can now consider this notion, the idea that every person who feels he has a vocation for scholarship ought somehow to be given an opportunity to exercise it, in a somewhat different light. For no one needs more than a lecture hall and a few students in order to disseminate his ideas, and these very requisites are to be always at his disposal from now on. If the initiative is put in the hands of the students, rather than the lecturer – for instance, twenty student signatures or a formal resolution by a student organization might be required to obtain the use of a hall – it is our assumption that everyone with something pressing to say will, in the end, be willingly heard by the student body. If he fails to find his twenty students, it will simply not be possible to grant him the use of a room, because of the limited number of lecture halls; let him then participate in student discussions until he finds enough disciples. It can happen, of course, in certain specialized subjects such as the philosophy of mathematics or the study of Indian poetry, that the material is worthy of presentation even though the number of participants is under twenty. In such cases, it must be left to the discretion and breadth of vision of the student leadership to arrange for the course and propose use of the lecture hall; the requirement of a minimum number is to be waived in the instance of such official student requests. Thus the proposition of open admissions for faculty is resolved into two components; payment of scholars for teaching and free use of university buildings by the student body.

Once the rector is stripped of his disciplinary powers and freedom of assembly is instituted, the rector will also lose his right to control student bulletin boards. Every student organization must be permitted to make such communications as it wishes. It nonetheless appears expeditious to confine the right to post notice to organizations, for if every student automatically had this right, it would become impossible to keep the bulletin boards in order. Given the large number of student organizations, this stipulation does not constitute a *de facto* restriction of rights.

The second administrative sphere is that of the faculty. Like the student body, the faculty must be granted self-government with respect to disciplinary and economic affairs. Here the problems are more easily managed, as the number of persons is not large. However, a major change must be made in the existing ranking of faculty members. Up to the present, the rights of individual faculty have varied enormously: the separation into full professors, associate professors, and lecturers, and the special status of the academic senate, the general assembly, and the rector have signified a wealth of injustices with respect to the broader mass of university teachers. These leftovers from the medieval status system must give way to the demands of justice. Setting out from the assumption that selection as a faculty member is a guarantee of academic qualifications, we demand equal rights for every faculty member, irrespective of position or age. The faculty assembly, in which every university teacher is to have an equal voice, without regard to faculty division, must serve as the last court of appeal in all matters concerning the faculty. (The division into various faculties is purely of pedagogical and scholastic significance and is not to play a role in administrative arrangements. Permitting it to do so could only result in giving smaller faculties undue influence.) Faculty members will elect from among their own numbers a five-to-ten member council headed by the rector, who is to be chosen by special election. The small size of the voting body permits the use of direct elections, rendering the formation of parties unnecessary; there is to be no system of rotation of academic subject. The council is to regulate the business affairs of the faculty members, for instance, to represent their salary demands *vis-à-vis* the state authorities; on the other hand, it will also administer the disciplinary code of the faculty. As with the student body, this amounts to a system of house rules. It is the council's job to see that the basic provisions of the university constitution are upheld by every faculty member and to issue directives for their execution. If a member, even after admonishment, repeatedly violates the constitution, the council will formally recommend expulsion to the provincial commission for the selection of university faculty. However, expulsion may not be carried out on grounds of adherence to an intellectual movement or questions of personal propriety or belief. The idea that questions of personal honor are subject to the judgment of society must be abandoned. Faculty position ought not to be influenced even by infringements of civil law or conviction by a court of law. We firmly maintain that civil punishment and its restrictions of freedom are quite sufficient. They should not result in further restrictions of civil rights, such as exclusion from the university faculty. On the whole, the social code represents a system of expediency, and infringements of it frequently have nothing to do with morality. Scholastic and scientific questions must be kept separate from those concerning the civil code. The sole ground for expulsion of a university teacher must be continual violation of the provisions of the university constitution. And even in these cases, the reason for expulsion is not the immorality of the violator's behavior, but rather the necessity of upholding a legal system within the university, lest its task be rendered impossible.

While the work of the faculty council among the faculty may be of a relatively small scope, it is faced with a considerable task in that part of the administration of the university that it shares with the student council. The powers of each must be restricted in such a way that the student council has freedom of decision in all purely student affairs, the faculty council in all purely faculty affairs; in those involving both groups, however, the student council shall have merely advisory powers, the faculty council having the decisive voice. In the interests of scholastic standards, the faculty must hold a position superior to that of the students in these matters. We pointed out earlier that democratic principles are inadequate for university administration, for the distinction between teacher and pupil is not a class distinction, but a rational separation of spheres. Accordingly, the scholastically superior group must bear the responsibility for the regulation of matters pertaining to the university as such.

The university government, that is, the combination of the two councils, is founded on the provisions of the university constitution, but is otherwise autonomous. Here its functions must be limited with respect to the state. How it will be organized in detail cannot be described straight off, for the form it takes will be related to the changes in the national constitution. It must in any event be borne in mind that the university faculty is a scholastic corporation and is not to be heavily burdened with administrative details for the regulation of which it possesses neither the interest nor the ability. The university corporation should be handed the reins of government only in those managerial and judicial affairs in which decisions have pedagogical or scholastic consequences. These decisions are no longer to be made under the direction of the civil government. In the past there was a procedure for appealing against decisions of the academic senate that led up to the minister of culture. This supremacy of political powers must come to an end; matters of justice should be handled by the regular courts. The appropriate judicial procedures must be worked out by the legislative bureaus.

VI. PEDAGOGICAL DEMANDS

The organization of the university under the category of society calls for certain economic and legal preconditions. However, the economic demands must themselves be put in legal form if they are to come to fruition; i.e., their contents must be expressed in the form of laws and be incorporated into the general provincial law. This is the only means of guaranteeing they will be carried out. Thus, in speaking specifically of legal demands, we were speaking of law in the narrow sense, having in mind the formal legal provisions which govern the basic freedoms, the admission requirements, and the administration of the university. Likewise, when we speak here of pedagogical demands, we are placing them over against legal demands in the narrower sense only. They are also legal demands in the broader sense, for they must take on a legal form if they are to be executed.

It is from just this point of view that we will select pedagogical demands.

For it is the nature of the case that the required social reform of the university cannot deal with material demands concerning the contents and methods of scholarship and its dissemination. To demand of society that it achieve goals with a certain material scientific content would be to misconstrue the concept of society as a union of diverse wills. It is not the task of a theory of society to decide, say, between various incompatible theories of knowledge or alternative methods of medical science, and our demands should not in any way rest upon judgments concerning these matters. We will limit ourselves to pedagogical demands that do not require one-sided scientific decisions, i.e., those that are capable of realization by a university, which is a society whose purpose is education.

Our demands set out from the fact that pedagogical principles have received only negligible attention at the university in the past. The lecture system left it up to every teacher to present his listeners with a wealth of material, never asking about the extent of the pupils' intellectual participation and involvement in the work. Thus our first demand is that instruction be set up so as to conform to pedagogical principles. We believe that the paths forged by the holding of seminars and working sessions point in the proper direction, and we demand that the whole of university instruction be fundamentally restructured in accordance with these methods. That the student body gain some influence over the forms of instruction seems to us essential, of course. For the teacher can never have a total perception of the position of the pupil with respect to instruction, which holds such significance for the suitability of the instruction and the interest of the students that it is imperative it find expression through face-to-face methodical discussion between teacher and pupils. In general, the pupil ought never to become the mere object of education, but should always, as its subject, retain the right to participate in its formation. This right will in essence have been upheld if teachers are considered superiors solely with respect to scholarship, while all other educational influences are left to the choice of the students themselves. Yet we must never forget that the dissemination of knowledge is the crux of education and must stand at the focal point of all pedagogical critique. And here it is particularly important to find channels for effective student influence. We recommend formation of special committees in each division of the university, in which the student body may express their will on matters of instruction. These special committees are to be kept completely separate from the student council. The council has administrative functions; its members will be largely students sophisticated in politics and rhetoric, and it would be pointless to transfer pedagogical tasks to a body with such a different orientation. Considering that the intellectually active portion of the student body is most frequently encountered in seminars, that these are really the scientifically productive workshops in the university, it would seem sensible to form the divisional committee from selected representatives from each seminar (cf. the writings of Hermann Schüller). It should have about the same number of faculty members as students. This body would work out plans for reforming the curriculum. Partial replacement of lectures by discussions; the dividing up of large classes into small groups, which, under the leadership of assistants or older students, discuss and debate the material presented; the establishing of special discussion groups who undertake to discuss any topic desired on the basis of their own theses – all these currently circulating ideas present fields for exploration by the divisional committee. At the same time, it would be thoroughly discussing the organization of class schedules and should be called upon in an advisory capacity when the schedule is set up for the semester. It could organize a curriculum for the students and present plans for a total program of study. Members could counsel students and offer helpful advice to their younger comrades. The committee would represent an organ giving concrete direction to the strong desire of the student body to participate personally in knowledge and to overcome the strangeness and coldness characteristic of the old system.

Of course, the divisional committee is to have purely advisory powers. Decisions concerning research and teachings must on principle be in the hands of scientific leaders, for they are the only ones who, on the average, have the maturity in scholarship that is presupposed by all forms of education. The increased degree of interest in and understanding of pedagogy to be observed among the student body today does not constitute grounds for a fundamental reorganization placing university teachers under their pupils. We must assume that, once the body of teachers has been somewhat restored and is really made up of scientific leaders, they will be similarly activated by the forces currently revolutionizing the student body. At that point, the faculty, deeply sympathetic with the student body but equally moved by scientific insight, will work out instructional methods which we cannot even imagine today. Forms of cooperation between teachers and pupils will develop of their own accord. The antimony posed by the freedom of the pupil under the direction of the teacher may undoubtedly be resolved in a variety of different ways; the pedagogical forms of an autonomous university, of academic society cannot be demonstrated *a priori*, but must be created anew each time on the basis of the empirical conditions. We do not wish to realize them by enforcing an artificial solution and expecting it to effect a spiritual reformation. For a start, we wish only to present the possibility of bringing attitudes closer together and taking the first steps towards cooperative efforts; the development of specific forms must be determined by the specific conditions.

Reform of instruction will also include changes in the examination system. Up to now, passing an examination has signified possession of a wealth of rote learning. Certainly, we cannot demand of an examination that it deliver a decisive judgment concerning the productivity and intellectual powers of a human being. Indeed, this is not the purpose for which examinations are instituted; they are simply supposed to pass judgment on concrete ability. But they ought to really do this, and should not confuse ability with knowledge. We know well how little regard past examinations have paid to this dictum. Even the dissertation, which as such is a good means of displaying an ability for scientific scholarship, has increasingly degenerated to a mechanical reproduction of the ideas of others or to a mere job of collecting information. The task of working out new methods of examination will fall to the faculty, jointly with the divisional committees and, in matters concerning state licensing, with representatives of the various professions. It goes without saying that historical vestiges such as the preferential position of ancient languages must be eliminated in the process of this reform.

The right to set examinations has up to the present been a monopoly in the hands of the examiners. Students have been compelled to attend their lectures, for experience has shown that examination questions are invariably drawn from the specialties of the examining professors. As a consequence, the lectures of older men are often heavily attended, while the rigorous lectures given by younger teachers with fresher minds are empty. In this way, the freedom to choose teachers is largely cancelled out. This monopoly of older professors can be broken by making a significant increase in the number of examiners. The student will thereby be granted freedom of choice and be enabled to pursue knowledge with all the enthusiasm of youth.

Another needed pedagogical measure is the supplementing of theoretical instruction by practical courses. Students in law and education, for instance, have been placed at a disadvantage in this respect, while medical students have experienced an extensive practical training. Laboratory work must likewise be increased. Scientific institutes and libraries are in need of enlargement, and reference libraries, which greatly facilitate specialized research, ought to be established for all subjects if possible. (Every mathematician is aware, for instance, of the great usefulness of the mathematical reading room in Göttingen in enlivening mathematical studies at that university.)

But all these improvements, which are only hinted at here and which must be worked out independently by teachers and students, can only be carried out within the framework of a new university, dedicated to science and learning. The past separation into four faculties [divisions], along with the distinction between university and technical institutes, has no justification in the systematic organization of knowledge. The nature of scientific research is expressed most clearly in the division of philosophy; the other three have become technical schools for doctors, lawyers, and priests. Surely the education of priests is not the job of the state. Once the socialist state has achieved the separation of church and state, the setting up of theological schools must be left to the Christian denominations, for no public interest is served by continuation of this profession, which has been outmoded by public enlightenment. Such genuinely scientific work as is performed in the theological division must be absorbed into history and philosophy; at the same time, it must be left to the discretion of each scholar, true to the principle of academic freedom, whether he chooses a theistic foundation for his philosophy. The university as such is to safeguard the level of scholarship only. The theological division may be altogether eliminated in the future; the other divisions, along with other institutions of higher learning, especially institutes of technology, have, in their turn, a need of basic reorganization. To start with, the position of research institutes and technical institutes requires clarification. This duality can not in principle be overcome, and although, being aware of the irreplaceable influence of pure scientific work on every form of technical work, we by no means demand complete separation of the two spheres, we must nonetheless insist that the organization of the university in some way accommodate itself to this division. The research institute must represent a narrower circle surrounded by the large circle of technical institutes, intertwined with it in innumerable ways. In order not to overburden the research scientists, the course of technical study will have to be separated from the research institute itself and will require a faculty particularly good at teaching. On the other hand, it is essential that the technical institutes not lose touch with pure research, for every professional intellectual ought to experience the fruitful and liberating spirit of true science at least once in his life. It is this relation alone that gives the university its central position in cultural life, making it the heart of the great flow of intellectual life that inspires the whole of the people; it must be linked with the adult education system, to ensure that, in the socialist state, the worker, too, is offered the opportunity of making contact with science. For a popular institute requires constant direction from the university if it is not to offer popular education in the derogatory sense; simplification of a theory for the untutored mind must never involve distortion of its contents. But all these are problems, the solutions of which lie without the scope of formal university organization, which cannot be adequately dealt with by the mere concept of a socialist society. Thus all we can do here is make the general request that a reorganization be undertaken of the conglomerate historically known as the university; how this is to be done must be left to the self-governing society of scholars, to an autonomous university. It will be the mission of all scientifically minded persons whose patterns of thought are not bound by history to formulate the necessary plans and to bring about their recognition within university circles. Once the socialist society has realized all of our demands – economic, legal, and socio-pedagogical – for the organization of the university, it will have created the legal forms needed for concrete collaboration. The special task of socialism will have been fulfilled; nothing but human reason, by virtue of its intrinsic powers of conviction, can actually bring about the organization of knowledge.

However, we cannot conclude a discussion of pedagogical demands without mentioning certain kinds of damage to the university that have resulted from the influence of the old class mentality. The exclusion of radical schools of thought has led to gross injustices so flagrant that we must insist upon their direct and immediate rectification. It is well known, for instance, that up to now Germany has had exactly one full professor of political economy lecturing on socialism, while in the world outside this doctrine has played an increasingly prominent role, its advocates constituting one-third of the population. Here this situation must be changed at once. There are quite enough leading intellects among the socialists with the academic qualifications to hold their own against any German professor of political economy. To name names is superfluous; let us simply demand that these well-known scholars be given university posts forthwith. The situation is the same in pedagogy. In the past few years, we have witnessed many attempts by educators to rebuild the schools according to a new system. Yet not one of these men is to be found among university teachers; the few existing chairs of pedagogy are occupied by historians of education who hold forth about Basedow and the Philanthropists while a whole generation languishes at the desks of outmoded public schools. In other specialties, too, we note the stultifying influence of caste exclusion. Philosophy, the systematic study of the sciences, has long since degenerated to an historical subject. To be sure, even in Kant's day there were professors "whose very philosophy was the history of philosophy." But one hundred years later, the model of this greatest of philosophers has not yet succeeded in deposing these eclectics. Indeed, many a systematic thinker has been silenced, those without means having been abandoned to commercial occupations in order that they might expend their energies in creating surplus value. The degeneration of scientific philosophy has not resulted from university teaching, but actually rests upon the dearth of systematic minds. Here we cannot demand appointments for famous men who have excluded heretofore, as we can do in sociology. The roots of the difficulty lie too deep: there are no such men. We can, however, demand that philosophy be developed in a manner favorable to original minds, that systematic thinkers always be given preference over historians, that the systematic talents be sought among the candidates of the younger generation independently of all outmoded prejudices. Reformation of the study of law follows naturally in this connection. Traditionally the law course has been dominated by descriptive and exegetical jurisprudence. This must be replaced by the philosophy of law; the critique of law, the science of the law as it *ought* to be. The same applies to philology and history, which must be raised from the dispersal of details to the status of systematic sciences.

That our socio-pedagogical demands reduce to demands regarding the selection of capable persons whenever we are dealing with primary questions of scientific organization is not to be regarded as a defect in our social and legal observations. Our purpose was to avoid taking any position on scientific questions and to search solely for the social conditions necessary for the university to achieve its ideal form. Since we defined society as the union of persons of different aims and outlooks, we may not include among our results demands that materially presuppose any one particular orientation. With respect to scholastic organization, which presupposes one or another concrete opinion concerning knowledge, our social demand must be reduced to a formal demand. But this can only be the demand that scientific qualifications alone be considered, which in turn becomes the demand for scientifically qualified persons. It must be said that, in our view, the differences of opinion regarding the organization of knowledge are not very great among those with scientific qualifications, and agreement on a definite organizational orientation could probably be reached. This presupposes, of course, the prior removal of the many insufficiently qualified teachers who are still at the university today. Thus the class of scholars will become mature enough to undertake pedagogical reform only several decades after the institution of the legal and economic measures outlined here. But our intention to wait for the actions of such suitable men, rather than expecting that a new edifice will result from organization as such, from bare reform, distinguishes our approach from schemes to improve the world by means of structure alone. In any case, we believe the realization of rationality is tied to the insights and deeds of *people*, and we consider it a point in our favor that we can find no better solution to the problem of the university than assembling leading minds and giving free rein to their activities.

8. REPORT OF THE SOCIALIST STUDENT PARTY, BERLIN

Winter Semester 1918/19

[1918c]

The Socialist Student Party was formed by a group of socialist students of all political directions who had come together in secret prior to the revolution, and some of whom had participated in the revolution. The Party made its first public appearance at a general student meeting one week after November 9th: its united stand and, even more, its very existence caused the greatest amazement, which soon was to change into open hostility. Our appearance caused students in Berlin who previously had been apolitical to organize themselves into parties as well, and thereby the election of the then non-existent Student Council became possible. Nine representatives were elected on the basis of general, equal, secret and proportional representation. The election returns (230 votes in our favor) permitted us to send at least one of our representatives to the Student Council.

According to the program, our activities are divided thus: (1) work of information and education about socialism which includes (2) collaboration in the proletarian youth movement, and (3) political work for the reformation of the university along socialist ideas.

The following offices were in charge of these activities:

1. *The Office for University Politics*

In addition to their work in the Student Council, our representatives are also members of the technical working committees of the entire student body which develop reform programs for the individual faculties and departments and organize lectures. Furthermore, we drafted a constitution for the student body which was submitted to the Student Council. As experience has shown, decisions made by the general student meetings almost never reflect the majority of the entire student body but are merely supported by chance majorities of a small number of students; therefore, we have proposed a student parliament consisting of approximately 20–30 members who are to be elected at the beginning of every semester by general, equal, and secret balloting. A central committee will be formed from the members of this parliament. Also, our representatives established direct contacts with the Ministry of Cultural Affairs in order to submit proposals for university reforms in accordance with socialist ideas as well as for the establishment of transition universities [*Übergangshochschulen*] for proletarians.

2. *Office for Scientific Courses*

(a) Dr. Hans Reichenbach conducted a course on the philosophy of socialism. This course focussed on the problems of the materialist conception of history and will later deal with the problems of ethics and socialism.

(b) Mr. Wittfogel conducted an introductory course about socialism. Lectures and discussions examined the question, "What really is socialism?", as well as different schools of socialist thought, and the basic theories of Marxist socialism.

(c) Further courses are being planned; e.g., a course by Otto Jensen entitled, 'From theory to practice'.

3. *Office for Lectures*

(a) Recently, evening discussion sessions of the entire Party take place weekly, introduced by lectures given by important politicians and theorists in which the principles behind the most urgent political questions such as 'Dictatorship or democracy', 'Bolshevism', etc., are examined. The following lectures have been planned: 'Bolshevism' by A. Stein, 'The materialist conception of history' by Prof. Cassirer, 'Geopolitics' by Dr. Engelbert-Graf, 'Problems of population politics' by Dr. Magnus Hirschfeld. Speakers who have accepted our invitation to present lectures include, among others: Kautsky, Bernstein, Gawronsky, Lederer, Schumpeter, Hilferding and Breitscheid.

(b) A series of public lectures has been planned on the question of guilt for the outbreak of the war. The speakers will be Helmut von Gerlach, Dr. Halpert, Wilhelm Herzog, Eduard Bernstein and Otto Jensen.

(c) Evening gatherings at which works of Leonard Frank, Karl Kraus and others were read aloud served to explore the ethical and artistic foundations of the socialist world-view. Further readings from Tolstoy and Dostoyevsky among others are planned.

(d) Prior to the Student Council elections, publicity rallies were organized for the students of this University and will be repeated frequently here as well as at other universities.

4. *Office for Party Work and the Youth Movement*

There are separate groups for the majority party and for the independents.

(a) Before the elections, the former group organized two large public propaganda meetings intended for academics. The speakers were Wally Zepler and Magnus Hirschfeld.

This group also obtained speakers for the young workers. It is planned to establish closer contact (through joint evening gatherings) between a larger group of students and leading personalities of the "Club of Young Workers" [*Verein 'Arbeiterjugend'*]. Through popular science lectures and discussions, we intend to introduce the young workers to all those fields of knowledge that are interesting to them.

(b) The independents set up a course for agitators under the direction of a party agitator. This course led to the formation of a group that tried to resolve the urgent question concerning the form and shape of the USPD [*Unabhängige Sozialistische Partei Deutschland* – Independent Socialist Party of Germany]. The group also began to compile material for a USPD-handbook and will continue that work. The group also assisted several of its members to join the USPD.

Some members who are affiliated with the USPD or the KPD [*Kommunistische Partei Deutschland* – Communist Party of Germany] worked for 'Free Youth' [*Freie Jugend*], the youth movement of the left socialist groups; they joined these individual groups and gave lectures about politics and the social sciences. There were also discussion evenings with leaders of 'Free Youth'.

5. *Office for External Propaganda*

Through the assistance of friends of members we were able to support the formation of socialist student groups at other universities. We circulated a letter with certain definite questions to established groups in Munich, Göttingen, Münster, Königsberg, Breslau, Leipzig, Dresden, Marburg, Heidelberg, Vienna and Cracow. In places where no groups yet existed, we encouraged the formation of local groups by sending information to socialist friends whose names were given to us by our members. We also contacted the International Socialist Conference in Bern and the Committee of Foreign Socialists in Berlin in order to establish relations with foreign socialist students in groups formed within the academic community. We considered at our particular duty to assist in the formation of new groups, to encourage unified action by existing groups and to establish a general socialist student association in Germany, something that could only be accomplished at a socialist student conference.

Therefore, we ask all friends of our cause to give us the names of socialist friends, especially those residing in places where no groups have yet been formed.

6. 7. *etc.*

The Office for Management was charged with the organization of our Party. For this purpose, it divided the party activities into separate functions, and relegated them to special offices (in addition to those mentioned before, there were: Finance Office, Secretariat, Office for Poster-Advertisement, Office for Convention Procedures, Press Office, Office for Propaganda in Berlin, Book-Exchange, Archives, and Housing Office). On the other hand, the Office for Management coordinates the activities of these offices by establishing a certain degree of centralization. In addition, the Office for Management was in charge of the Party's budget and funding.

By February 10th, we had 250 members.

Furthermore, it should be mentioned that we sent a delegation to Karl Liebknecht's funeral. When our delegation gathered inside the University, it was thrown out by reactionary students. We sent a letter in reply to Henri Barbusse's appeal in the *Republic* and also issued a note of protest to the press regarding the court-trial in Moabit of members of the 'Spartakus' Organization.

The Socialist Student Party, Berlin
Dr. Hans Reichenbach, Anna Bresser,
Ernst Leitritz, Alfred Seidel.

NOTES ON THE PROGRAM

We chose the name 'Socialist Student *Party*' because we wanted to make clear that we are a university party, i.e., a group which aims at reforming the university in a definite way. We also want to overcome the aversion which exists in the student body against the term 'Party'; furthermore, we wanted to make clear to the unpolitical and reactionary members of the student body that our group is politically aggressive in character.

The first sentence of the program states that the Party as such, when all groups agree, will involve itself in politics only on questions related to university matters.

For we consider it to be our chief task, as stated in point (2) to contribute to scientific enlightenment on problems of socialism. Individuals will be politically active within their parties and youth movements for which we shall serve as a coordinator. Furthermore, we intend to accept members of all socialist persuasions, from majority socialists to communists to anarchists; this would be impossible if we were to get involved in political questions of the day. Unfortunately, some communist students have left our Party and have founded a communist student party.

The Socialist Student Party of Berlin consists not only of students of the University as such, but also of students from the Technische Hochschule and the Business School. The Party is projected as the united organization of all socialist students in Berlin. It also accepts a limited number of older academics and intellectuals as members, and proletarians are always welcome to its events.

We understand ourselves as a local group of a yet to be formed Socialist Student Association, which all socialist student groups in Germany should join. This General Socialist Student Association of Germany will then become a member of an International Socialist Organization of Students and Academics.

PART II

POPULAR SCIENTIFIC ARTICLES

9. THE NOBEL PRIZE FOR EINSTEIN

[1922a]

Albert Einstein has been awarded the Nobel Prize. After the enormous popularity that the creation of the theory of relativity has brought to this great man, this distinction will hardly come as a surprise to the general public. What is surprising is that the citation from the Swedish Academy of the Sciences does *not* mention the theory of relativity, but specifically honors Einstein for a different achievement: "for his investigations in the field of theoretical physics, and especially for his discovery of the law of the photoelectric effect". It would seem that the Swedish Academy wished to avoid taking a position on the question of the theory of relativity – in itself a display of justifiable caution, since the Nobel Prize ought not to render a decision in scientific disputes. Nonetheless, the Nobel Committee has thereby rejected the historic opportunity to do public honor to one of the greatest physical theories of all time.

That the prize has been justly awarded for the reason cited, that the physicist Einstein would rank among the most significant present-day scientific investigators even if the theory of relativity were erased from among his life's achievements, will certainly not be disputed by anyone familiar with modern physics. Yet the public may not realize this, and a brief description of Einstein's work on other subjects may therefore be in order.

Einstein's first work appeared in the year 1900. In it the scientist, then only twenty-one, discusses consequences of the phenomena of capillarity and calculates the surface tension of a liquid by means of the molecules' powers of attraction. But 1905 proved to be the most fruitful year of his early period; three major works date from that year. First came the basic exposition of the special theory of relativity, then the theory concerning so-called Brownian motion, and finally the study of the generation of light from light quanta, which has now won him the Nobel Prize.

The extraordinary boldness of Einstein's gift for making connections was apparent even in his work concerning Brownian motion. It had long been assumed that the smallest particles of matter are in a state of motion that we experience as heat. Yet these particles, called molecules, were thought to be too small ever to be seen directly, even through the most powerful microscope. Einstein calculated that small particles of dust, which are just barely visible under the microscope, are not, after all, very big in relation to molecules and that they consequently cannot withstand the impact of the molecules of a liquid in which they are floating around, which forces them to move back and forth at random. The scientist Brown had already noted this phenomenon, but no explanation had been found for it. It was Einstein who came up with the precise theory of Brownian motion and calculated, for example, the average velocity of the dust particles. Investigations with the microscope were subsequently renewed and led, in the event, to a ringing confirmation of the Einsteinian calculations. Thus the phenomenon is regarded today as one of the strongest pieces of evidence for molecular theory. The sight of the particles swirling about under the field of the microscope is indeed quite wonderful, giving us the feeling of taking a look deep into the structure of matter.

We come now to the work on light quanta. It bears the modest title, 'An Heuristic View Concerning the Origin and Transformation of Light' [*Annalen der Physik* **17**: 132–148 (1905)], and is dated March 17, 1905, at Bern. Einstein was at that time an official in the patent office in Bern; hence it is that the city of Bern is named in the work which is now being accorded such great honour by the Swedish Academy of Science. In this work Einstein attempts to explain the connection between *light rays* and *corpuscle rays*, i.e., rays consisting of the most minute particle (electrons). Previously it had always been thought that light rays were propagated evenly (continuously) through space. In connection with Planck's quantum theory, Einstein proposed the hypothesis that light, like matter, is made up of tiny elementary particles, consisting, however, solely of light energy, rather than having a corporeal character. He called these elementary structures *light quanta*. This hypothesis offered a superior explanation of the interaction between bodies and light. According to it, an oscillating electron emits light quanta at intervals and, conversely, the individual light quanta are absorbed by the individual electrons when they strike a body. In this way the electrons receive an impulse and are able to leave the body in the form of corpuscular rays. This final effect is known as *photoelectric*, because electricity is released by means of light. The modest manner in which Einstein communicates his deeply significant ideas is quite remarkable. In the work under discussion, we find the following passage.

> It does indeed now appear to me that the observations that have been made of . . . the generation of cathode rays by means of ultraviolet light . . . become more comprehensible if it is assumed that the energy of light is distributed discontinuously in space. According to the assumption upon which we must now concentrate our attention . . .

the energy of the light ray consists of a finite number of energy quanta, located at space points, which move without splitting up and which can only be absorbed and generated as a whole. In what follows, I wish to reproduce the argument and enumerate the facts which led me into this way of thinking in the hope that the view to be presented may prove useful to some scientists in their research.

This hope has been abundantly fulfilled. The Einsteinian Law of the photoelectric effect has been firmly established by experiment and has been the stimulus for numerous studies in this field. Einstein himself published further investigations of the quantum theory of radiation. In 1909 he addressed the Congress of Natural Scientists in Salzburg on 'The Constitution and the Nature of Radiation'. The most important of his subsequent works on this subject appeared in 1917, only two years after the conclusion of the general theory of relativity – a proof that alongside his work on relativity, to which he has dedicated the greater part of his energy, Einstein still found enough time to carry out other work which in itself has made him worthy of the Nobel Prize.

Leafing through Einstein's works, we are amazed by the wealth of subjects of which they treat. We find 'A Thermodynamic Foundation of the Photochemical Law of Equivalence' [*Annalen der Physik* **37**: 832–38 and **38**: 881–84 (1912)], 'The Theory of Opalescence' [*Annalen der Physik* **33**: 1275–98 (1910)] (which explains why the sky is blue), 'Periodic Variations in the Size of the Moon', 'Propagation of Sound in Partially Dissociated Gases' [*Preussische Akad. der Wissen., Sitzungsberichte*, 1920, pp. 380–85], along with various comprehensive presentations, e.g., his article on atomism for the physics volume in the collection, *Kultur der Gegenwart* (*Contemporary Culture*) [Part 3, Section 3, Vol. 1, pp. 251–63 (Teubner, Leipzig, 1915)]. And all this has been, as it were, a secondary occupation, for his principal occupation has been the theory of relativity, which he has discussed in numerous publications and which has also induced him to present various philosophical ideas.

Truly, here is a physicist who has earned the Nobel Prize. He has been called the Newton of our century, a comparison that is doubly appropriate. Newton's primary achievement lay, like Einstein's, in the field of gravitation, but in addition to this Newton made significant contributions in optics which in themselves would have justified his fame. Newton, too, would have been awarded the Nobel Prize for his optical work alone, had this honour existed in his day. Einstein may view the citation of the Swedish Academy not as belittling his theory of relativity but as a very special tribute to his comprehensive scientific achievements.

10. RELATIVITY THEORY IN A MATCHBOX

A Philosophical Dialogue

[1922b]

Characters

A. A man of common sense
B. A physicist

A: Come on now, Mr. Physicist, isn't the theory of relativity sheer nonsense? What are its claims: Time is the fourth dimension, and my walking stick will get shorter when I'm riding on the train, and the older of my twins will soon become the younger because my wife has brought him along on the trip . . . ?

B: If you'll just permit me to . . .

A: No, I won't. It contradicts common sense, and therefore I don't believe it.

B: Well, give me a chance anyway. Your common sense is superb; just allow it to listen to me for ten minutes. Your common sense can tell you, for instance, exactly how to strike a match, not so?

A: Yes, of course, but . . .

B: No interruptions, please. This is just a simple experiment, by means of which I'm going to give you a quick demonstration of Einstein's theory.

A: What? Right here, sitting in the café? Are you going to arrange for an eclipse of the sun? Do you keep the speed of light in your pocket?

B: More likely the latter, for I can certainly strike a match. But I don't need the speed of light for the moment. As I was about to say, it is my opinion that I can give you an experimental proof of the entire theory of relativity in two minutes.

A: Now, look, my good man, I must ask you to . . .

B: No, I must ask *you* for something, namely, your attention. Now, just look this way. I'm holding the box in my left hand and moving the match along it with my right. You see, it lights.

A: Yes, of course.

B: All right. Now I'm holding a fresh match in my right hand, but this time I keep it still and move the box instead. You notice that this match lights, too.

A: Well naturally!

B: See, now you think it's natural too. And that is exactly what Einstein says; what I've just shown you is almost the whole of the contents of his great theory. There's just one more small detail . . .

A: And what's that?

B: Well, as I say, it's just a detail, just a single amendment. Namely: the way it is with the match and the matchbox is the way it always is, with everything that exists.

A: And that's *all* that Einstein says? Isn't he interested in anything else besides matchboxes?

B: Let me remind you, please, that you promised me your attention. What I said was, with everything, not just with matchboxes.

A: All right then, maybe he includes a few other kinds of incendiary devices.

B: Now, now! As I was saying, from my experiment with the matchbox and that one small amendment I can prove to you that Copernicus was no smarter than old Ptolemy, that light rays are deflected at the sun, that spectral lines at the sun are 'redder' than on the earth . . .

A: You must be joking!

B: Please do listen. Let's take Copernicus first. He said that the earth does not stand still, but turns around. We can tell this from its so-called centrifugal force. You're familiar with that, aren't you? If you tie a stone to a thread and whirl it around, the thread becomes taut. The earth, too, becomes taut from spinning, and that is why, as is well known, it has become too thick at the equator; the poles have flattened out.

A: Goodness, don't you think I know what every schoolchild knows?

B: Yes, certainly I do. That's because of your common sense. Now, here's where the matchbox comes in. Suppose I think that the earth stands still and all the stars rotate around it. That's just the opposite, and is what Ptolemy believed. Now, what must happen, according to the matchbox model? The earth will still flatten out.

A: How is that supposed to happen, if the earth is standing still?

B: "How?" is quite another question. You didn't ask me *why* the match caught fire the second time around as well as the first. It just does. Similarly, in this case, the rotating heavens will produce an attractive force such that the earth flattens out.

A: But I have to be able to imagine that.

B: Imagine it? You're very hard to please, my friend. Are you under the impression that you can imagine why the match burns?

A: Of course I can imagine why the match burns!

B: Look, you are in the habit of lighting twenty cigarettes a day, and therefore you observe twenty times daily that the match burns. Suppose you were so big that your feet rested on Sirius, and you habitually made the earth spin twenty times a day – and, just for a change, the fixed stars, too: then you could also "imagine" what I'm telling you.

A: (after a moment's thought) You could be right. But why does it follow from the effect produced by the stars that Copernicus is wrong?

B: When the postman comes to my door, the bell rings. When you come to pay me a visit, the bell rings. Therefore I cannot conclude from the fact that the bell is ringing whether it is you or the postman at the door. It's the same way with the earth: the flattening takes place no matter what, whether Copernicus is right or Ptolemy is right. I can't figure out who is right from the fact of the flattening.

A: That's very clever. But you've forgotten one thing: even if I can't figure out anything from the ringing of the bell, I can go to the door and see who's there.

B: Right, that's just the difference between Copernicus and the postman. Poor Copernicus would have to be able to observe other effects, in addition to the flattening at the poles, in order to conclude that the earth is in motion. There *are*, indeed, other such effects. But we come to my small amendment to the business of the matchbox: it is the same with all other effects. They occur if Ptolomy is right, too. That is why Copernicus has, in principle, no means of knowing that the earth is in motion.

A: Aha, now I understand. You trapped me with your little amendment. There is no means . . . Umm, yes, I see. The effect will remain the same, whether the carrousel goes around or the world goes around . . . But then both possibilities amount to the same thing!

B: You really *have* understood. It is all one; we only draw a distinction because we haven't thought the matter out to its conclusions. Einstein was the first to think it out to its conclusion – and that's all.

A: Fine, we should always think things out to their conclusion. Yes, indeed, though that doesn't look so easy. But, tell me, didn't Einstein go on thinking even beyond the conclusion? I mean, the business with the light rays – that doesn't have anything to do with motion, does it?

B: We must not think at every stop that the trip is over; that way, we'll keep on making more discoveries. Watch me drop this stone; it falls to the ground in a straight line. The man over there in the train that has just left the station and is constantly accelerating also saw the stone, through the window, but for him the line it described would be curved, if he were to draw it on the windowpane.

A: Because he is moving along with the train.

B: That's right. But according to the principle of the matchbox, he could also say that he is at rest, while we are moving past him, along with the whole city and the whole world. He would then call the force that presses him against the wall of the compartment as the train moves a gravitational field, and say that gravity causes the stone's path to be curved. But for him, light rays are curved as well, for the same reason, so that he would also say that gravitation makes the light rays curved.

A: What about the sun?

B: There, too, we say that a gravitational field exists, and therefore we must also say that the light rays are curved. If we crashed into the sun on an airplane, there would be for us no gravitational field, and the light rays would be straight for us.

A: That isn't so easy for me to understand. I'll have to think about it some more.

B: It's the very same point, but you do have to think it out very carefully.

A: Too bad that I can't grasp it right away with my common sense.

B: Look, common sense is a very good thing, but considering a good tip for the stock market is one thing and mulling over light rays on the sun is another. What did Kant say: "A chisel and mallet may serve very well for carving a piece of wood, but engraving requires an engraving needle." Anyone who wishes to understand Einstein's theory thoroughly must be able to draw with the engraving needle in his brain.

A: That may well be true; but I'd rather stick with the good market tip. But I do understand one thing now: Einstein must be able to think fantastically well, right to the conclusion.

B: And you will never again say the relativity theory is nonsense?

A: No. I've got just enough common sense to notice when someone else has thought something through further than I have – to its conclusion.

B: Look: a matchbox, a simple statement, and ten minutes' listening time. If only everyone with common sense would listen for ten minutes! (*Says a friendly goodbye and leaves.*)

A: It's a good thing after all that not everyone thinks matters out to their conclusion. If they did, I don't think we'd get any more good market tips! (*He lights his cigarette, carefully holding his matchbox steady in his left hand: then he sticks it furtively into his pocket and walks away.*)

11. TYCHO BRAHE'S SEXTANTS

[1926a]

The city of Prague houses two artifacts that serve as a landmark in the history of civilization: two sextants of the famous Danish astronomer, Tycho Brahe. Tycho, who lived in the latter half of the sixteenth century and who was widely famed throughout the world as an astronomer, made his home on the Danish island Hveen, where King Frederick the Second of Denmark had a castle, the Uranienborg, built for him. Here, by using various newly constructed mechanical devices, he prepared his instruments, which superseded all their predecessors in their eminent precision and with which he himself undertook very exact observations of the planets. But under Frederick's successor, Christian, Tycho's luck turned. Some jealous courtiers were able to turn the king against him, and after 21 years of successful activity, the 51-year-old Tycho had to flee from Uranienborg. He came to Germany, a country with which he had become familiar as a student, and spent two years in Wandsbek as a guest of Count Rantzau. From thence he was called to Prague, having received an appointment from Emperor Rudolph the Second, and he succeeded in having his instruments from Uranienborg sent after him. In Prague he continued his observations with the support of the young Kepler, but he died of a sudden illness after only two years. Kepler took over his instruments and continued his observations. Fate was unkind to these valuable treasures, for almost all of them were destroyed during the Battle of White Mountain. Only two sextants remained intact, and they are still to be found today in the Sternwarte Museum in Prague.

One of these sextants is seen in Figure 1. It was constructed for the purpose of making astronomical measurements of angle and has a radius of about $1\frac{1}{2}$ metres. When measurements were to be made, the instrument was placed on a stand, supported by the point seen in the foreground. The angles were marked off exactly on the large arc on the right; the transverse arm, the pivotal point of which is in the left-hand corner, can be moved along the arc. In this same corner is a sight-hole, or dioptre, through which the astronomer sights the rim at the other end of the arm. The sight-hole itself is shown in Figure 2. It is a few centimetres across. Through the small slit in the lower iron plate the astronomer had to sight the upper position of the rim, also keeping in his view, of course, the star at which the sight-hole was directed.

Fig. 1. One of Tycho Brahe's sextants. Radius about 1½ metres.
(From the Archives of Institut für Zeitungsforschung der Stadt Dortmund.)

Today we are amazed at what primitive arrangements the astronomers of Tycho's time had to work with. The present-day astronomer can watch the stars in comfort through a telescope and can focus on them very exactly, as they are magnified many times. The telescope was unknown to Tycho, yet he must have been able to manage very well with his sextants; no modern astronomer with Tycho's relatively crude instruments for observation could reach the precision he was able to attain.

For what was learned from the stars with this historical instrument was nothing less than the elliptical form of the course of the planets. Kepler was the first to make the discovery, but he did so on the basis of Tycho's measurements. Even Copernicus had still taught that the planets rotated in a

Fig. 2. The sight-hole, measuring a few centimetres across.
(From the Archives of Institut für Zeitungsforschung der Stadt Dortmund.)

circle, but Tycho's observations were so precise that Kepler was able to recognize from them the slight deviation from a circular to an elliptical orbit. Kepler then formulated this results in the famous laws bearing his name which in turn constituted the most important basis for Newton's laws of gravity and, thereby, of modern mechanics.

Tycho, incidentally, was not able to agree altogether with the Copernican system of the universe. According to the old, Ptolemaic view of the universe, the world is stationary and the sun and planets revolve around it; on the Copernican view, on the other hand, the sun remains stationary, the earth and planets revolving around the sun. Tycho's system might be said to lie between the two. On his view, the earth is stationary and the sun revolves around it, but the other planets revolve, not around the earth, but around the sun, which thus revolves around the earth, drawing the entire cluster of planets along with it. In Figure 3 we see an old clock, now in the Sternwarte Museum in Prague, that was constructed by a Jesuit father around 1780 and demonstrates Tycho's system of the universe. In the centre of the dial is the earth, below it and to the right the sun, which in turn stands in the centre of a cluster representing all the other planets. When it was still in operation, this old clock demonstrated correctly the relative motions of the planetary system. It is interesting that the very latest conception of the problem of planetary motion, as present in the theory of relativity, gives credence equally to

Fig. 3. An old clock, now in the Sternwarte Museum in Prague, constructed by a Jesuit father around 1780, demonstrating Tycho's system of the universe.
(From the Archives of Institut für Zeitungsforschung der Stadt Dortmund.)

all three of these systems. According to this theory, Tycho's system, too, is an admissible description of the movements in the universe, of which it is possible to give a number of logically equivalent presentations.

Finally, Figure 4 depicts the Tein Church in Prague, the final resting-place of the remains of the famous Danish astronomer.

Fig. 4. The Tein church in Prague, the final resting-place of the remains of Tycho Brahe. (From the Archives of Institut für Zeitungsforschung der Stadt Dortmund.)

12. THE EFFECTS OF EINSTEIN'S THEORY

[1926d]

The great upsurge of public interest that Einstein's theory of relativity attracted several years ago has subsided. If we now wish to weigh the net results of this movement, we must first try to comprehend the roots from which it sprang.

Discoveries that signal a turning point in science invariably begin to exercise a public influence only after a certain lapse of time. Even if occasional public notice is taken quite early on, the great bulk of explanatory literature will not appear until the new theory has established itself in professional circles, until it has ceased to be a subject for academic discussion and has already given rise to results that place its scientific significance beyond doubt. The public should feel nothing but gratitude for this delay in their opportunity to participate in the life of science, for the checkered and circuitous course of science would prove simply too much for them; it is a rough path, full of extensive detours and dramatic situations unsuspected by those who enshrine science as a lofty goddess of unerring certainty. The course followed by relativity theory has been no different; it is just that the hardest part lay far behind by the time the public heard about it. By 1905, Einstein had already published the special theory of relativity, but only in 1919 did the public begin to take an interest in it. During the interim years, physicists occupied themselves intensively with the new theory. Shortly after Einstein's first publication, which transfigured the scientific world like a magic spell, Minkowski published his fundamental mathematical reformulations of the theory, thus supplying it with the equipment indispensible for its subsequent development. While, from that time forward, the effect of the theory began to be felt publicly under the impetus of the slogan (much misunderstood) that time is the fourth dimension, the majority of leading physicists busied themselves with the greatest problem then still outstanding, the integration of gravitational phenomena into a more inclusive theory of relativity. We may safely conjecture that every physicist and mathematician of any standing was considering this question at that time; the direction of future developments was suggested by the mathematical account itself, which contained, as it were, the impetus towards generalization. A particularly interesting approach – and by far the most abstract and formal – was constructed by the Swede,

Nordström, who proposed the introduction of a fifth dimension. But it became apparent that the goal was not to be reached by following mathematical instincts alone. Einstein's physical genius was called upon to contribute another *physical* idea before the apparatus of mathematical calculation was able to come into play. Einstein had announced the basic elements of this idea as far back as 1906, but he himself had to try a whole series of approaches – including one in which he erroneously believed himself to have demonstrated the impossibility of a completely general relativity theory – before he discovered the final form, in 1915. It was in 1919, when his new theory found astronomical confirmation for the first time, that it began its triumphal march through the public as a theory that had long since had its rough spots smoothed off and could be acceptably presented to the public as a complete and certain thesis.

But why was there such strong and widespread interest in it? Its influence on public thinking is best seen in the statistics on publications put together by the Belgian, M. Lerat. According to his findings, an average of 120 publications about relativity theory appeared annually during the years 1913–8; the number then shot up, reaching a maximum of 550 in 1922, and finally subsided after that. The great increase beginning in 1919 is principally due to the popular accounts, which at that point far exceeded the scientific publications in number. Yet how was it possible for such a difficult scientific theory to command the interest of the most diverse segments of the public?

To understand this phenomenon, it is necessary to grasp the line of historical development to which the theory of relativity belongs. Two hundred years back, at the time of Newton, Leibniz, and Huyghens, a dispute concerning relativity theory had taken place in connection with the Newtonian theory of space; at that time, the public took as lively an interest in the matter as they do in Einstein's theory today. The advocates of relativity, Leibniz and Huyghens, did not, of course, discover the decisive physical arguments of Einstein: thus they were unable to block general acceptance of Newtonian absolutism, to which Kant later assented in its essentials. But the repeated reappearance of the idea of relativity, most recently in the 1880's, shows that there lies at the core of the relativity idea an element that perpetually fascinates the human mind quite independently of any scientific basis for it, that convinces us of the correctness of the conception of relativity and is so immediately evident that it finds defenders even among the general public.

This element is the relativity of motion. Everyone who has ridden a train has doubtless experienced the illusion of believing himself to be in motion, when in fact the neighbouring train has started up. This experience is the intuitive seed from which all relativity theories grow. Is it correct to speak of an illusion here at all? Is the 'false' conception not equally justified? Is not motion a matter of the reference point one selects, a relative assertion? Anyone who has but once seriously entertained this doubt has been, as it were, infected by relativity theory and will not rest content until he has found an answer to this question, which is so simply posed and which appears to be quite independent of complex scientific investigations. Here lies the well from which springs the public interest in relativity theory, extending, as it does, so far beyond professional scientific circles. And this well was replenished by a stream from a new and deep-lying source when, with its astronomical confirmation, the news of Einstein's relativity theory came into public view.

Professional scientists tend to display a certain scorn for the popularization of scientific ideas. Again and again, we hear expressed the opinion that scientific investigators have no interest in forcing their findings upon the public, that science itself is not advanced by their doing so, and that it behooves scientists to display a detachment becoming to their station. The popularization of relativity theory in particular imparted new strength to this standpoint, for it was attended by so many undesirable side-effects that it often really did seem best to look away – especially for those of us who were ourselves relativity theorists.

What was the source of these phenomena? Suddenly a whole host of new 'inventors' and 'philosophers' appeared on the scene, claiming either to have refuted the theory of relativity or to have been preaching it for years. Some even proclaimed that they have done both. People who had taken no interest in the theory from 1905 to 1919, the period of its scientific development, felt themselves called upon to take a position in the face of the public show of interest. Thus began that noisy quarreling, that back and forth of public battle, which even reached a stage at which positions were taken up along political party lines, so that the theory of relativity took on the appearance of a political idea, some supporting it as a matter of course and others rejecting it, equally as a matter of course. As always in such struggles, the least attractive products were the most visible and the loudest screamers the least appealing. In all this, many people forgot that good seed lay at the core even of the popularization, quite aside from the high scientific quality of the theory itself. For among the popular publications are unquestionably to be found some displaying admirable artistry of presentation in the best spirit of popularization.[1] This highly sophisticated idea of a mature science has been rendered accessible with most remarkable success. It is equally beyond doubt that, despite the loud shriekers, a goodly amount of genuine interest in the foundations of knowledge was present among even the least sophisticated readers. Precisely because the logical nub of the theory was so attractive and so easy to grasp, the public was able to participate with a degree of involvement not possible with other scientific disputes. Primitive as they may have been, the resulting logical discussions achieved some measure of mass education: they taught clear thinking, the formulation of ideas, the making of a firm personal decision. The connection between the theory of knowledge and the education of the people was strengthened in the process; once again, a slice of scientific culture penetrated to the masses and schooled their minds – even if this entailed a simplification of the scientific content and even if the benefits accrued only to ordinary, highly unscientific minds. Is science supposed to be above all this? Quite the contrary: it should be delighted with such results, which may have a more profound effect upon humanity than does sophisticated culture among the educated classes.

But what is the situation today? Does the waning of the interest of the public signal the end of relativity theory itself? To draw this conclusion would constitute a serious misunderstanding of mass psychology. Once the passionate opponents of the theory grew quieter – at last! – a number of grave-diggers were already standing in waiting, eager to give it a hasty burial.[2] They make appeal to an experiment carried out recently on Mount Wilson by the American scientist Miller, which, while its findings indeed conflict with the foundations of relativity theory, is so dubious in its quantitative results that the majority of German publications do not acknowledge it. And they fail to mention that equally basic experiments were performed at the same time by German scientists on a still higher mountain, the Jungfraujoch, with results that beautifully confirm the theory and make Miller's result unacceptable. They further fail to mention that recent measurements concerning the satellite of Sirius have agreed beautifully with one of Einstein's predictions. These would-be heralds of death interpret the waning of public interest as signifying the decline of relativity theory – quite wrongly, for this scientific theory has never stood on firmer ground than it does today.

Leaving aside isolated opponents, notorious from the years of open battle, marked disagreements are scarcely to be found among physicists today. Relativity theory has become an unshakable constituent of physics; the younger generation of physicists is already scarcely able to imagine how physics could ever have been possible without relativity theory. The logical and epistemological foundations of the theory – which are precisely the aspects that still required clarification after Einstein completed his physical account – have been considerably extended and now form a solid groundwork from which the genuinely physical nucleus of Einstein's work can be surveyed – from which, indeed, a deeper understanding of the pre-Einstein space–time theory has become possible for the first time. In short, the theory of relativity has become *classic* – and that is why we now hear so little about it.

Especially, by the way, among physicists themselves. They have involved themselves with all conceivable intensity in the problem of the atom, which has undergone marked advances in the past few years. Work is being done on spectral lines, on the internal differences between chemical elements; efforts are being made to construct a model of the atom, combining mathematical acuity with experimental skill. Decisive progress was made just last year; the idea of the atom as a small-scale planetary system had to submit to a further modification, according to which it consists of a positively charged nucleus surrounded by a vibrating cloud of negative electricity. The greater consequences of this modification are still hazy: the interest of the investigators themselves still flutters, much like an electric cloud, in the vicinity of this problem, the fundamental nature of which is still unknown to the public. Relativity theory is mentioned in this connection only when, as a matter of course, the relativity corrections are added to the model for purposes of precision.

"To truth there is granted only a brief victory celebration between the two lengthy periods in which it is denounced as paradoxical and dismissed as trivial" – thus spoke Schopenhauer. The theory of relativity has now entered the third of these stages.

NOTES

[1] I shall list just a few of these publications here. In *Die Idee der Relativitätstheorie* (Springer, Berlin, 1921) [trans. by R. A. B. Russell as *The Ideas of Einstein's Theory*, Methuen, London, 1921], H. Thirring offers a thorough and very clear presentation (especially as regards the principles), rendering even the more difficult general theory of relativity comprehensible; note particularly the tables at the back, which present schematically the logical relations among the conceptual foundations. F. Beer's *Die Einsteinsche Relativitätstheorie* [Einstein's Relativity Theory] (M. Perles, Vienna) is distinguished by its intuitive style, comprehensible even to the untutored. The book by G. Mie, *Die Einsteinsche Gravitationstheorie* [Einstein's Theory of Gravitation] (G. Hirzel, Leipzig) is carefully worked out from the pedagogical standpoint; however, it covers only the general theory of relativity. While the aforementioned accounts manage to avoid mathematical formulas, the following two make use of elementary mathematical calculations without, however, presupposing the more difficult relativity mathematics: W. Bloch, *Einführung in die Relativitätstheorie* [Introduction to Relativity Theory] (Teubner, Leipzig, 1920) and M. Born, *Die Relativitätstheorie Einsteins* (Springer, Berlin, 1920)

[trans. by H. L. Brose as *Einstein's Theory of Relativity*, Methuen, London, 1924; reprinted by Dover Books, 1962]. A. Eddington's *Space, Time, and Gravitation* (Cambridge University Press, 1920) offers a lively presentation in a conversational vein, though requiring some mental agility on the part of the layman. Finally, let me mention a few of the philosophically oriented works, which, again, are completely free of mathematics. M. Schlick's *Raum und Zeit in der gegenwärtigen Physik* (Springer, Berlin) [trans. by H. L. Brose as *Space and Time in Contemporary Physics*, Oxford, 1920] increases our understanding of the theory specifically from the philosophical viewpoint. E. Cassirer presents the theory of relativity in the context of the historical evolution of modern epistemology in *Zur Einsteinischen Relativitätstheorie* (Cassirer, Berlin, 1921) [trans. by Swabey and Swabey in *Substance and Function and Einstein's Theory of Relativity*, Open Court, Chicago and London, 1923; reprinted by Dover]. I myself discuss the relation of Einstein's theory to Kant's philosophy in *Relativitätstheorie und Erkenntnis apriori* (Springer, Berlin, 1920) [trans. by M. Reichenbach as *The Theory of Relativity and A Priori Knowledge*, Univ. of California Press, Berkeley and Los Angeles, 1965] [1920f and 1965a]. A further discussion of the philosophical literature on the subject is found in my article "Der gegenwärtige Stand der Relativitätsdiskussion" [trans. in this volume as "The Present State of the Discussion on Relativity"] [1922f].

[2] Namely, H. Dingler (*Süddeutsche Monatshefte,* **23**, 210) and K. Lichtenecker (*Kunstwart* 7, 63 (April, 1926)).

13. AN OPEN LETTER TO THE BERLIN FUNKSTUNDE CORPORATION

[1926e]

Gentlemen:

On New Year's Eve, we heard on many stations the announcement that German radio is going into its third year. We also heard you, along with many other broadcasting companies, take the opportunity of proffering thanks to your announcers and artists – as was, indeed, most appropriate. For when, yet again, we come to take stock of the year just drawing to a close, it behooves us to acknowledge the many fine achievements accomplished within its span.

Many solid, average programmes were offered over the past year. We heard good German music – sung, played on solo instruments, played by orchestras. We heard sound scholars deliver lectures, we were stimulated or amused by readings of short stories or humorous sketches. We also heard records and dance music – well, that was to be expected; it is as much a part of radio broadcasting as a merry-go-round is a part of popular amusements. I have always resented the way in which some people, with a great show of pathos and cultural morality, accuse the radio of lowering the standards of civilization or of fostering cheap pleasures. We must take for granted that the level of culture maintained by the radio will not be any higher than the hodgepodge of intellectual demands and needs for recreation determining the cultural level of the vast millions who are caught in the daily treadmill of stultifying labour and cannot fairly be judged by the standard of the happy few who possess the advantages of education and a capacity for critical discrimination. My usual response in such discussions on this topic is that infinite gains are made so long as good programmes are *also* available, so long as the loftier spiritual demands are *also* satisfied. This argument is most particularly applicable in the case of the radio. If you don't want to hear the shimmy, you can turn the dial to Mozart; if you don't care for music, you can seek out an educational programme. And if you prefer to sit by yourself in your own cosy corner, you can turn the radio off altogether. That we have the choice, that we can select according to our pleasure among programmes from all over Germany – indeed, from all over Europe – is one of the unrivalled advantages of radio, making it a serious component in our civilization.

But – everything hinges upon this 'also'. Programmes must also be offered for people of refined tastes; programmes must also be offered which can be made available to a wider public only by means of radio. We are all familiar with quartets, with Wagnerian arias, with the short stories of Marie von Ebner-Eschenbach, and we can hear them elsewhere. This fact should not deprive you of the joy of holding concerts. Please, go on broadcasting quartets and arias, and we shall be delighted to hear them once in a while, sitting over our cup of tea. But we won't rest content with hearing nothing but opera singers and actors, nothing but edifying gentlemen or chitchat for the housewife. Radio has other possibilities, for there exists no other institution with the same capacity as the electric wings of the radio for bringing the voice of a single person to the general population.

A few weeks ago, you see, I listened to five New Year's Eve broadcasts at the same time. I simply turned the dial on my radio, and one after the other I heard Berlin, London, Paris, the Danish station Ryvang-Radio, and my local station. What I heard was not entirely cheering. Berlin, at least, played ballet music until ten o'clock, whereas Stuttgart treated the world to a kind of mindless buffoonery that was too much for even my tolerant ears. Paris was middling; at least there was some decent piano playing, and also a good violinist. Ryvang-Radio sounded very solemn. Unfortunately I don't know any Danish; perhaps that is just the way the Danes celebrate. England tried hard. Their main wave carried first London and then Birmingham, and there were serious lectures alongside blues. After ten o'clock Berlin, too, broke out into the shrill racket which we Germans evidently feel is called for in seeing out the old year. That's all very well: the inability to take a serious look back over the old year as it draws to a close appears to be so widespread that the radio must be expected to join in, too. But as I listened, something else became clear to me.

I became, once again, vividly aware of *how many people* hear radio announcers, and of what it means for the spoken word to be audible throughout the continent. The announcers must have experienced a similar feeling, for they used the occasion to make ceremonial speeches. Berlin declared with a bit of a swagger that hundreds of thousands of listeners were tuned in at that moment, thanked their broadcasting staff – quite rightly, as I have remarked – and then gave a New Year's toast to their board of directors in Zehlendorf. With all due respect to the directors, was it necessary to do that within the hearing of the entire continent of Europe? The only really tasteful touch came from London. In closing, the announcer said, in his polished English manner:

Good night, ladies and gentlemen!

Bonsoir, mesdames et messieurs!

Gute Nacht, meine Damen und Herren!

He said it in three languages, and his foreign accent was marked as he pronounced the unaccustomed words. I cannot think of a more elegant way of expressing the feeling that the ear of all Europe is open to the radio announcer. Those words, with the foreign ring to them, were for me on New Year's Eve, the epitome of radio.

Please, don't you imitate what they did! That won't work. By the second time it's already a bit stale, and by the third time it's an empty phrase. The suggestion I am going to make to you is quite different.

Keep in mind, as did the English announcer, that your audience is Europe. Don't just tell us you have a hundred thousand listeners; act as though you do. Be careful that you do not always offer these hundreds of thousands a reflection of their average level; give them, instead, something worthy of being heard by hundreds of thousands. You are the very ones who can do that. Yes, you gentlemen! You stand at the hub of Germany; in addition to various short waves, the vast powers of Königswusterhausen are at your disposal, and by means of their long wave you can broadcast anywhere without difficulty and without the interference of local stations. With these extraordinary technological facilities you can offer what no one else in Germany is able to offer: you can lend your microphone to the elite of the German world and give them the opportunity to address every German, every European.

After all, there are some really fine intellects among us Germans. There are men whose books are read by everyone, men who have really had an influence upon the life of the mind. Up to now, these men have only been able to use the language of the printed word. Put radio waves at their disposal, so that they may spread the spoken word in their own accents, with their own breath! We would like, for once, to *hear* these men, not just to read them.

Which men do I have in mind? I mean those who are above average, not those who simply repeat and present the thoughts of others or those whose sole accomplishment is that the state has adorned them with office and honours. Nor do I mean politicians: they talk far too much already. I am thinking of the creative spirits from whom we otherwise hear only indirectly.

But, you may reply, opinion is divided as to who is to be counted among these men. Everyone who has experienced some success thinks himself a great man, and some are held to be great by one faction while their opponents consider them insignificant. That is certainly true. I therefore advise you, whenever you have a choice of two, to take both. For the point is simply to include the great men *as well*. If a few more average voices are heard on the radio it will do no harm. What is harmful is the total absence of great ones – and that is what we are faced with at present.

Simply take the prominent people – it isn't so hard to establish who they are. Yes, yes, I know: some prominent people are forgotten three years after their death while others are still known one hundred years after their death, though they were not famous during their lifetime. It won't be your fault if you don't find those people. But at least take the people who are prominent. We would like a chance to hear them, for we would like to be able to judge for ourselves whether they are as great as they are famous, and we can do this much better from their spoken words than from their works. It is an eery feeling of intimacy that a good clear voice over the radio engenders. It is exciting to hear the timbre of the voice, the passion of a speech, the forming of the words, the flow of which was formerly known to us only from written works. Here lie possibilities of a kind of influence that calls to mind the era of the ancient rhetoricians and philosophers. In those days the circle of the cultivated was so small that people could go to it. Today we are grateful to the art of printing, which at least delivers books to our door. Give these prominent men a voice, and the resulting leap will be no smaller than that taken in Gutenberg's time. For it will be, as well as a leap forward, a leap two thousand years back, to the time when teachings were handed on directly from one person to the next by means of the spoken word.

If you want me to mention names: I can imagine Gerhart Hauptmann reading from his works, or Thomas or Heinrich Mann. Hermann Hesse, too, would have something to contribute. Albert Einstein is a unique phenomenon in science – as any scientist will tell you. Sigmund Freud has forged new paths in psychology. Gustav Wyneken is as hated as he is important. Picking at random from the brilliant array, I could further name Hermann Keyserling, Oswald Spengler, Hans Driesch, Max Scheler, Max Planck, Käte Kollwitz, Franz Werfel, Ferdinand Damaschke, Max Weber, Flettner, Eckener . . .

Do I hear you saying those are the people who happen to interest me? You are mistaken: some of them don't interest me at all, others a great deal. I know of some others, too, whom I won't mention, even though I hold them to be very important. I know you wouldn't take them, for they are not famous. But you yourselves must know of many others to whom I, and everyone else with me, would cheerfully agree. Please, then, add to these names a list three times as long. But leave out those promoted only by factions – I am thinking of minds, not movements.

This would keep you busy for about a year. For I imagine this institution as being effective only at weekly intervals: one evening a week, an hour is reserved for a great figure. Every German station receives the Berlin broadcast and relays it to its own provincial vicinity. During this hour, all Germany listens to the words of one man, and you may be sure that the rest of Europe will also listen to this 'German hour'.

You will excuse me, gentlemen of the Berlin Funkstunde Corporation, for burdening you with my proposal. But I am convinced that you are so busy with the trying task of putting together your daily program that you will welcome the suggestion of someone who is detached from it. The question of costs will surely prove manageable. As the programs will be transmitted to other stations, their allotment will be released for the joint project. And I am further convinced that the people in question will regard their talks not as a chance to increase their earnings, but as an opportunity to have some public influence. An invitation to address the nation would be regarded as an honour.

If you carry out this proposal, you will be assured, gentlemen, of the thanks of the hundreds and thousands of listeners whom you mentioned on New Year's Eve.

Very truly yours,
Dr. Hans Reichenbach

14. LAYING THE FOUNDATIONS OF CHEMISTRY: THE WORK OF MARCELLIN BERTHELOT

[1927a]

From October 23 to October 27, the scientific world will be celebrating the one-hundredth birthday of Marcellin Berthelot. This great French chemist has not yet achieved the status of a remote historical personality. It was a mere twenty years ago that he died, and some of the leading scholars of today had occasion to engage with him in scientific dispute. Nonetheless, his lifetime is sufficiently removed from the present for us to survey his works and his influence and to construct a clear picture of his significance.

Marcellin Berthelot was one of the most prominent scholars of the past century. He was one of those men who are permitted to witness the success of their work in full flower and who reap from this experience the strength for further undertakings. The wealth of his scholarly publications demonstrates the breadth of his scientific activity and his interest in the most varied subjects, not only within the realm of chemistry but far beyond its boundaries. It was his early success that placed him in a position to perform such a vast amount of scientific work. Thus he became the embodiment of the scientific spirit of an age and the symbol of a century of science.

His brilliance became apparent when he was yet a child. That he was the best pupil at the Collège Henri IV demonstrates that, for those with certain kinds of talent, the frequently alleged contrast between performance in school and thereafter does not exist. He distinguished himself as a student and soon attained fame and respect. He became a professor at the École Supérieure de Pharmacie when he was just thirty-two years old. Yet it was not teaching, but scientific research to which he was most devoted. He fought against the burdening of scholars with examinations, declaring it to be "one of the most foolish of ideas to use the time of inventive and distinguished men in order to give beginners an opportunity to utter nonsense". Such words express the most prominent strain in his character. For him, it was important above all to produce, to generate a maximum amount of scientific knowledge extending over the greatest possible sphere. The professorship at the Collège de France, which he attained soon thereafter, well suited his aspirations, for it occupied him almost entirely with research. He retained this post until he was an old man.

Berthelot's influence, then, did not arise chiefly from his attraction of disciples. He never had many assistants, and those he did have were engaged solely in carrying out his own ideas. His influence derives mainly from his publications, through which his impressive array of scientific findings became generally known. It was through this means that he set a new course for the development of chemistry. He began by working on the synthetic production of organic materials and laid, in the process, the foundations of organic chemistry. This structure led him, in turn, to important investigations into the speed, the direction, and the limits of chemical reactions, which he published in conjunction with Péan de Saint-Gilles in their classic works on esters. Most famous of all is his work in thermochemistry, which we will be discussing in greater detail presently. Berthelot did most of this work between the ages of forty and sixty, when his creative powers had matured. In his later years he turned his attention to agricultural chemistry on the one hand and historical investigation on the other. He founded in Meudon the Institute of Phytochemistry, which is currently directed by his son, the physicist Daniel Berthelot. A trip to Egypt, in response to an invitation to attend the opening of the Suez Canal, stimulated his work on the history of chemistry. He was particularly interested in the historical origins of alchemy and concluded from his study of Theban papyrii that the desire to produce gold synthetically arose from the ancient Egyptian recipes for forging precious metals, which were supposed incorrectly to be recipes for producing gold.

If, in the development of science, we can regard the nineteenth century as the century of expansion, of the amassing and ordering of information, we can regard Berthelot as one of the most brilliant representatives of the period. The accomplishments of an extraordinary memory and inexhaustible energy for work, which he utilized fully during the course of each highly regimented day, enabled him to embrace and to master the entire field of chemistry, which in his day was growing by leaps and bounds. Once, on an official occasion, Emil Fischer called him "the only living chemist who could withstand the shattering onslaught of the towering masses of material". Berthelot imparted his great knowledge to his contemporaries in a magnificant series of textbooks. But ambition would not let him stop until he had pushed beyond the collection of facts to a more penetrating comprehension of the universal laws that lie behind them. However, those same characteristics that were the basis of his greatness – his ambition and his self-confidence – drove him on to form theories that could not withstand closer scrutiny. His principle of maximum work did perhaps more to promote his fame than his more successful efforts. In observing that most chemical reactions take place by means of the spontaneous generation of heat, he believed that he had discovered a general law of chemical transformations determining the direction of chemical phenomena. The direction was to be determined by means of the striving to produce a maximum degree of heat, to perform a maximum amount of work. This hypothesis was vigorously denied, especially by German scientists, and led to an extended scientific controversy. Contemporary insights into the basic principles of the transformation of matter can leave no doubt that Berthelot's hypothesis was wrong. In his later years he himself found it necessary to place strict limits on the application of his theory. Only with the establishments of Nernst's heat theorem did the kernel of truth embodied in Berthelot's theory come to light, and then admittedly in a substantially altered form and on a sounder basis.

Berthelot's tenacity in holding to obviously mistaken views fits in only too well with the image we must form of this outstanding personality. In some areas his authority was a positive hindrance, particularly in the sphere of atomic theory. He opposed the theory of atoms and still believed in the importance of equivalent weights when virtually the whole of the scientific world had taken the next step forward. He considered the belief in atoms to be a religion, to be opposed on the grounds that it was not scientific. It was not until he was very old that he backed down and at least made the concession of writing his formulas in atomic symbols.

The multiplicity of interests that distinguished Berthelot also led him into political affairs. He was both a free-thinker and a republican and numbered himself among the opponents of the Second Empire. At the time of the 1870 siege of Paris he belonged to the *Comité scientifique pour la défense de Paris*, whose members developed incredible plans for defending the city. They hoped to use the waters of the Seine to conduct electrical signals from outside, and Berthelot figured that a great amount of saltpeter could be extracted from the refuse within the city for the purpose of making gunpowder. On the other hand, such notions inspired his important investigations into the phenomenon of explosion. After the rise of the Republic he even became a cabinet minister, holding office as late as the eighteen nineties: yet even then he did not set aside his scientific work.

Extraordinary honors were awarded this man, who had such a full and successful life behind him. In 1902, on the occasion of the fiftieth anniversary of his first scientific work, representatives from every scientific institute and academy in France and abroad foregathered in Paris to honour Berthelot under the aegis of the President of France. After his death, in 1907, he was buried in the Pantheon along with his wife, whose death so shocked him that he died a few minutes after she did.

It was Berthelot's aim in life to become the Lavoisier of the nineteenth century; he wished to found a new science of chemistry upon the ideas of synthesis, chemical mechanics, and thermochemistry. Fate permitted him to attain his goal in a different way. To hold a position parallel to Lavoisier's proved impossible, for the development of science does not follow the same course in each era. The period in which a new science arose had been replaced by a period in which it underwent expansion. Such, too, was Berthelot's destiny. In both his fertility and his final unfulfilled plans, he is one of the most brilliant representatives of the era of scientific expansion.

15. MEMORIES OF SVANTE ARRHENIUS

[1927b]

With the death of Svante Arrhenius, the world has lost a scholar whose personality and work became the embodiment of a new era in the exact sciences: the era of the fusion of two basic sciences, physics and chemistry. If the great development that took place in both sciences in the first three-quarters of the nineteenth century can be credited to their separation and the concentration of efforts within a narrower sphere that it made possible, then the blossoming experienced by both sciences in the last quarter of the century is to be attributed equally to the fact of their coming together. This unification was given visible expression in the concept of physical chemistry, the name given to the common border area. If we were to assign to Arrhenius a professional classification, we would have to call him a physical chemist.

The dual character of this area of study was clearly displayed in the life work of this great Swede. He studied chemistry and physics simultaneously and took his degree in physics. When he wished to qualify as a lecturer in physics at the University of Stockholm, it was the chemist Wilhelm Ostwald who interceded for the significance of the young scholar's work and who first succeeded in conquering the resistance to the recognition of his ideas. Later on Arrhenius spent several years working with Ostwald, a chemist, but he also did work with the physicist Boltzmann and the physical chemist van 't Hoff. When, at thirty-two, he became eligible for an appointment, he was offered simultaneously a chair in chemistry and a chair in physics. He chose the latter, principally, no doubt, because it took him back to his native Stockholm. If we examine his works, we find, intermingled, themes from pure physics, from astrophysics, from chemistry, and from medical physiology: a multiplicity of achievements probably attained by no-one else except Nernst, who was just a few years younger.

Arrhenius did not divide his work between the two basic disciplines of exact science merely because of his personal inclinations. Rather, the necessity to do so arose from the very state of the disciplines themselves, for chemistry became compelled to seek in physics theoretical explanations of chemical laws while, conversely, physics looked to chemistry for factual material that made possible for the first time the completion of physical laws. At that time, great attention was directed to the phenomena that occur in diluted salt solutions, which Van't Hoff sought to clarify with his theory of osmotic pressure. According to this theory, the dissolved salt is comparable to a gas the molecules of which swirl about freely in the solution – let us say, water – and are restrained only by the surface of the liquid, as if by a membrane. Like other gases, this salt-gas exerts pressure, and it can be verified with the help of semi-permeable membranes made from a fish's bladder, through which the solution, but not the salt, can pass and which become inflated by the pressure of the salt-gas. Van't Hoff also discovered that the salt-gas followed the well-known Boyle–Mariotte law for the relation between the pressure, volume, and temperature of gases, but he found a noteworthy irregularity when, in addition, he applied Avogadro's law for calculating the number of atoms or molecules of gas. Application of this law indicates, for instance, that the osmotic pressure of hydrochloric acid dissolved in water is almost twice what would be predicted by the law for gases.

The scientific development initiated by these discoveries was enriched, several decades later, by an influx of material from an entirely new source; with the discovery of radioactivity, chemical research and physical explanation once again sought guidance from one another. Today, when this source of material, along with the mathematical analysis of precise physical–optical observations, has given rise to Bohr's theory of atoms, we can no longer imagine away the marriage between physics and chemistry. After all, it has now become possible to explain the chemical characteristics of the elements in physical terms, so that chemistry has become the physics of protons and electrons and blends into physics in much the same way as other subsidiary subjects such as mechanics and thermodynamics.

In contrast, Arrhenius' interest in pure physics pertained to quite different areas. He spent a good deal of time working on cosmic physics, being particularly attracted to the phenomenon of radiant pressure. To be sure, the notion that a beam of light exerts pressure when it strikes an object is quite alien to the layman, and, indeed, these extraordinarily weak forces are imperceptible in normal observation. Yet this force becomes proportionately very much greater when it acts upon small particles of dust, as their surface area is large in proportion to their mass. Such particles can even be blown away by a beam of light. Anyone who has watched this phenomenon with a strong microscope is amazed at the mechanical force exerted by a beam of sunlight. Arrhenius figured that cosmic dust takes long journeys under the pressure of such forces, and he came up with the bold hypothesis that the most minute germs of life can be transported from planet to planet at the same time, giving rise to an 'international' form of life freed from the accidental and cosmically admittedly unpromising fortunes of the planet earth – a most happy prospect from the biological point of view.

Distinguished also by his personal quality, by his kindness towards others, by his ability to write books for the general public, and by his gift for speaking languages, Arrhenius often spoke at German conferences, and all who heard his lectures, delivered, as they were, in a deep voice, took delight in the ease with which he "replaced grammar with courage", as Ostwald once put it. To quote further his dear friend Wilhelm Ostwald, he confirmed

> the proposition that early dissemination of the great new discoveries the guiding spirit gives to the world; that early entry into relations with enthusiastic disciples and colleagues who take upon their shoulders a portion of the burden of working out ideas and, most importantly, protect the creative organism against premature weariness by offering a sympathetic atmosphere during his periods of greatest strain; that all these elements, which lighten both physically and mentally the burden of performing great work, do in truth also have the effect of minimizing the dangerous repercussions to which the productive organism is liable and of preserving in the organism the energy that fosters further creation.

In his autobiography, *Aus meiner Jugendzeit* (Akademische Verlagsgesellschaft, Leipzig), which is dedicated to "my dear friend Wilhelm Ostwald", Arrhenius says,

> When I look back upon the stage of development of the natural sciences during the time I was a student (1878–83) and compare it with the present stage, I must admit that the progress has been enormous.

He modestly forbore to mention that he himself was a major cause of this progress.

The man who has now departed from among us was able to look back upon a life that was not only fruitful but also beautiful and rich, and perhaps that is the most comforting comment that can be made at the bier of a man of science. For it is not often that science is able to bestow the gifts of fortune upon one of its own.

16. A NEW MODEL OF THE ATOM

[1927c]

The historical springs of the idea of the atom lie as far back as classical antiquity. The Greek philosophers Democritus and Epicurus were the originators of atomic theory; from them, an unbroken line can be traced through the Middle Ages right down to our own day. Over this long span of time, the theory has not only taken on a basically different appearance, but has also become susceptible to far more precise confirmation than could be carried out by the speculative philosophers of the past. Nonetheless, the modern theory must be viewed as an historical continuation of past theories. For the basic idea giving rise to the conception of the atom was the same in ancient times as it is today.

The idea of atoms is occasioned by the phenomena of compressibility and capacity for change that are displayed by matter. Even the ancient atomic theorists adduced as evidence that these phenomena were otherwise unintelligible. That the self-same substance sometimes fills a smaller, sometimes a larger space would be incomprehensible if compression were supposed to affect the matter itself. Thus it must be assumed that the compression simply reduced the interstices between the individual particles, leaving the particle unaltered. Chemical conversion of matter likewise becomes understandable if we visualize it as occurring through a mixing of essentially immutable atoms.

While the atomic thesis has been confirmed by modern science, the picture of the internal structure of the atom has changed altogether. The ancients thought of an atom as a tiny piece of compact matter equipped with hooks by means of which it was attached to other tiny pieces of matter. This very primitive notion was not overcome until Newtonian forces of attraction became familiar, making the hooks superfluous. This was the first great step beyond the old theory of the atom. The second important step was the forfeiting of the idea of uniform material particles. The atom became recognized as a complex system made up of yet smaller individual particles, each chemical element possessing its own characteristic atomic structure, to be established only by individual investigation. The credit for translating this theory into fact is due principally to the English physicist Rutherford. He adapted to this model the ideas that had in the meantime been developed concerning the atomistic structure of electricity and thereby created the planetary model of the atom, according to which negatively charged electrons, the atoms of electricity, revolve around a positively charged nucleus, which makes up the bulk of the atom. Chemical elements are differentiated according to the mass of the nucleus, on the one hand, and the number of rotating planets on the other. As was eventually shown in detail, this model makes the chemical characteristics of the atom comprehensible; chemical compounds could be explained as fusions of two planetary systems into a single system. Rutherford himself based his model on his deflection experiments, in which he impelled alpha-rays, that is, small, positively charged particles, through atoms. These experiments showed that the atom was of a highly porous structure and also revealed the existence of extraordinarily small, heavy, strongly deflecting central bodies.

The third step of fundamental significance was taken by Bohr, who linked Rutherford's model of the atom with the results obtained through investigation of the atom with the results obtained through investigation of the spectral lines of the elements. If a chemical element is made to emit light through 'excitement', e.g., by means of the force contained in electric sparks, it sends out light rays of a definite colour: potassium produces a violet light, sodium a yellow, hydrogen a red, and so forth. In general, each element emits not just a single characteristic colour, but a particular colour compound. Each colour corresponds to a particular number of oscillations, or *frequency*; the frequency, in turn, corresponds to a particular wavelength for the light: thus the yellow sodium rays have a wavelength of 589 $\mu\mu$ (1 $\mu\mu$ = 1 millionth of a millimetre), a violet potassium ray a wavelength of 405 $\mu\mu$, and so on. If we write down one after another the wavelengths of the colour compound emitted by a certain element, a so-called spectral series, their distribution at first appears completely random. However, closer inspection reveals a characteristic regularity; like a telegram written in code, these random numbers can be decyphered with a key. It was Bohr's great achievement that he was able to use this key to unlock the door of the story below, so to speak. It was he who realized that the key to the spectral series was at the same time the key to the construction of the Rutherford atom.

Somehow, it must be possible to explain the emission of light by the atom through the revolution of the electrons. A possible explanation is, in fact, to be found in the electromagnetic theory of light: an electrically charged particle in revolution produces in its vicinity a periodically alternating electric state which is propagated as an electromagnetic wave and can be directly conceived as a light wave. However, this explanation involves one great difficulty. The emission of the light wave signifies a loss of energy, and the revolution of the electron must therefore follow an orbit of an ever-decreasing radius. There is nothing impossible about this attenuation as such; indeed, we must assume that each atom can emit only a certain store of light and will ultimately exhaust its reserves. Its 'excitement' by external electrical forces will then consist precisely in the electron again moving further away from the nucleus and recommencing its revolution. But another consequence of continuous emission leads to a contradiction. The smaller orbits will entail a different period of revolution, and therefore the frequency of the emitted light ought to change continuously. Since at any one time we observe in the spectral lines many atoms, in a variety of different states, we would also have to observe a continuous colour spectrum if this hypothesis were correct. The existence of distinct spectral lines with their regular series would become quite incomprehensible.

At this point, Bohr's new idea enters in. He makes a strange assumption, viz., that, contrary to electromagnetic theory, the revolving electron does not emit radiation, but instead describes its orbit with a constant unattenuated frequency. Nevertheless, not just any conceivable orbits are possible for the electron. Stable orbits exist only at certain distances from the nucleus, and an electron situated in any of the intermediate areas will move until it falls once again into a stable orbit. The electron may, for one reason or another, occasionally leave its stable orbit and fall until it reaches a narrower orbit. During the period of its fall – and this is Bohr's other assumption – it emits the light wave. Each transition from one orbit to another corresponds to a definite frequency of the emitted wave. Consequently, we find repeated in the series of spectral lines the series of stable orbits, from which the planetary model of the atom is constructed.

This so-called Bohr model of the atom displaced the preceding physical theories. It was worked out by the most intricate mathematical methods, and was established on the basis of Planck's concept of an elementary quantum of energy; the stable orbits showed themselves to be those orbits the energy of which could be reduced to an integral multiple of the quantum of energy. Spectacular agreement was achieved with experimental observations of spectral lines. Nonetheless, this theory was never entirely satisfactory to physicists.

There were two unsatisfactory aspects. First, Bohr's account contradicted electromagnetic theory. Although this theory was to be regarded as generally valid, an exception was to be admitted in a single point: a revolving electron was supposed not to radiate light. But a second difficulty was also involved. Although, on Bohr's theory, light radiation is emitted during transitions from one orbit to another, it is the orbit itself that is said to determine the light

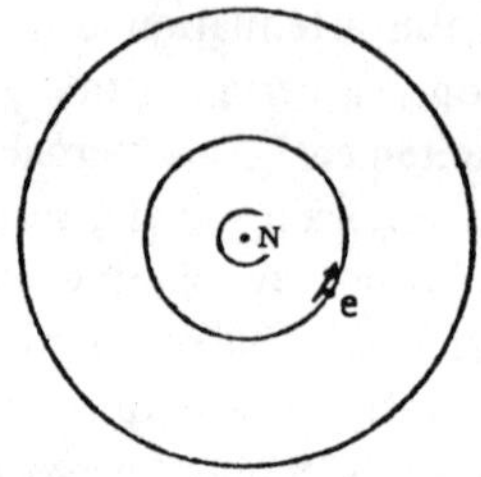

Fig. 1. *The hydrogen atom on the Bohr model.* A negative electron revolves around the positively charged nucleus N on one of the stable orbits shown. In the intervening areas, no orbit is possible, but there are continual further stable orbits outside those shown here.

emitted; the frequency of revolution is held not to be a direct characteristic of the emitted light, yet the nature of the light is supposed, in its essentials, to be determined by this frequency. Bohr formulated this idea as the *correspondence principle*: according to this principle, the emitted frequencies were supposed in a particular way to be coordinated with (to correspond to) the periodic frequencies of their upper harmonics (as shown by an harmonic analysis of the orbit according to Fourier's principles) and to possess the corresponding intensity.

This law, which curiously enough received experimental confirmation, made far-reaching demands of nature. The orbits of the electrons were to have an influence on an event not in any way caused by them. Explanation of this fact would have required conception of a mechanism such that the form of the orbit determines the electron jump, but no clues as to how to go about this were at hand. Scientists began to suspect that Bohr's planetary model, while it produced the correct results, was nonetheless completely inadequate in essential respects.

It seemed, indeed, as though the Bohr model were a sort of magic key to the correct laws of the atom, while not itself possessing direct physical significance. The suspicion arose that the image of an electrically charged particle revolving in planetary orbits around a nucleus was altogether wrong; that the real state of the atom had a quite different appearance; that the Bohr model was admissible only as a fiction that, while it leads to the correct result, cannot constitute a valid physical assertion. But then what did the atom look like? Scientists faced the task of constructing a model entailing the same laws of radiation as the Bohr model while avoiding its contradictions and violations.

This problem, which has hovered before physicists for a number of years, has recently entered a new stage with the help of some recent research. In fact, such great progress has been made that we may begin to hope that the right solution is finally at hand.

The first of two recent yet independent advances was made by the physicist Heisenberg, of Göttingen, who then undertook its further mathematical development in collaboration with Born and Jordan. These three scientists ultimately abandoned the Bohr model and proposed the principle that the laws of the radiation mechanism should simply be formulated without any preceding explanation by means of some sort of model. Through their own particular approach, they achieved a presentation of great concision and rigor by using a special calculus, the calculus of matrices. In the logical economy of its account, which asserted no more than was demonstrably true, this theory certainly signified substantial progress. This work, along with a mathematical essay by the Englishman Dirac, was published in the autumn of 1925.

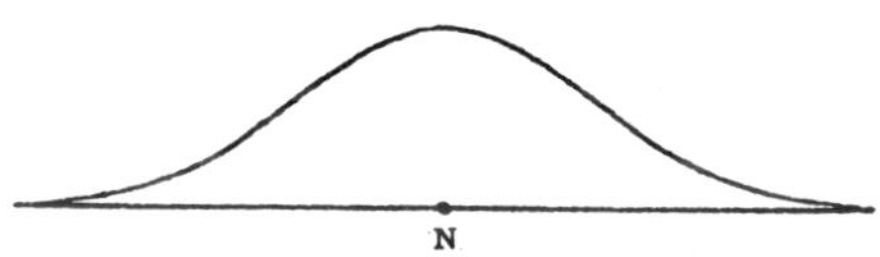

Fig. 2. *The hydrogen atom on the Schrödinger model.* A negatively charged electric state of oscillation surrounds the positively charged nucleus N. The distribution of the density along a cross-section is shown schematically; an antinode occurs in the middle. Upper harmonics are superimposed, but not shown here.

The second advance, which involved more fundamental insights, was made during the past year by Schrödinger, a physicist from Zürich. Under the impetus of an ingenious attempt by de Broglie to introduce the idea of waves into mass-point mechanics, Schrödinger constructed a wave theory, displaying admirable sensitivity towards the reason for the failure of all past wave theories. His account goes beyond that of Heisenberg in that it not only contains a formulation of the laws of radiation but also offers, once again, an intuitive model from which the characteristics of radiation are derived: a model that certainly bears no resemblance to Bohr's, yet is like it in developing an intuitive spatial mechanism. We shall attempt to give a simple description of this model, which must be considered the latest phase in the development of the idea of the atom.

The surprising feature of the new model is its rejection of the idea that the electron is a small electrically charged sphere. Schrödinger assumes instead that the negative charge is distributed continuously over the whole of the space surrounding the positive nucleus. This charge, however, is not at rest, but executes oscillations, namely, so-called standing waves, in which nodes occur at certain points and antinodes at others, just as a vibrating violin string has nodes at the ends and an antinode in the middle. The atom, then, is not a planetary system, but a positive nucleus with a surrounding vibrating cloud of negative electricity.

This system, like all other oscillating systems, displays not only the fundamental vibration but also upper harmonics. The overall state results from the superposition of all the individual states of oscillation; it varies according to the number of harmonics present. At the same time, the electrical state present in a given space results from the addition of all the individual states, but since we are in this case dealing with waves, it is quite probable that an interference phenomenon will occur, making the density of electricity zero in certain places. And now comes the remarkable twist: If a large number of upper harmonics develop, the electrical density is destroyed through interference over almost the whole of the space; yet in one single place there remains a small 'packet' of energy which does not stand still, but instead revolves in a planetary orbit. The Bohr–Rutherford model becomes a limiting case reached by the development of numerous upper harmonics, i.e., for higher quantum values; the electron appears as the result of an interference phenomenon, which singles out from the electrical density distributed continuously over the entire space a particle-like area in which alone the contributions of the individual states of oscillation are re-enforced.

We can now understand why the Bohr model was, despite its deficiencies, always able to produce so many positive results: it is a limiting case which the Schrödinger model achieves for high quantum values. For low quantum values, however, the Bohr model must be totally abandoned, for in these the image of a planetary system loses all meaning. Schrödinger illustrates this transition by means of a comparison with optics. For large spatial areas we may speak of light rays and employ the laws of geometrical optics; for small areas, however, we must conceive of light as a wave motion following quite different laws and leading, for instance, to observations involving diffraction phenomena that would be incomprehensible from the standpoint of geometrical optics. This analogy is the principal theme of Schrödinger's mathematical account, which is admirably elegant: it is also capable of explaining the distinctive regularity of quantum states, reducible to the condition that they be integrals, which appeared in the Bohr theory as the law of stable orbits. Schrödinger is able to show that the integrity follows necessarily from his mathematical account, from the differential equation of oscillation, just as the frequencies progagated by a vibrating string must stand in integral relations to one another. There are, of course, certain disadvantages standing over against these advantages of the Schrödinger model, but it would appear that the principal features of matter are accounted for by Schrödinger's theory.

The Schrödinger model, then, is to be regarded as the fourth decisive step in our knowledge of the atom. The atom, originally thought of as a particle of solid matter equipped with hooks, later supplied with Newtonian forces of attraction, and finally analyzed into a planetary system, is presented today as an area of space with no clear boundaries which is filled with complex electrical states of oscillation. Thus the theory of the atom, which replaced continuous matter with an aggregate of particles and empty space, has in the end led us back to a continuum theory, for the atom itself is now conceived of as an electrical field.

17. ON THE DEATH OF H.A. LORENTZ

[1928a]

The death, in Haarlem, of Holland's greatest physicist, Professor Hendrik Antoon Lorentz, was reported yesterday in the morning paper.

The importance of this man extends far beyond the boundaries of his fatherland, which has already given the world a number of leading minds. H.A. Lorentz ranks among the fathers of modern physics; we are indebted to him for knowledge vital to physics as we know it today. He is rightly acknowledged to be the true originator of the theory of electrons, the theory of the atomism of electricity, for he was the first who was able to construct the mathematical formulas for the laws of this theory. According to this conception, an electric current consists of the rapid motion of the tiniest particles, the atoms of electricity, for which the term 'electron' was introduced. The next problem was to deduce the known laws of electric currents from this model. To this end, Lorentz had to calculate the influence of magnetic fields upon electrical charges in a state of agitation. This he was able to do with the help of the concept of 'Lorentz-power', which is usually formulated by using a special form of mathematical calculation, vector computation. His ideas attained their greatest triumph, however, when, using these concepts, Lorentz was able to explain the phenomenon of the splitting of spectral lines in a magnetic field, discovered by his fellow Dutchman, Zeeman. Ever since then, the electron theory has been firmly established, and anyone familiar with physics today knows what a powerful system it has become.

Lorentz' name has become still better known through its appearance in two other terms of physics, 'Lorentz contraction' and 'Lorentz transformation', for these concepts figure in the theory that ultimately gave rise to Einstein's theory of relativity.

Lorentz attempted to explain, by means of a contraction of all rigid bodies resulting from their motion, the negative result of the famous Michelson experiment that was to determine whether the earth really moved through the alleged luminiferous ether. In the course of this effort, he hit upon mathematical formulas remarkably similar to those of the special theory of relativity later developed by Einstein.

Lorentz differed from Einstein only in his interpretation of these formulas. Thus there developed a singular situation in which scientists possessed two physical theories differing not at all in their physical significance, but only in their philosophical aspects. Even in later years, Lorentz was unable to fully accept Einstein's theory. One is tempted to conjecture that no one has such grave reservations about embracing a thorough-going innovation as he who was once very close to discovering it himself.

But the growth of friendly relations between Lorentz and the great German physicist was not disrupted by their opposing viewpoints – more a matter of sentiment than otherwise – concerning whether the ether was more or less of a material nature. After all, they had enough in common. The personal kindness of this great man, Lorentz, his warmth and interest in the private concerns of every person with whom he had dealings, be it a colleague from abroad or the cleaning-woman at his institute, won friends for him everywhere. In addition, he possessed a great talent for organizing; he had the gift of recognizing a person's productive abilities and placing him in the right position. Thus he came to be appointed as head of the Commission on Intellectual Affairs of the League of Nations, to which Einstein was a German representative. His name, then, is permanently inscribed in the annals of science not only as one of the great discoverers, but also as one of the great organizers.

An intellectual power of such magnitude is always a teacher and a transmitter of a legacy of knowledge at the same time. Lorentz disseminated great intellectual riches in the world through academic instructions and, still more, through numerous lectures held throughout the world, and through his textbooks, which enjoyed a wide distribution. We cannot but admire the gift for presentation displayed in these works, whereby Lorentz is able to explicate clearly complex subjects, and we must admire equally his ability to master German, French, and English as well as his native Dutch. Just last year, a lecture tour took him to California, where, fresh as ever at seventy-five, he delivered a course of lectures in conjunction with a famous, younger colleague. The hardships of the trip and a bad winter could not daunt him. The news of his death was thus all the more shocking to those who are familiar with his heartiness and who had believed his scientific endeavours to be not yet ended. With the death of Lorentz, a significant chapter of modern physics has lost its founder.

18. PHILOSOPHY OF THE NATURAL SCIENCES

[1928c]

There are those who have called our era the century of the natural sciences and yet, while fully acknowledging its scientific and technological achievements, have made the accusation that there is no appreciation of intellectual or philosophical problems nowadays. This judgment, however, reflects a highly superficial characterization. Indeed, it is difficult to imagine a greater misunderstanding of the spirit of our age than this short-sighted rejection of the positive forces in which technology and industry, science and objectivity, are grounded; it is misplaced humanism. On the one hand, such a judgment does an injustice to the humanities, for in spite of technology and industrialization, humanistic endeavours are flourishing; this is clear to anyone who has a broad conception of the liberal arts and who makes an unbiased attempt to understand the variety of approaches within the literary, aesthetic, and sociological spheres, even where they do not wear the professional cloak. On the other hand, and more significantly, calling modern science unphilosophical displays a critical lack of understanding. Quite the contrary: there has seldom been an era in which philosophical ideas have been so deeply at work in the natural sciences as they are today.

To be sure, the concept of a philosophy of the natural sciences has fallen into disrepute since the latter half of the past century, when the materialistic ideas of such men as Büchner and Haeckel usurped the name philosophy of science. In the public mind, philosophy of nature has not yet recovered from this blow. There can remain no doubt that their primitive interpretation of the theory of biological development is philosophically untenable, and we must also remember that it always played a far smaller role in science itself than in the popular mind. We must further recall that this popular philosophy owes its strength precisely to a reaction against certain intellectual and philosophical tendencies which represented an equally naive encroachment from the opposite direction, for no further proof is required that the efforts of Schelling and Hegel in natural philosophy have no value as philosophy of science. Quite the contrary: the deep misunderstanding with which the humanistic philosopher has repeatedly approached the world of the natural scientist has seldom been so clearly displayed as in their speculative philosophy of nature. And one can hardly resist the feeling that it is precisely in this respect that the present-day opponents of materialism have failed to make any progress – namely, that the construction of scientific concepts is just as alien to them as to those whom the materialists justifiably rejected as philosophers of science.

In fact, it is striking how little understanding the humanistic philosophy of today brings to the sciences. Ever since Kant, in building his grand system, fitted the concept of scientific knowledge into his framework of a universal rationalist philosophy, the image of scientific knowledge has remained fixed in the philosophy of the academies; that science has undergone a fundamental transformation in the intervening century and a half is completely forgotten. Insofar as it touches upon natural science at all, the current course in epistemology is based upon the Copernican–Newtonian model of the universe. Certainly, Kant's construction of the philosophy of this classic epoch was an historic achievement, yet it is entirely too antiquated when reproduced with minor variations in the era of the Einstein–Bohr model of the universe. This remark is not intended to deny present-day academic philosophy its significance. It has, indeed, made significant contributions to phenomenological analysis, to the comprehension of historical connections, and to humanistic psychology. Yet it is an unalterable fact that it has lost all relation to science and that, indeed, the scientist has given up all hope of obtaining from any such philosophy answers to the epistemological questions that confront him in the course of his research.

Thus it is that natural science has created its own philosophy – a philosophy, to be sure, that is not so much a codified science as it is a mental attitude, an immanent, active tendency, recognizable only by its effects upon scientific theories. This mode of thinking has as yet scarcely been written down in books, but it appears in the special proceedings in technical journals, plays its part in discussions at professional meetings, and becomes at most apparent when a professional scientist endeavours to address his scientific remarks to a lay audience beyond his own professional circle. The urge towards clarity is at the same time an urge towards a philosophical frame of mind, towards an explication of the fundamental concepts used in the sciences. This force, which otherwise is at work only at a subconscious level in the solution of narrowly circumscribed technical problems, is released into the sphere of consciousness and takes on a life of its own through precise formulation.

Anyone who has made closer enquiry is astonished at the wealth of epistemological ideas at work here. Consider, for example, Einstein's theory of relativity, in the foundation of which epistemological ideas played a vital role. Or examine the most recent attempts at a solution of the quantum problem, in which epistemological assertions stand side by side with complex mathematical equations, by means of which scientists have sought to give an exact formulation of these ideas. Thus the theory of relativity has given rise to the philosophy of space and time that superseded Kant's theory of space, while thermodynamic statistics has produced the conception of probability laws, which has gained new perspectives today within the framework of quantum theory and is perhaps destined to replace the traditional concept of strict causality. Out of mathematics there has finally grown a new logic that must be regarded as the first genuine step forward in logic since Aristotle; a logic put into mathematical form, in which calculations are made by using formulas and which is able, with this technique, to tackle problems that formerly eluded solution.

In later articles we intend to provide some insight into this world of precise philosophical thinking, and we hope, at the same time, to refute the facile accusation that these discoveries in natural philosophy are unphilosophical. For they are responses to just the same problems that classical philosophy tried in vain to solve with its systems. The only difference is that the modern solutions do no longer appear in the guise of speculative systems, but within the bounds of positive science, in the form of specific results rigorously based on evidence and claiming to be true independently of all viewpoints. Arising out of the communal effort of a group of individual investigators, this exact philosophy of nature is itself a social good; like other sciences, it is capable of being communicated, verified and expanded.

Thus it is not a failure of philosophical thinking that leads to the separation of natural science from humanistic studies, but rather an abundance of epistemological insights that the natural scientist knows to be in his possession and the humanistic philosopher refuses to acknowledge. To be sure, we do not believe that a combination of the two directions will be possible in the near future. The field has already grown too large; anyone wishing to participate in the philosophy of natural science requires an acquaintance with science nearly equal to that of a scientist, and a division of philosophy into two branches, one embracing philosophy of science, the other humanistic philosophy, appears unavoidable today. In any case, these few humanistic philosophers to whom we owe a debt of gratitude for their earnest efforts to reach results in philosophy of science have by their example demonstrated precisely the point that productive work within the sphere of natural philosophy is no longer possible as a sideline, that philosophy of science as it is today requires a full-time commitment. On the other hand, as this field is also too large to enable natural scientists to do work in it as a sideline, there is an increasingly well-defined group of investigators who dedicate all of their effort to the borderline area between science and philosophy. If our era is to be called the century of the exact sciences, a new philosophy must also be included among these sciences: the philosophy of the exact sciences.

19. SPACE AND TIME

From Kant to Einstein

[1928d]

The philosophical content of modern science has found perhaps its clearest expression in the widespread interest of the public in Einstein's theory of relativity. While this interest set in, of course, only after the theory had taken firm root in physical circles, and while it is certainly riddled with misunderstandings of the physical theory, it is nonetheless only to be explained by the fact that the theory resolves more than merely technical problems, that it touches upon genuinely philosophical issues the intellectual consequences of which are of a significance similar to that of Copernicus' discovery in his day.

Indeed, the problem of space and time has always played a vital role in philosophy. For in discussing the spatio-temporal forms, we are considering very general characteristics of reality, pertaining equally to all sciences. Whereas such phenomena as mechanical force, electricity, and the ability to sustain life are characteristics of particular objects in nature and therefore become the subjects of particular individual sciences, space and time appear to be characteristics pertaining to all natural objects; an analysis of these concepts ought, accordingly, to lead to insights having philosophical import. The philosophical systems of such men as Leibniz and Kant concerned themselves extensively with the space–time problem, and Kant's space–time doctrine, in particular, was long thought to display the most profound philosophical wisdom.

But today we may declare that Kant's theory of space and time was viable only within the framework of a stage in the development of mathematics and physics at which non-Euclidean geometry and relativity theory was unheard of. As long as there was only one known geometry and one known time, it was possible to take the philosophical position that this geometry and this time possessed absolute significance and were to be interpreted as being *a priori* certain. For purposes of clarification, we might point to analogous phenomena in the ethical sphere: as long as the people of Europe believed their ethical views to be the only ones possible and were unacquainted with, e.g., the general tenor of the Chinese outlook, to which the concept of progress is so alien, they held their own ethic to be the only one thinkable, to be absolute. In mathematics, the evolution to a multiplicity of geometries had taken place by the beginning of the last century, and the leading mathematicians of the day, such as Gauss and Helmholtz, had already concluded that the Kantian theory was untenable.

The various geometrical systems are differentiated according to their propositions regarding parallelism and also to those regarding congruence in different places, i.e., assertions about the metric. In themselves the systems are all of equal value; that is, it is impossible to declare one system or the other especially true on purely mathematical grounds. But the question is quite different for the geometry of real space. Which of these geometrical system *fits reality* is a question that can only be answered empirically. We may therefore say that the development of mathematics led to the recognition of a *special physical problem of space* alongside the mathematical problem. Thus the development of the technical field has led to far more than specialized knowledge; it gave rise to a philosophical problem at the very time when philosophy was dissipating its strength in unfounded speculations or in fruitless exegesis of the Kantian system.

It might be supposed that, once it had been discovered, further investigation of this problem would be taken over by the professional philosophers – but nothing of the sort took place. It was, instead, the physicists themselves that carried the problem a step further. Helmholtz and Einstein are the two physicists who recognized the particular problem regarding the relation between theory and reality that lies in the question – Helmholtz with respect to space, Einstein with respect to time. For it can be shown that the coordination of a particular geometry with reality cannot be carried out without some arbitrary stipulations. Certain assumptions about rigid bodies as a standard of measurement for distant spatial segments must be laid down before the question as to the geometry of real space takes on a definite meaning. Hence the *relativity of geometry*; in physics, too, no geometry is true in itself, but only in connection with a particular definition of congruence. To this idea of Helmholtz, Einstein added an analogous idea respecting time; simultaneity (and uniformity) can only be discussed upon the assumption of a certain definition, viz., the Einsteinian definition of simultaneity, which employs light signals. Thus Einstein's famous doctrine of the relativity of time is an extension of Helmholtzian conventionalism from space to time. This is one of the reasons that space and time appear in modern physics as such closely related forms that they can be treated as a single four-dimensional manifold.

This is not the place to argue this claim in detail; we merely wish to indicate the philosophical problems involved, or, more precisely, the problems in their philosophical aspect. For the questions are truly philosophical, and if many academic philosophers in our time have expressed the opinion that relativity theory is a purely internal concern of physics, they have thereby displayed a profound ignorance of the nature of epistemological problems in general. It is the relation between *knowledge and reality*, and thus the most central problem of epistemology – think, for instance, of Kant's doctrine – that comes under discussion in connection with the relativity of space and time and takes on an entirely new character. Once the irrefutable insights of Helmholtz and Einstein became known, there fell to philosophy the task, not of defending its past untenable position with artfully contrived arguments, but of carrying out systematically within the framework of the new insights the process of epistemological analysis for which Helmholtz and Einstein simply offered a particular target, as it were. For the physicist, of course, is not able to carry out systematic philosophical investigations; he pursues philosophical questions only to the extent that they free him from those conceptual restrictions which can prove a positive hindrance to physical knowledge. This becomes particularly clear in the work of Einstein, who, displaying admirable instinct, has been able to uncover such restrictive philosophical conceptions and yet consciously limits himself solely to physical evaluation, thus giving his physical theories their immense impact.

But again, we do not wish to discuss here the results of a systematic investigation of the theory of space and time, which has led to a peculiar coupling of space and time with the principle of causality, to *a causal theory of space and time*. Instead, we will look briefly into the consequences for the scientific concept of knowledge to which such a turn of events gives rise. Dependence upon certain stipulations, called coordinative definitions, is thereby shown to be a very general basic feature of knowledge as such, in accordance with which making assertions about nature is meaningless unless we provide at the same time the conditions relative to which we assert them to be true. It could be said, then, that the naive view of knowledge has lost its final round to the theory of relativity in as much as the relativity of all assertions has been acknowledged, and yet we realize immediately that this relativism is not in any way tantamount to subjectivism. It does not signify a *relativism of truth*, but only the shifting of truth to a *higher level of formulation*. It is indeed necessary to give a *system of reference* in order to establish an assertion about nature, but the very *combination* of reference system and assertion about nature constitutes in turn an assertion possessing objective significance. The meaning of the physical assertion is recognized to be an assertion about relations; there are no absolute specifications of location, length and time, but only data about the *relations of bodies to other bodies*. What was formerly regarded as an absolute position in space, as an absolute length, or as an absolute time, likewise turns out to be, in fact, information about a relation between bodies, namely, the relation of the body under observation to certain standard bodies, or measuring instruments, that appear in all propositions about nature as a connecting link. It is they which constitute the arbitrary element in knowledge, but as they are, after all, only a link, the assertion of objective connections does not depend upon them in the least. No matter which link is chosen, the same objective connection results. Thus release from the arbitrary aspect of the description of nature is not achieved by maintaining naive absolutism in the face of them, but only by recognizing and formulating them as arbitrary: the road to objective knowledge must proceed by way of a realization of the subjectivity of our methods of learning.

Thus the philosophy of space and time results in an epistemological thesis in the prudent moderation of which is to be found the reason that the theory of relativity has become an essential feature of the modern view of the world.

20. CAUSALITY OR PROBABILITY?

[1928e]

The idea of *strict causality* has long been regarded as the most visible hallmark of modern natural science, and it is indeed a concept through which modern science has taken a long step forward from the science of antiquity. While the ancient Greeks found strict regularity only in mathematics – that is, in a science dealing solely with abstract entities – and contented themselves with merely approximate rules in their knowledge of nature, the modern age has embraced the idea that mathematical strictness is to be found in every natural event, that the "book of nature is written in mathematical symbols", and since the time of Galileo scientists have learned how to implement this idea systematically in the form of a *mathematical physics*.

It is the mark of distinction of classical physics that it coordinates with every natural event a mathematical equation which, as it were, reflects conceptually, by means of its strict functional form, the strict relation between quantities in nature. The conceptually necessary dependence of the function upon the physical arguments is the true model of the naturally necessary dependence of a natural event upon its determinant parameters. This thesis is the ground of the predictability of events; we learn by observation the determining factors – say, the characteristics of a planet's path – and calculate from these its position at a future time. That this method has such immense success – that the subsequent observations virtually always agree with the calculated result – constitutes the justification of the underlying idea. And if here and there an exception turns up, we nevertheless do not give up the basic idea but hold instead that the error is the result of inadequate knowledge of the determining factors and that a strict agreement between calculation and observation would assert itself after all, were *all* relevant conditions taken into account.

This conception gives rise to *determinism*, the doctrine of the strict predetermination of all events. The course of events ticks away like clockwork; we cannot interfere with this clockwork, for any attempts at intervention on our part are themselves a part of the one all-inclusive course of events and are therefore harnessed to the clockwork. Whether such attempts will take place or not is also predetermined, and our belief that our will is free is an illusion stemming solely from our ignorance of the ultimate determining factors of our own behaviour. A being with a perfect mind, possessing a knowledge of all natural laws and being, at the same time, totally and precisely acquainted with the state of the universe at a single moment, would be able to calculate its state at all future times. He would know in advance not only the mechanical sequences, such as geological and astronomical developments, but also all the sequences of events involving living creatures; he would be able to give, say, the political conditions in the year 2000 as well as the individual actions of any given person at that time.

The doctrine of strict causality is still regarded by many people as an unshakeable verity. Philosophers in particular present it as the most certain fact in natural science, as an *a priori* truth. Yet we must report here a development that has called this doctrine into serious doubt. It is precisely that strictest of all natural sciences, mathematical physics, that has led to a fundamentally different conception of regularity in nature, and it can no longer be said today that the deterministic doctrine can hold its ground with a reasonable degree of certainty.

The shock emanated from thermodynamics, a discipline that developed into an exact science in the middle of the last century and at first appeared to signify the ultimate triumph of classical physics. It turned out that one of the basic laws of the theory, the so-called second law of thermodynamics, was not to be conceived as a strict law in the old sense but as a law of fundamentally different type, as a *law of probability*. The second law asserts a certain *direction* of all regular events in nature; it asserts, for instance, that heat can only flow from a warmer into a colder body and that gases can only become mixed and cannot reverse the process. It was Boltzmann's great discovery that this law only becomes truly comprehensible if we substitute for causality a second basic concept of regularity in nature, the concept of *probability*. Suppose, for instance, that two gases have been put into the two compartments of a container and the partition separating them is then removed; the molecules of the gases will rapidly intermingle. A state will eventually develop in which molecules of both sorts of gas will be located throughout, at every place in the container, and despite continued intermingling of the molecules the mixture will maintain the same proportions. Yet we can only assert it to be highly probable that the mixture will be maintained as it is, for it is not, after all, impossible that on some occasion all the molecules of one of the gases will just happen to collect on one side of the container and all the molecules of the other on the other side. In a particular card game it could likewise happen, through an extraordinary coincidence, that the cards turn out to be shuffled in such a way that one partner gets all the red cards and the other all the black (supposing there to be only two partners). It is not impossible for this to occur; it is merely extraordinarily improbable. The corresponding case for the two gases is far less probable, but we cannot speak of impossibility even there.

This very important thermodynamic principle thereby takes a singular turn. Instead of being a sure and strict causal law, it has only the significance of a *statistical law* and possesses the rank of a mere *rule*, such as the statistical rule that male and female births occur in approximately the same proportions or that there are a certain number of traffic accidents in Berlin every year. Thus the Boltzmannian idea proved as treacherous to physics as the Greeks' gift to the Trojans; while offering a most striking explanation of the second law of thermodynamics, it at the same time deprived it of the rank of a strict law of nature and introduced the peculiar uncertainty of the concept of probability into a basic principle of physics.

The physicists found this result most disturbing and attempted to escape by looking upon the use of the concept of probability as nothing more than an expedient that is to be employed only so long as the acquisition of more complete information remains impossible for technical reasons. In their view, the precise determination of the position of each separate gas molecule and the prediction of its path, which is highly complicated because of constant collision with other molecules, fails through the inadequacy of human instruments of observation. If, however, we were capable of making such calculations, the probability principle would be superfluous, and we could calculate in advance the course of the mixing process of the two gases just as strictly as we can predict the course of the planets. The uncertainty, they claimed, lies not in nature, but only in our *knowledge* of nature; it is, so to speak, a merely human failing, not a failing of objective reality.

But this subterfuge cannot withstand epistemological scrutiny. On the one hand, it is highly artificial to maintain a belief in strict regularity when it can no longer be carried out in individual cases. And on the other hand, there is no getting rid of the fact that the probability laws, despite all their uncertainty, nonetheless express an objective state of affairs; even the superman with the perfect mind, were he to pursue natural events in detail, would in the end come to the conclusion that in the vast majority of cases the probability law is correct. Thus this so-called expedient must, after all, give expression to an essential feature of nature. It must itself be a law of nature, but of a second variety.

In a very curious sense, this led to an extension of the concept of regularity in nature. We must not form such a narrow concept of a natural law that it can only correspond to the concept of a univocal coordination of cause and effect. Instead, we must admit as laws of nature also such functional relations as are found, for instance, in games of chance and are known as *laws of chance*. Indeed, more penetrating investigation shows that without the concept of probability law every law of nature, not merely exceptional laws, would be completely untenable; all laws of nature would lose their meaning. This is to be explained by the fact that the number of determining factors in every natural event is infinitely large and a strict law can only take account of a finite number of factors, while a probability hypothesis must be formulated for the remainder. The concept of probability thereby takes on an objective meaning and takes precedence over even the concept of causality. Accordingly, it no longer makes sense to refer to a strict predetermination of natural events which is simply unknown to us. Instead, we must speak of an *objective indeterminacy* in nature; the course of the universe is comparable, not to a clockwork, but rather to a perpetual dice game in which each stage in the event corresponds to another throw of the dice. The reason that we nonetheless are acquainted with relatively strict laws such as the classical laws of nature is that in such cases, as with the second law of thermodynamics, a compilation of instances achieves such a high probability that they may be regarded as virtually certain.

This philosophical perspective has recently been significantly strengthened through the development of the very latest physical theory, the *quantum theory* of the atom. Arising from the basic research done by Planck and Bohr, quantum mechanics arrived, under the influence of the theories of Heisenberg and Schrödinger, at the daring conjecture that events within the atom can no longer be comprehended by laws of the strictly causal type and that only the concept of probability is applicable to them, the experimental results being inexplicable in any other way. Readers of the *Vossiche Zeitung* have already been instructed as to this development through the discussion between Lasker and Born, which demonstrated clearly the necessity of the physicists defending themselves against excessively dogmatic demands on the part of philosophers. We need only add here that this new development in quantum mechanics is not even so very astonishing for the discerning, who will see that it simply means a logical extension of a fundamental epistemological principle worked out some time ago in the philosophy of science.

The consequences for philosophy of such a modification in the concept of regularity in nature cannot yet be conclusively surveyed. But bear in mind that all epistemological systems have always set out upon the assumption that strict causality was a certain fact; if mathematical physics itself now throws this concept overboard, philosophy is faced with a novel state of affairs compared with which all previous transformations of our world view, such as that which occurred under the influence of the theory of relativity, must appear as nought. It would seem that we are standing at the very beginning of a development; the goal that hovers before us is not only new *physical* knowledge but also new *philosophical* insight into the deepest relations in nature.

21. THE WORLD VIEW OF THE EXACT SCIENCES

[1928f]

If we wish to comprehend the vast expansion of science during the past few centuries from the aspect of its influence upon our world view, we must bear in mind that the endeavor cannot consist of casting about for tangible signs, for clearly visible manifestations of a newly constructed conception of the universe. In all its tendencies, science has a positivistic orientation; scientists are concerned with research and knowledge and do not contemplate directly the effect of their work upon attitudes towards the universe. If, nonetheless, something like a world view characteristic of the exact sciences has evolved, it is because of the very nature of intellectual activity, which is such that conceptual knowledge, pursued seriously to its utmost consequences, comes in time to be accompanied by a gradual redirection of the emotions. During the course of such an extensive intellectual process as the growth of modern science, this tendency may flower into the all-embracing basic approach that constitutes a world view. To be sure, this world view will not exactly be a formulated system: it will not, that is, be yet another explicit conceptual structure. It will, rather, be given in the sense of a basic attitude, an intellectual stance, a habit of mind enveloping the edifice of science like an invisible cloud felt only by those who circulate within its precincts and inhale its atmosphere. Yet it is present nonetheless, and its actual influence will be all the greater for the fact that it is as yet not systematized and has not grown into dogma. I shall attempt, in what follows, to sketch briefly the general outlines of the scientific approach to the universe.

The first basic characteristic appears to me to be a belief in the unconditional attainability of the scientific goal. Often as it may have been pronounced in theological circles, the idea that knowledge might some day come up against limits is totally alien to science. The tremendous quantity of scientific production, and its undeniable success, have formed the basis of this belief that its method can overcome all obstacles and have hammered it into our heads more firmly than any philosophical theory would ever possibly be able to do. The precedence of experience over intellectual construction asserts itself in this conviction, and the intensity of scientific research and the boldness and tenacity that characterize the conquest of even the greatest difficulties posed by the object of study spring essentially from this source.

Thus it is all the more remarkable that in the most exact discipline in this whole cluster of knowledge, in mathematics, there have recently arisen doubts as to the reliability of its results. Unbounded faith in the most elementary tool of scientific thinking, logic, has been shaken through the paradoxes of set theory, discussion of which has, in conjunction with the development of axiomatic methods, resulted in the crisis of the foundations of mathematics, occasional word of which has already reached the general public. The issue under discussion is the problem of the consistency of mathematical methods, and as yet it is by no means resolved. Discoveries have likewise turned up in mathematical physics that appear to point to the existence of limits to a rigorous description of elementary physical events and to the motion of electrons and light quanta. It almost seems as though Dubois-Reymond's cry *Ignorabimus* [we shall never know] is to be taken up in a novel form and on a sounder basis – yet the discerning eye can recognize in this crisis more clearly than ever before the belief in the certainty of scientific methods. For it is typical of modern science – and here we come to the second basic feature of the modern scientific world view – that it no longer views itself as having goals firmly established from the outset which it is to encircle, approaching ever nearer to the target. Rather, it has become steadily clearer that the goals of scientific research crystallize gradually during the course of research, that they show a different aspect when viewed from the higher plane of advanced knowledge, and that their material formulation itself becomes the object of a gradually maturing knowledge. The squaring of the circle hovered before the ancients as a goal of mathematic research; in its proof of the insolubility of this problem, modern mathematics did not simply declare *Ignorabimus*, but rather recognized the meaningless of the former goal and substituted for it, through this proof, a new and meaningful goal which it subsequently actually succeeded in achieving. Thus the crisis in the foundations of mathematics to which we have referred is to be regarded as displaying not a mistrust of the power of mathematical research, but a critical discrimination in the setting of mathematical goals itself. The manner in which this problem is tackled by great mathematicians displays most clearly the faith in the power of rational thinking and how it achieves ultimate certainty only by means of unsparing critique of its own methods.

In conjunction with this development, we find an extraordinary independence from scientific tradition. It is safe to say that the goals could not be modified in this way within the framework of traditional concepts. Thus the ability to modify long-standing habitual concepts and views is a vital condition of modern science. Mathematical set theory, modern logic, Einsteinian relativity theory, the most recent development in quantum mechanics, would all have been impossible had the mathematical scientist not possessed that ability for reorientation which finds its most visible expression in an unconditional rejection of the Kantian *a priori*. Faith in the eternally unchanging categories of reason is replaced by a faith in their capacity for change; the conviction that the forms of human thought are valid for the mastery of nature is replaced by a conviction that these forms are adaptable – indeed, the greatest victory ever achieved by the force of scientific thinking may well consist in this change in orientation, the success of which now stands unchallenged.

Alongside this belief in its own methods comes a belief in the regularity of natural events: the third basic feature of the scientific world view. That all events possess an intrinsic order, that science must succeed in establishing with ever-increasing accuracy the law governing that great machine which is nature, is a principle that has grown up upon the basis of so many successes that it is the tacit presupposition underlying all scientific activity. The ancient conception of an order determined by ends has finally been laid to rest in science, although it constantly reappears in philosophy. The idea of an order determined by ends has given way to that of causal order, and events in nature are seen as blind, i.e., as running a course determined solely by the past and present, in contrast to the course of purposeful human behavior, which, possessing sight, shapes its activities with an eye to future successes. It may, perhaps, appear bold to propose such a thesis at a time when investigations of matter at the micro-level have strengthened earlier suspicions that statistical laws, rather than causal laws on the classical conception, form the true nucleus of all regularity in nature. He who, nonetheless, would see such a modification as a rejection of the concept of a law would simply be making the mistake of confusing an abandonment of traditional goals with a rejection of knowledge. For the afore-mentioned modification of physical knowledge offers an instance of just such a modification of goals as we have discussed: an advance in knowledge, not a renunciation of it. On the contrary, it is our belief that, through the critique of the causal concept that has been realized with this complex of problems and through the insight into the fundamental place of the concept of probability, the essence of all regularity in nature has become more profoundly understood and the concept of natural events will be formulated all the more distinctly, precisely in its function as the embodiment of a 'becoming'.

There may be those who will refer to our sketch of a world view of the exact sciences as rationalist and will regret that it leaves so little room for the needs of mysticism and speculation, of feeling and fantasy. But they ought at least to keep clearly in mind that this concept of rationalism is different from that belonging to the eighteenth century, that rational knowledge in our sense is not tantamount to categorization within the pre-established cubby-holes of a reason that governs *a priori*, but simply amounts to unconditional faith in the power of the human capacity for knowledge – within the framework of a critique of its own goals. Thus the rational element is itself subject to change; and it emerges with increasing clarity that the basic stance of science is a faith more akin to an instinct than to rational insight, to will than to knowledge. Thus the will, the tenacious, malleable, indefatigable, and yet eternally modifiable will, is probably the basic element that truly represents the world view standing behind the scientific investigation of nature. Nature is blind, and faith in its benevolence and meaning is unfounded. But it is approached ever more closely through unceasing effort, surrounded and caught in the mesh of the finest conceptual net – in the sure consciousness that a knowledge of nature is one of the highest goods man is capable of attaining.

22. NEW APPROACHES IN SCIENCE: PHYSICAL RESEARCH

[1929a]

The extensive development undergone by the most exact of all the sciences during the past century, and particularly during the last quarter of a century, has found expression not only in a wealth of new knowledge but also, and equally, in a transformation of its working methods. It may be this transfiguration that holds particular attractions for the observer interested more in the picture as a whole than in the individual details. It is, indeed, sometimes possible to gain deeper insights into a science through a glance at its methods than through a summary of its material results, for such insights reveal the evolution of the very mechanisms of human thought.

Around the turn of the century, that separation between theoretical and experimental physics which today's physicist takes for granted had already become an accomplished fact. It is a separation based on the distinction between the abstract and the concrete: the theoretical physicist occupies himself with abstract thinking about natural phenomena, while the experimental physicist produces or observes natural phenomena through active intervention. While the tool of the experimental physicist is manual technology itself – he must be able to blow glass, make electrical connections, and service machines – the tool of the theoretician is mathematics, which he must master in all its subtle ramifications. This great difference in instruments has necessitated the distribution of the functions among different persons. Yet let us expressly emphasize that no division between the sciences is thereby intended. On the contrary, theoretical and experimental physics are intimately related as parts of a single science and belong together. This division of labor is of a solely practical nature, arising from human limitations.

Meanwhile, intellectual progress has continued; within the separate spheres of operation, special working methods have developed that were unknown to physics in its former undivided state. Let us first look into the experimental side. Here the most prominent feature of the working procedure is the penetration of modern technology. The purely manual work, necessary as it is, is of little significance in comparison to the use of modern machines and refined mechanical apparatus. Through the blossoming of physics, technology has experienced an unanticipated increase in its assets. It is now rendering its thanks to the mother science by providing physics, in return, with technologically refined equipment which enables physical research to press forward to fresh results. The first X-ray tube was constructed by a physicist in his laboratory; today, every radiological physicist uses factory-made X-ray tubes, constructed in accordance with the greatest refinements in precision engineering, glass blowing, and vacuum technology, in producing X-rays for research purposes. Corresponding developments have occurred in all spheres. Galvanometers and vacuum pumps, amplifying valves, optical apparatus, and so forth are today produced in factories and assembled by the experimental physicist in his laboratory according to the directions. Thus an up-to-date instruction manual for experiments is indispensable for anyone lacking specialized knowledge, and success in experiments will accrue only to the physicist who is able to make skillful use of industrially produced apparatus at the right point in the course of an experiment.

The theoretical physicist does not have these special abilities: the tool he requires instead is a mastery of modern mathematical methods. The refinement of mathematical thinking in our time is, indeed, closely comparable to the refinement of the art of experimentation. Even in these ultra-rational spheres, we find something along the lines of sociological parallelism. The physical problems of today cannot be solved without being harnessed to the conceptual apparatus of modern mathematics. The contemporary theoretical physicist must display consummate mastery of complex differential equations and higher-dimensional geometries of the most general type if he wishes to gain insight into the propagation of electric waves or the internal structure of the atom. This has led to the development of a special conceptual apparatus, to a technicalization of thinking, so to speak. Just as the experimental physicist places technically constructed machines at the service of his course of experiments, the theoretical physicist uses the methods of modern analysis as highly subtle calculation machines. He is indebted to their versatility for results that would have remained unavailable forever to naive intuitive thought. Indeed, it has turned out in more than a few instances that the conceptual machinery man has created for himself has come to be, as it were, cleverer than its creator, that it has gone on calculating on its own and has automatically produced the result which the human mind, thinking in terms of particulars, was unable to discover. Such trends have been at work in the most significant theoretical achievements of the past decades. For instance, the general theory of relativity was able to progress from the status of a programmatic approach to that of a theory only after Minkowski created a special mathematical apparatus for it. And the quantum mechanics of the past few years is likewise indebted for its success to the development of a special mathematical apparatus. In this case, the process whereby the conceptual machinery has become independent may be traced with remarkable clarity: on the basis of purely mathematical considerations, Schrödinger created a new wave theory of matter, the physical interpretation of which could only follow later on. A species of mathematical experimentation has become common in theoretical physics. Its method of trial and error is not so very different from that of the experimental physicist, who frequently comes to understand his results only after he has reached them. This may seem unsatisfactory from the standpoint of the classical approach to physical work, but, given the complexity of the problems faced today, a technical approach of this kind may be unavoidable. Furthermore, it still affords adequate leeway for intellectual creativity, for it is precisely in the production of new mathematical mechanisms that it may blossom.

A further differentiation of working spheres goes hand in hand with the refinement of working methods. Today we have already seen the last of the universal physicists, the free-ranging spirits such as Helmholtz, who was able to combine experimental and theoretical physics. And among the newest generation we can scarcely find an experimental physicist who is acquainted with the whole of experimental physics or a theoretical physicist acquainted with the whole of theoretical physics. For the time being, the specialization that this has entailed has not received much official recognition. The technical physicists alone have had some success in this direction; backed by the needs of industry, they have brought about the creation of a few special chairs for technical physics. In the theoretical sphere, interest in the complex of quantum-mechanical problems is at present so great that it overshadows the investigation of all other special spheres. In particular, the need for special work on conceptual philosophical problems involved in physical knowledge has as yet received little acknowledgment. Physicists believe that an occasional general remark on the part of outstanding representatives of theoretical physics will suffice in this area, but here they display the same kind of shortsightedness concerning conceptual problems that the experimental physicists showed some decades ago in espousing the view that theory could be dealt with quite adequately by experimentalists on the side, as it were. But in the long run this state of affairs will prove untenable, and treatment of these special problems will have to be given over to specially trained physicists, particularly as this work possesses great significance within the framework of yet another science, philosophy.

It would be quite wrong-headed to undertake criticism of such a scientific development on the basis of value judgments which compare it to other scientific epochs. Science, too, even in its most abstract form, is tied to the step-by-step progress of time and displays in its development such universal social tendencies as abstraction and differentiation. All we can do is gain a clear understanding of this development. For only within the framework of a critical approach to the overall course of the ages will we be able to make a final evaluation of the science of today as well as the science of yesterday.

23. NEW APPROACHES IN SCIENCE: PHILOSOPHICAL RESEARCH

[1929b]

That it is possible today to speak of philosophical research can itself be regarded as a characteristic of modern philosophy. For the very word 'research' designates a method. In admitting that it engages in research, philosophy classifies itself, from the methodological standpoint, among the special sciences, which progress step by step by solving particular problems and attain a systematization of the whole only by assembling the various components of knowledge embodied in these situations.

This is indeed a new approach to philosophy, at least compared to that of the past century. For past philosophical generations, it was the overall picture that stood in the foreground, and philosophy set out to construct a system in which knowledge and consciousness were analysed according to certain categories, and world and man were viewed together in an all-encompassing perspective. For a long time, the presentation of a world view was held to be the only permissible form of true philosophy, and in developing in this article the idea that the approach taken by contemporary philosophy is fundamentally different, we do not mean to claim that this view is already generally accepted. The ideal of a system still plays a certain role today, and the older generation is still endeavouring to defend it before the younger generation. Nonetheless, we do not believe that the process of transformation can be stayed by these efforts.

On the contrary, we hold that the unmistakeable lag in the development of philosophy compared to that of the individual sciences, its obvious stagnation, results from clinging far too long to the ideal of a system. Once it had been recognized that the attempt to construct a philosophical system suited to our times was bound to fail in view of the wealth of factual material, the systematic philosophical approach was necessarily transfigured into an unproductive historicism. It appeared that the great men who were capable of projecting a philosophical system belonged in the past, and the fact that such systems could succeed only in the presence of a relatively meagre store of material in the individual sciences, particularly in the natural sciences, was completely forgotten. A return to the great systems of the past was preached, in the belief that there could be no other task for philosophy in the present day than the modification and extension of the classical systems so as to make them conform to modern science. The exaggerated stature attributed to the great men of the past could not but cripple initiative, could not but turn the attention of present thinkers from present possibilities and produce that unproductive epigonism the sight of which must repel anyone who works in the stimulating atmosphere of the sciences and who has even once looked at contemporary philosophy from this vantage point. We need ask only one prominent natural scientist his personal opinion as to the value of philosophy in order to encounter the most astonishing condemnation. This attitude of mistrust can be explained only partially by the difference in the subjects. It is, for the most part, the tendency towards historicism, the uncritical retreading of the paths of old systems, that turns the natural scientist from the philosophy of our time.

Yet we cannot claim that historicism was present in the works of those philosophers of the past who are now held up to us as models. Quite the contrary: the great philosophical systematizers, with the possible exception of Hegel, have been of a decidedly unhistorical cast of mind, and we are somewhat justified in expressing a suspicion that this very fact is the ground of their effectiveness. The great achievement of these philosophers consists in their original treatment of the knowledge possessed in their time, and they would have refused to ponder so endlessly over an outmoded system, twisting it until it fits in with a new state of knowledge. On the other hand, we cannot accuse the unhistorical philosophers of today's younger generation – which is really unhistorical – of undervaluing the intellectual achievement of the classical philosophers. Indeed, it is precisely in the original thinkers that we observe a marked appreciation of the intellectual achievement of past generations; they reject only that crippling glorification which obscures insight into the special character of their own problems.

Contemporary philosophy can be quite uniformly characterized as possessing the negative feature of rejecting all ties to systems; this makes it all the harder to look for common positive features. More than other sciences, philosophy today is split up into different movements, and in scarcely any other field is the battle between the factions waged with greater passion. While this may show that we do not possess today a universally recognized scientific philosophy, we can by no means take this state of disunity to be a sign of decline. The freeing of a science from the fetters of tradition will always be heralded by its splintering into factions. At present philosophy finds itself in a transitional period, leading from the era of systems into the era of positive research, and only by looking backwards can we fail to recognize in this a process of convalescence. It is possible to be totally convinced of the greatness of a Plato or a Kant and yet advocate the view that the way of the future in philosophy lies not in a new exposition of the systems of these giants but in steering a path out of the chaos of current opinions in accordance with one's own laws.

Opposition to systematizing is the first basic feature of modern philosophy; it necessarily merges with a second basic feature which is now becoming more and more clearly delineated. This is the close dependency of philosophy on the sciences, which is particularly to be observed in philosophers of a scientific and psychological orientation. For the course of development in philosophy that we have just described is none other than that found in natural science. The transition from the construction of systems to the investigation of individual problems, from synthesis couched in obscure imagery to conceptually clear differentiation, from speculation to concentration, has already taken place in science, and it is the powers released in philosophical minds by contact with productive branches of science that are at work in the forefront of the overhauling of contemporary philosophy. Indeed, anyone who takes a closer look at science finds that philosophical forces are already on the upswing, that certain problems have already been perceived and partially solved which were formerly to be found in a less precise form among the philosophers, where they never progressed beyond a state of conceptual speculation. We need think only of the discussion of the concept of an atom, of space–time problems, of the problem of life; in these scientific discussions, so much conceptual material has been assembled and digested that it would be a short-sighted philosopher indeed who would not gladly become a pupil of the sciences. It has been said that there is more by way of an art for our times to be found in the static constructions of modern technology than in the works of contemporary artists. A parallel to this idea can be discerned in the relation between the sciences and philosophy: more philosophy is contained in the scientific discoveries of the present century than in the intentionally philosophical productions. Certainly, acknowledgement of this fact does not entail a refusal to engage consciously in philosophical activity, for the present state of affairs need not remain unaltered. On the contrary, this period appears to be coming to an end – if only because we are becoming conscious of the philosophical significance of the various sciences. Just as architecture and fine art are borrowing new forms from the purely functional approach of technology, so, too, a new philosophy is coming into being, a philosophy that is itself no longer a part of natural science even though it has taken over from it a wealth of problems and conceptual instruments.

Finally, a third basic feature necessarily accompanies the described transformation. The change in emphasis from construction to detailed investigation brings, along with specialization, an attendant differentiation. Concentration upon a specific problem requires such a precise knowledge of details and working methods that the individual is no longer able to embrace an entire field. Today we already find specialists in the philosophy of mathematics, the philosophy of physics, the philosophy of psychology, and the philosophy of biology, and anyone who has carefully scrutinized matters knows that the differentiation is going to continue further. It is regarded as a rejection of true philosophical work in some circles, where philosophy is conceived as essentially dealing with the whole of knowledge, with a totality. But this verdict appears to us to miss the mark entirely. A phenomenon that has its roots in the ever-increasing domain of intellectual stores relative to the capacity of the individual brain cannot be disposed of by denying to it the name true philosophy; rather, the concept of philosophy will have to be altered in the face of such facts. The name, incidentally, is irrelevant. Philosophy has spawned many sciences, and today, as in the past, new disciplines will continue through specialization to become detached from philosophy. We may use for them such names as natural philosophy, epistemology, or the theory of knowledge, but we must keep clearly in mind that in so doing we are removing vital appendages from the body that we still wish to salvage as so-called pure philosophy. The vast problems connected with knowledge and the external world that play the leading role in the classical systems are just the ones that in the future will no longer be found in that 'pure philosophy' but rather in the new special disciplines, where they are resolved with the help of new conceptual instruments. There is simply no disputing the fact that, e.g., the problems Kant treated in the Transcendental Aesthetic in his *Critique of Pure Reason* have found an entirely new solution through the philosophical discussion of the theory of relativity, and decisive strides have also been made towards resolving the problems included in his theory of categories through the philosophical discussion of modern physics. Or consider how the logical problems of classical philosophy, and especially the Eleatic paradoxes, have taken on an entirely new look at the hands of mathematical set theorists. This time, then, we are faced with a differentiation that should be described preferably not as a splitting off of new disciplines but rather as a subdividing of philosophy, if we wish to avoid excluding some of the essential problems from membership in philosophy.

We do not mean to claim that our mode of viewing contemporary philosophy is shared by all philosophers. On the contrary, our position divides us from several currently widespread philosophical movements that would take up metaphysics and speculation once again as the proper domain of philosophy and are prophesying the decline of scientific rationalism. But these prophets of a new philosophical mysticism forget that the opponent they are fighting is merely the rationalism of the last half of the past century; that the critical rationalism which today is developing out of natural philosophy is, in contrast, unscathed by their weapons, for though it faces problems just as serious, the precision of its conceptualization is superior. The path to philosophical knowledge we see before us lies not in renewed reflection upon the essence of things or in introspection and the analysis of simple experiences, but solely in disciplined investigation and in rendering scientific methods fruitful for epistemological ends. Technicalization, not dialectic, is the instrument of modern philosophy, although it may consequently prove more sober than the metaphysician's philosophy of world views. What counts is to follow the way of truth, and nothing more.

24. NEW APPROACHES IN SCIENCE: MATHEMATICAL RESEARCH

[1929c]

Inquiry into the paths pursued by contemporary mathematical research leads naturally to the question of the extent to which the expansion of technology, which has been of such far-reaching significance for the natural sciences, has penetrated into the development of this most abstract of sciences, for mathematics has always been seen as closely parallel to the natural sciences. Certainly, mathematics cannot be designated a natural science in the strict sense, for its subject matter is taken not from natural objects but from abstract conceptual constructs, and it might therefore be more properly described as a humanistic discipline. Yet granting it membership in the natural sciences is nonetheless fully justified, for the conceptual world treated by the mathematician is precisely the one required by the natural scientist, and particularly the physicist, in order that he may gain a knowledge of nature. For this reason, mathematics has always developed hand in hand with the exact sciences and an exchange of problems and methods of solution has grown up through reciprocal inspiration.

However, we must not therefore suppose that mathematical thinking merely stands in a close relation to the theoretical side of natural science. On the contrary, mathematical methods have been required and developed within the other side of natural science, in the area bordering on technology. This has led to the establishment of a special technical branch of mathematics, so-called applied mathematics, the continuous growth of which we may regard as an expression of the movement towards technicalization which permeates contemporary science. The idea of applied mathematics is certainly not new. Quite the opposite: practical mathematics is the older sibling, for the mathematics of the ancients was an outgrowth of practical, i.e., geodetic and astronomic problems. Thus it is the conscious setting apart of applied mathematics as a special discipline that is more clearly a distinguishing feature of the present. Indeed, differentiation and specialization are the characteristic mark of modern scientific pursuits, which are as thoroughly permeated with the division of labor and the attendant realization of maximum output as are industry and technology. This tendency is likewise discernible in the purely theoretical branch of mathematics; and the ideal personified in a mathematician like Gauss, who with his free-ranging mind was productive in such diverse fields as algebra, analysis, the theory of numbers, geometry, astronomy, and probability calculus, is not to be looked for in our time. Today, a mathematician who is able to make significant contributions in even two of these basically very different spheres is hailed for his extraordinary achievement.

The movement towards technicalization has found expression in yet another phenomenon. It is discernible in the extraordinary virtuousity of the conceptual instruments with which the modern mathematician tackles his problems. If we may appeal to a metaphor, mathematical work is no longer done by hand, but done instead by employing the machinery of a conceptual apparatus that automatically produces a series of results which would have remained hidden from the view of the thinker using immediate intuition. Connected with this is the fact that a great amount of material must be absorbed before mathematical production can begin. Thus there are scarcely any more outsiders to be found in mathematics today. He who has not undergone the regular course of training cannot substitute his own productivity for this versatile conceptual tool.

Parallel to the technicalization of the conceptual apparatus runs another tendency which has been released by the increasingly technical powers. We have in mind the conceptual creativity expressed in the wealth of newly created concepts, in the superabundance of intellectual constructs. The transformation of Euclidean geometry into the more general form of Riemannian geometry and the modern extension of this trend; the introduction of higher-dimensional spaces, which did not cease until it reached an infinite number of dimensions; the handling of the problem of infinity by means of a ranking of infinite sets, ordered according to powers – all these intellectual constructs bear witness to an abundant conceptual creativity which now apparently knows no bounds. Former generations believed themselves to possess, in abstract mathematical constructs, the most general model of the possibilities of the human intellect. Subsequently, through the imprint of this superabundance of conceptual constructs, which have become substantially more general and abstract, there has emerged with increasing clarity the idea that there is no limit to the degree of generality, that the possibilities for human thought are unbounded, and our knowledge of conceptual possibilities already greatly exceeds what is required for the comprehension of nature. Thus Einstein's relativization of time and his combining of space and gravitation found mathematical expression in a four-dimensional geometry, and similarly, the peculiar state of affairs existing in modern quantum mechanics has already found its special mathematical apparatus by borrowing from existing mathematical systems.

The uninitiated, hearing of these abstract conceptual constructs for the first time, usually feels oppressed by their alien quality. It seems to him as though the intuitive clear thinking characteristic of ancient and classical mathematics has no place in this world of abstract speculations; indeed, even mathematicians have been heard to express the opinion that intuitiveness can no longer be guaranteed throughout mathematics. This view may, of course, represent a merely temporary phenomenon. In an era of great change and expansion, anything contrasting to the preceding, more naive conceptual world will invariably be received as a violation of the requirements of intuitive clarity. Only gradually, through repeated use, will the new conceptual apparatus become furnished with intuitive contents; eventually it will come to appear as familiar and intuitive as the old one. The insight that the earth is a small sphere and revolves around that far greater sphere, the sun, must have appeared to those who experienced this revolution in conceptual thinking to be a denial of all possible forms of intuitive depictability, difficult as it may be, in turn, for us today to see any problem in picturing this proposition in our minds. This same sequence is to be observed in other fields. Today it can no longer be said that non-Euclidean geometry is less intuitive than Euclidean, even though the outsider may find this claim incredible, and it is impossible to see any reason why the human capacity for intuitive visualization should not keep pace with the human power of conceptual productivity, eventually rendering contemporary mathematics once again intuitive, although in a higher sense than heretofore.

Alongside the technicalization of methods and the creative power of conceptual instruments, there comes yet a third feature, stemming from a completely different basic attitude and yet equally characteristic of the current era: this is the critical approach, the logical-epistemological analysis of the foundations of mathematical thinking, which in our time has undergone particularly extensive growth. This form of critique began in connection with the creation of set theory, and thus in a particularly fruitful period. It arose as a reaction to excessively self-confident productivity and pursued a 'very dramatic' course, as Hilbert once put it. In their delight over new and fruitful results, mathematicians had obviously become too uncritical in their procedures with respect to the admissibility of arguments, for contradictions began to appear through the mere application of certain conceptual constructs and modes of argument that had gradually become customary. At first instances occurred in isolation, but over the course of time these so-called paradoxes of set theory became more distinct and more serious. The most exact of all science, long held to be the prototype of logical thought, suddenly found itself threatened in the territory most peculiarly its own; it became entangled in contradictions and thereby did violence to the criterion which alone may issue a verdict regarding the truth of mathematical creations: logic. Yes, logic itself appeared threatened, for at this level mathematics and logic are scarcely to be separated, and a revision of mathematical methods invariably signified a revision of the logical foundations. Russell's theory of types and Hilbert's axiomatic methods were attempts to find a way out of this situation, and Hilbert's plan for a proof of the consistency of arithmetic in particular represents a magnificient effort to establish logic and mathematics on a new basis. The attacks of intuitionism [in mathematics] are aimed in a similar direction, subjecting logic and mathematics to still sharper criticism and excluding certain old and apparently self-evident conceptual constructs; it introduces a positively Spartan rigor into its critique of mathematical foundations and is not afraid to trample underfoot the most precious possessions of mathematical knowledge if they fail to withstand scrutiny.

Thus we end our brief survey of mathematical research on a philosophical note. Indeed, it is characteristic of contemporary thought that, along with its turn towards the technical, along with the technicalization of its methods and the resulting superabundant productivity, it simultaneously turns its gaze towards the philosophical, that it reflects upon itself and attempts to set up, alongside the naive positivism of productivity, a critique of its own activity, of its own significance. This is the same basic trend that is visible elsewhere, in science and in art. Perhaps it will one day come to be understood that a philosophical attitude need not be conceived of as the antithesis of highly technical productivity, however much they may initially appear to pull in opposite directions – that, rather, critique and technicalization are merely different expressions of one and the same basic attitude that has borne fruit in our time and the sociological tool of which is our brain.

25. THE NEW PHILOSOPHY OF SCIENCE

[1929d]

It is the lot of philosophy not to enjoy, as do other sciences, a more or less independent existence, but to grow and flourish in tandem with the development of the special sciences. Never has there existed philosophical investigation 'as such', resolving philosophical problems through its own springs of knowledge during periods of stagnation in the sciences. Rather, philosophical problems are not seen with all their ramifications, philosophical puzzles not really thoroughly thought out, until investigations in the individual sciences have encompassed a broad range of knowledge and the process of learning has reached a certain maturity, which forms the only possible basis for philosophical activity. We need only recall the philosophy of Plato, which evolved out of a serious concern for mathematics, or the grand system of Kant, which, while purporting to be a system of pure reason, actually represents a thoroughgoing reworking of the mathematical physics of the eighteenth century, which accounts for its overwhelming importance.

This is not to say that philosophical investigation automatically accompanies work in the special sciences, that the scientist is, as such, also a philosopher. On the contrary, a philosophical frame of mind gives rise to a fundamentally different approach from that operative in scientific work; it involves, not direct reflection upon nature and its laws, but rather reflection upon the methods of knowledge and learning themselves. Our intention was rather to express the idea that, in addition to the individual sciences, there exists a mode of treating and evaluating its subject matter that is itself not scientific but philosophical instead, that examines not the content but the methods and form of scientific knowledge and consequently leads to insights into the essence of processes of knowledge as such.

It is this type of philosophy of science about which we wish to report at present. We are not talking about mere theoretical possibilities, but may point to a growing philosophical movement that actually exists today and that has as its theme the ideas set forth here. The exact sciences in particular, mathematics and mathematical physics, have reached in our time a point of maturity positively demanding philosophical evaluation. That it is now no longer possible to bring about developments such as Einstein's relativity theory and non-Euclidean geometry in the peace and quiet of discussions limited to professional scientific circles stems solely from the much more widespread interest in the philosophical significance inherent in such intellectual creations. Scientific details, modifications or increases in precision of the methodology will leave the greater public unmoved, but more is at stake during scientific crises. The issue is, in truth, the ultimate insight into the essence of nature and knowledge as such.

In contrast, it is less widely known that as a result of these crises in scientific development there has sprung up a special method of philosophical investigation, and with it a special group of younger philosophers. Working closely together with science, they are attempting to establish a new form of scientific philosophy. Upheavals such as Einstein's relativity theory and quantum mechanics are, in their eyes, just starting points, not the end of philosophical investigations. These thinkers are primarily interested in grasping the common core of such scientific revolutions and subsequently using to the full the epistemological forces lying behind them for the advancement of general philosophical inquiries. While the strength of this approach lies in its intimate connection with science, this connection also accounts for the extraordinary difficulties involved in its working methods. It requires not only a precise mastery of the sciences but also a philosophical training, which makes possible extrication of the proper epistemological question even if it is buried under masses of mathematical and philosophical material.

A short time ago, this movement made its first public appearance by sponsoring quite a large conference. The conference, entitled 'The Epistemology of the Exact Sciences', was held in the middle of September in Prague, and questions of scientific epistemology were discussed in the presence of a large circle of interested listeners. The interest felt by the audience was particularly evident in the discussions following the papers. Not often are such sophisticated discussions, including so many personal contributions by the participants, to be heard at a philosophical conference. The conference took on a particular tone through being held immediately after the conference of mathematicians and physicists; through this organizational connection, both the material and personal relation between the two kinds of work found overt expression.

A whole series of epistemological questions and complexes lie at the forefront of current research. A new set of problems regarding regularity in nature grew out of physics, once investigation of space–time problems in conjunction with Einstein's relativity theory had in a sense been concluded; these problems are visibly expressed in the contrast between causal and statistical regularity. From the mathematical direction, the so-called problems of the foundations of mathematics are still pressing to the fore; they focus upon a critique of logic, and have found a powerful tool through the construction of a notation for logical concepts, similar to the mathematical notation for formulae. But they are also related to questions of a general epistemological sort. After all, discussion of all these individual questions culminates in nothing less than the question as to the general nature of consciousness and reality.

In contrast to other philosophical movements, which set out from the totality of a system and approach all individual questions from that standpoint, it is the mark of this new scientific movement in philosophy that it begins by asking questions regarding particular problems and does not make bold to ascend to the systematical level until it has resolved them. This may seem a defect to old-style philosophers, but we regard this conscious restraint as the very strength of the approach. When, some three hundred years ago, physics set out to become an exact science, it too began by conscious specialization, and it owes its splendid system to following this path, which leads via specialization and differentiation to a higher synthesis only at an advanced state of maturity.

Thus modern philosophers of science, while they reject for the present the establishment of a system, do not wish thereby to rule out the possibility that their work, too, may some day grow up into a philosophical system. But for the time being this aim appears to them far too remote for them to dare to regard their work in this light. They also do not wish to have the scope of their work determined. Although the mathematical sciences stand at the forefront of their interests at present, they nonetheless cherish the hope that critical work on biology, sociology, and the humanistic sciences in general will get under way, so that in this more or less inductive manner a comprehensive system will finally be attained.

But comprehensiveness is only a goal; it does not belong at the beginning. The beginning calls for concrete problems and a firm grasp of ideas. Such are the forces that stand behind this new philosophy of science.

26. EINSTEIN'S NEW THEORY

[1929f]

[The *Vossische Zeitung* reported recently the news that Einstein had succeeded in deepening the significance of the theory of relativity by means of a new hypothesis. We have asked Professor Reichenbach to give us his views on Einstein's new work, which appears in the *Berichte der Preussischen Akademie der Wissenschaften* (*Reports of the Prussian Academy of the Sciences*) – if only to prevent the emergence of public misunderstanding about the contents of the theory, which is above all of purely factual interest. – Ed., *Voss. Z.*]

The goal of the new theory is not so new at all – it has been pursued with great tenacity by a number of mathematicians and physicists for about ten years now. The great achievement of the theory of relativity was the combining of a series of physical facts under a single law of great scope, yet it had to interrupt its progress upon encountering one last barrier; it could not unite the theory of electricity with the theory of mechanics. Thus two vast bodies of laws stood at the pinnacle of physics. On the one hand, there was the complex of mechanical laws, which happily brought about a union between the phenomena of gravitation and the facts of space–time theory, culminating in Einstein's gravitational field equations; on the other, there was the complex of electrical laws, which had already been formulated in essentials by the Englishman Maxwell by the end of the last century. These two sets of laws had, as it were, nothing to do with each other; the world of physics was divided into two kingdoms, one ruled by Einstein, and the other by Maxwell. The temptation to attempt a supreme union was irresistible: however, nature proved to be more stubborn than had been anticipated.

The initial attempt to construct a unified field theory for gravitation and electricity was made by the mathematician Weyl. For this purpose he created a mathematical aid, a geometry of even greater generality than the known non-Euclidean geometries. In so doing, he created the apparatus of which others subsequently made use, including, finally, Einstein himself. A series of approaches were attempted along these lines, most notably by the English astronomer Eddington. Einstein tried out a number of theories, all tending in this same direction. The difficulty of the problem was so great that he had to withdraw one hypothesis after another. In 1925 he wrote, "After an unrelenting search of two years, I believe I have now found the real solution", yet later in the same year he pessimistically admitted that "the extension of the foundations of geometry proposed by Weyl and Eddington is unable to assist the progress of physical knowledge." But today, Einstein has taken a new step, deviating somewhat in its mathematical apparatus from his previous accounts, and this time he is firmly convinced of its significance.

And indeed, the new theory succeeds in uniting the fundamental laws of relativity mechanics and the fundamental laws of electricity into a single formula. According to this formula, there is only one substance, the 'field', and only one law of the universe; the field is made up of electrical and gravitational components, and all these components are united under a single formula. Einstein is able to show that the previously known laws can be derived from this formula, so that it signifies the subordination of the two formerly divided realms under a higher law. Yet the new formula achieves still more; it represents the older theory of two systems as a special case and makes new assertions concerning the relation between gravitation and electricity in relatively complicated fields. Thus the new theory is of more than merely formal significance, for it asserts the existence of an effect of gravitation upon electrical events and vice versa. To be sure, it is not yet possible to form a picture of how this connection will work out in detail from a physical standpoint. Specifically, it remains an open question whether the new theory will enable scientists to solve the puzzle of quantum theory, which itself represents a peculiar combination of mechanics and electricity theory.

For the time being, no pronouncement can be made concerning the physical significance of the theory. The situation was different when the general theory of relativity made its first public appearance. What was presented at that time was a complete theory, worked out in all its consequences, which had already passed its first great empirical test and therefore justly deserved widespread public interest. In contrast, the latest expansion of the theory is only a first draft, lacking the persuasive powers of the original relativity theory because of the very formal method by which it is established. The theoretical physicist is now faced with the task of thinking through the new theory in all its ramifications. This does not deprive it of its value, for all great discoveries start out in a similar state; when, in addition, the hypothesis is presented by a man of the experience and theoretical insight of Einstein, it must be taken completely seriously as science. Yet until the new theory has been thoroughly worked over, no leads are available for public discussion of this matter, which is primarily of a purely factual nature.

27. JOHANNES KEPLER

[1930a]

On November 15, three hundred years will have passed since Johannes Kepler, one of the founders of modern astronomy and of modern science as a whole, closed his eyes for the last time. Measured against the brevity of our own individual lives, three hundred years is a long time, but measured with an eye to the laborious progress of science as a social phenomenon of civilization, it is but a short span of time, encompassing such a closely related set of intellectual approaches that, from this stand-point, Johannes Kepler is regarded as a member of the present epoch of civilization – indeed, as one of those leading minds whose thoughts continue to be applied in new ways by scientists.

If modern science is distinguished from ancient by its rejection of natural laws invented by the imagination and their replacement by exact measurements of quantities in nature that are then set forth in laws expressed in formulas, Kepler is to be designated as one of the first who made use of its methods. His precise calculation of astronomical measurements, which found their overt expression in Kepler's famous three laws of planetary motion, are still accounted a model of quantitative description of natural events, and we can see today what a great influence Kepler's discovery in particular had upon the subsequent course of astronomy and physics. Although Copernicus, through his great discovery that the planets revolved around the sun, laid the groundwork for establishing the laws of planetary motion, he nonetheless made no substantial progress towards carrying out the necessary investigations. With his recognition that planetary paths take an elliptical form, Kepler became the first to establish the physical laws of planetary motion, and he himself always regarded this relation of astronomy to physics as highly significant.

And yet the mathematical physicist Kepler is not to be thought of as one of those sober scholars who live out their lives between numbers and formulas, leaving no space in their minds for ideas of more general significance. To be sure, he had to maintain an unceasing tenacity in working through quantities of numerical material in order to establish his laws, but he did this merely out of necessity, as the only available route to an answer to his more general questions. By nature he tended to be romantic and mystic rather than dry and scholarly.

In *Mysterium Cosmographicum* (1596), his first work, Kepler reveals himself as a still whole-heartedly religious thinker, imbued with the belief that there is a mystical harmony between the structure of the universe and the basic mathematical forms at that time, he was a twenty-five-year-old professor at the University of Graz, having completed his schooling at monastery and cathedral schools such as Maulbronn and Tübingen, as was then customary in his native Swabia, and subsequently pursued his studies at the University of Tübingen. Even then, the question of the laws of planetary motion left him no peace, and he constructed a theory according to which the diameter of the path of a planet bears a particular systematic relation to the dimensions of regular polyhedrous geometrical forms.

Fig. 1. Johannes Kepler (astronomer, 1571–1630) as a young man. After a painting (probably retouched picture). Photograph Deutsches Museum, München.

The most noteworthy fact about this theory is that it offered him at the same time a rule for establishing the number of planets. Geometricians had long known that there are only five possible regular polyhedrons, and as it was a consequence of Kepler's theory that the number of planets had to be greater by one than the number of polyhedrons, the theory justified the belief that there were six planets, the number known at that time. Today, it is easy for us to recognize that the establishment of planetary laws on a mystical basis of this kind is untenable, for in the meantime, the telescope has revealed two, and recently even a third, further planets. We know now that the number of planets is an arbitrary fact, the law of the planets revealing itself soley through their paths.

But, in truth, we have Kepler's research to thank for the fact that we know this today, for as the pioneer he had himself to tread this thorny path to knowledge. "A prior speculation must not be permitted to run counter to manifest experience; rather, it must be brought into conformity with it." This comment was written by Kepler with reference to his own early work. It was his special destiny to overcome the mystic ideals of his youth in the course of his pursuit of systematic investigation, replacing them with exact scientific knowledge. Thus he travelled within the confines of his own mind the course traversed by science, from its irrational sources to its strictly rational modern form. It is no coincidence that the man who had to conquer within himself a belief in the harmony of the spheres in favour of the impartial interrogation of the factual belongs to the founders of modern science.

Kepler was helped in the pursuit of his course by the peculiar twistings and turnings of his personal destiny. The agitation stemming from the religious wars profoundly affected the life of this free spirit. As a Würtemberg Protestant, he suffered from the hostility of the Catholic church. With other Protestants he had to leave Graz, and he and his new family had thus to give up any outside means of subsistence. Thereupon the great Danish astronomer, Tycho Brahe, who was then living at the court of Emperor Rudolph in Prague, invited him to join him at Castle Benatek.

Tycho's entry into Kepler's life was responsible not only for providing him with a fresh existence from the economic standpoint; his work with Tycho offered him for the first time the purely scientific opportunity to carry out his great work. For Tycho Brahe was the greatest technician of astronomical measurement of his day. He was the first to construct precise astronomical instruments, and he himself spent years making highly precise observations of the positions of the planets. Needless to say, even Tycho's observations are quite primitive compared to those made with precise modern instruments; as no telescope was at his disposal for use with his sextants, the stars had to be viewed with the naked eye, the viewer sighting them through a slit along a

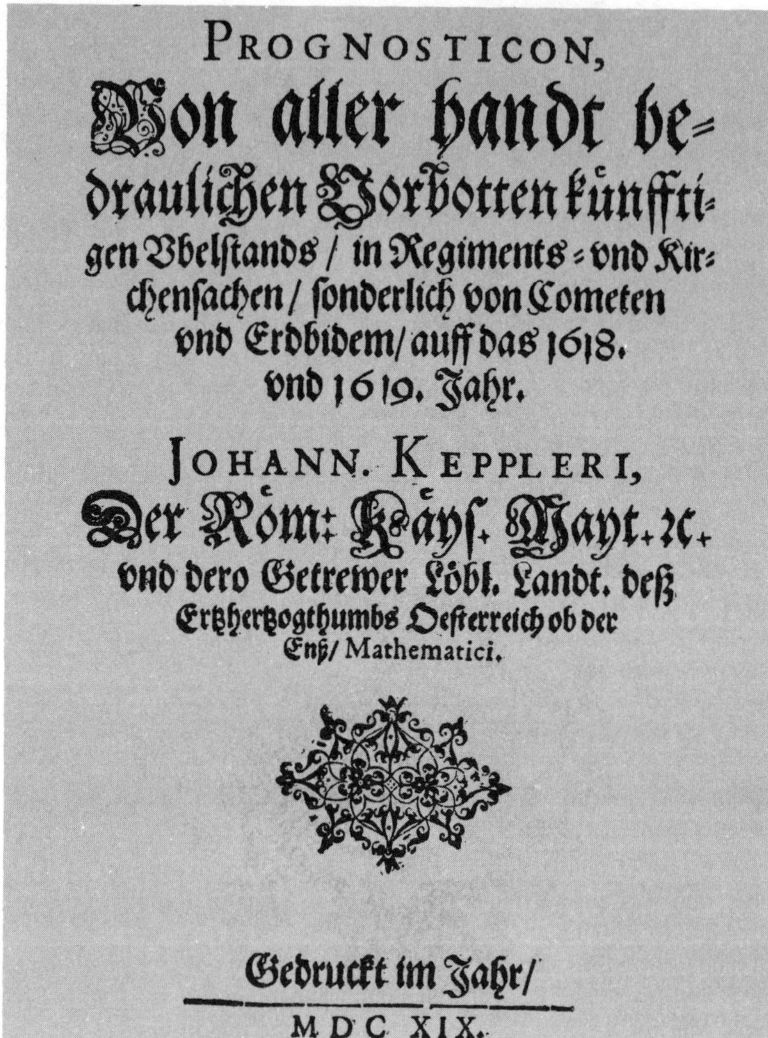
PROGNOSTICON,
Von aller handt be-
draulichen Vorbotten künffti-
gen Vbelstands / in Regiments- vnd Kir-
chensachen / sonderlich von Cometen
vnd Erdbidem / auff das 1618.
vnd 1619. Jahr.

JOHANN. KEPPLERI,
Der Röm: Käys. Mayt. ꝛc.
vnd dero Getrewer Löbl. Landt. deß
Ertzhertzogthumbs Oesterreich ob der
Enß / Mathematici.

Gedruckt im Jahr /
MDC XIX.

Fig. 2. Title page of *Prognosticon* from Johannes Kepler (dated 1619). Vorlage: Kepler-Kommission Fr. List.

movable arm towards the rim. Nonetheless, Tycho attained an accuracy of within eight degrees in his measurement of angles.

Making his own measurements with Tycho's instruments, Kepler completed Tycho's extensive records and then used them as the basis for a laborious job of calculation. Kepler used the path of Mars as the subject of his investigations, probably finding it of particular interest because it did not accord with the theories of Copernicus. Copernicus had still believed that the planets followed a circular path; in order to explain the minor deviations revealed by observation, he had to make the assumption that the sun was situated slightly to one side of the centre of the circle, rather than directly in the centre. Kepler realized that the paths that would result could not possibly be reconciled with the empirical observations, yet he did not find it easy to establish the ellipse in place of the circle as the form of the planetary paths. For a time he believed in an egg-shaped path, and this hypothesis alone delayed him for a year, until he recognized its inaccuracy. It was in 1605 that Kepler first recognized the ellipse as the correct form. In his principal work, *Astronomia nova* (The New Astronomy), he gave a thorough presentation of his discovery and also established his second law of planetary motion, which gives a formula for the area covered by the radius vector of a planet in motion. However, he did not establish his third law, which concerns the periodic times of the planets, until some years later, and it is presented in another work.

Kepler's name is associated with a number of other physical mathematical discoveries in addition to his great discovery in astronomy. He had already conceived the idea that the motion of the planets is to be explained physically by a force emanating from the sun, yet he did not conceive these relations quite properly, and thus their discovery had to be reserved for another thinker. But Isaac Newton was able to formulate his law of gravity only because he knew Kepler's laws as facts of astronomy and could presuppose them. Incidentally, Kepler laid the groundwork for another of Newton's innovations, the infinitesimal calculus, in his *Cubic Calculation*, which deals with investigations of the cubic capacity of bodies.

Kepler also did work in the field of optics; he freed the laws of refraction from the erroneous conceptions of the ancients and invented a special kind of telescope that still bears the name Kepler's telescope.

Although such discoveries early established Kepler's fame, making him a highly respected astronomer even within his lifetime, the external circumstances of his life certainly returned him no thanks for his work. His life was a constant struggle for subsistence, and it was only with difficulty that he was able to spare some time for his genuinely scientific work. Economic necessity compelled him to deliver himself of astrological prophecies and to produce calendars forecasting the year's weather and political events. Kepler himself was quite clear as to the value of such activities. As he once wrote,

This astrology business is certainly a foolish offspring, but the Lord only knows what would happen to its mother, the highly rational *Astronomia*, without this foolish daughter . . . Stipends for mathematicians are also so small, and so few and far between, that the mother would certainly go hungry if the daughter were not earning something . . . "

Moreover, he and his family suffered a number of misfortunes. His children and his first wife died of smallpox in Prague, and he himself often suffered from illness; he had been a seven-month baby, and was never strong in body. He had to make a trip to Würtemberg to rescue his mother from a terrible fate in her old age; she had been accused of being a witch and had already been condemned to torture upon the rack.

The famous son was able to save the old woman through his influence, yet

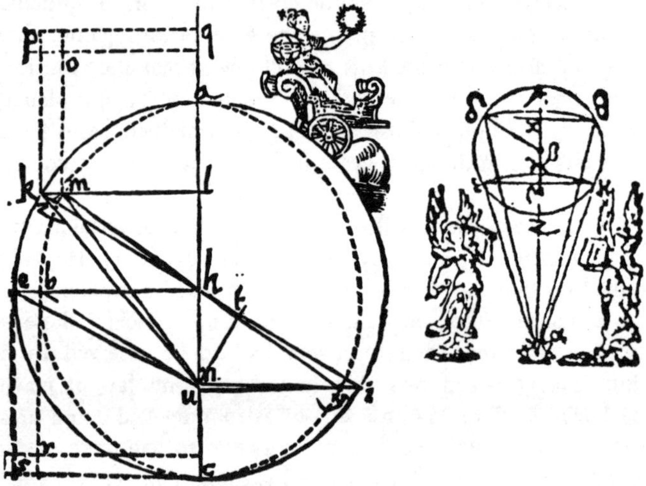

Fig. 3. Illustration from Kepler's magnum opus *Astronomia nova* (Heidelberg, 1609) which contains the calculation of the orbits of the planets. The part at left: Photograph Deutsches Museum, München. The part at right: From *Die Woche* (November 1930).

this influence was never great enough to secure Kepler a position for himself at his native university at Tübingen. The faculty at Tübingen considered him far too free-thinking to be received into their midst.

Thus Kepler had to subsist upon the favour of princes and was forced to squander a great portion of his energy on the intrigues of court officials.

In Prague, where he succeeded Tycho Brahe, after his death, as the emperor's court mathematician, Kepler was entitled to a good salary. Yet he never received more than a small portion of it, and the outstanding debt owing to him finally reached the imposing sum of 12,000 guilders. Emperor Rudolph's successor, Mathias, had even less thought of paying the debt and struck a bargain with Wallenstein, according to which Kepler was to enter his service and Wallenstein, in return, was to pay off the outstanding salary to Kepler. Kepler, who had already given up Prague for Linz, where he taught at the high school and pursued his investigations of measurement on the side, and who had then moved on from Linz to Regensburg, now betook himself and his whole family to Wallenstein's court in Sagan.

But Wallenstein, who had little interest in Kepler's scientific work and who gave the great scholar a position only for the sake of his tinkerings with astrology, had no intention of paying off the emperor's debt. Kepler resolved, therefore, to go to the Imperial Diet in Regensburg, which was just then convening, in order to establish his claims before the representatives. He made the trip on horseback, taking the circuitous route via Leipzig, and arrived in Regensburg on November 2, 1630. But the strain of the journey was so great for the fifty-nine-year-old man that he was stricken with fever and died thirteen days later, on November 15, 1630.

He was buried in Regensburg, but the cemetery that housed his grave was totally destroyed in the upheavels of war that filled the succeeding decades. Not even in death was this great man granted the peace that was always denied him in life. That he was able to give the world such a wealth of ideas in spite of his misfortunes bears testimony to boundless energy for work, and also to a passionate devotion to that task for which fate had chosen him. "I came into the world destined to occupy myself with matters from the difficulties of which others retreat." – Today, three hundred years after the day of his death, we can give no more profound expression to the significance of his life than he himself set forth in these words.

28. THE PRESENT STATE OF THE SCIENCES: THE EXACT NATURAL SCIENCES

[1930b]

A description of the contemporary state of the exact sciences has not always presented as difficult a task as it does in our time. Over the course of the development of a science there are periods of ascent and periods of lingering at a plateau. Both are essential for the progress of science, for although the ascent constitutes the period of actual growth, the time of quiet elaboration, of the gradual maturation of newly made discoveries, is equally necessary for the internal consolidation of the whole developmental process. The distinctive rhythm of all life, the step-by-step progress, in which ascent and a time for maturing alternate with one another, appears to be intrinsically necessary for the over-all development of sociological phenomena, to which the growth of the natural sciences bears a living testimony. After the steep ascent accomplished in the era of such scientists as Galileo and Newton, the succeeding century coincided with a period of undisturbed consolidation, but about one hundred years ago a new ascent was begun which even today has not yet flattened out into a horizontal curve. On the contrary, over approximately the last three decades we may observe a substantially steeper ascent of the curve than was to be seen before, for our epoch is characterized by overly hasty development, by the appearance of totally new and revolutionary theories of fundamental significance, by the amassing of new discoveries, so that the contemporary observer barely recalls the impression of steady progress and finds the image of growth by fits and starts to accord far better with his experience.

This epoch of highly rapid development is no longer played out in the quiet of the scientific institute. Today the public takes an active interest in the discoveries of scientific research, and they became aware some time ago that developments in contemporary physics have been unfolding at a downright revolutionary pace. And yet only a handful of those observers who are not personally involved have any inkling of what concentration and intensity are at work in physics today or of what deep internal consistency lies in the new turn taken by physical thinking. When they hear of the two fundamentally new theories of modern physics, relativity theory and quantum theory, today's public probably attributes the fresh impetus to powers of bold speculation or creative imagination; little do they suspect how closely the existence of these forces is associated with the refinement of experimental methods, with the process of technicalization that has taken hold as firmly in the sciences as in technology and in social organization. For the new physical theories have sprung primarily from that penetrating and extremely precise interrogation of nature achieved by the physical experiment. Much as it is to be attributed to breadth of conceptual thinking, Einstein's relativity theory would nonetheless never have come into being, or at least not have achieved such remarkable recognition, if it had not been able to appeal to the experimental results which displayed clearly for the first time the great precision of space–time measurement in modern experimentation. Similarly, the quantum theory of such scientists as Planck and Bohr progressed only in the closest relation with the experimental investigation of the minutest particles of matter. The yield of theoretical reflection is bound up quite essentially with the increasing precision of experimental methods. Indeed, it is, for the most part, these methods that have demanded the modifications and critical revisions of older theories that we regard today as the hallmark of contemporary physics.

And yet there is a great difference between the two theories. Relativity theory is complete, having by now already reached a certain stage of maturity, whereas quantum theory, although temporally some five years its senior, is still in a very fluid state and has taken a different direction in the last four years, the ultimate outcome of which cannot yet be foreseen. Consequently this theory is, to a far greater extent than relativity theory, the object of scientific investigation for the contemporary generation of physicists. A further reason is that the quantity of experimental material encompassed by quantum theory greatly exceeds that affected by relativity theory, and a far greater number of collaborators was required to develop from Planck's discovery of the quantum-like, i.e., atomistic, nature of radiation a theory of matter of quantum-like behaviour. The new orientation resulting from this theory entailed far more than the assertion that matter possesses an atomistic structure, for in contrast to this older notion, quantum theory asserts that the regular mechanism of the atoms, both in their reciprocal reactions and in their internal structural constitution, also conforms to quantum-like laws. Matter, particularly within atoms, is restricted to a number of discrete states, and natural events consist of discrete transitions from one of these states to another. This signifies an abrupt break with all past conceptions of regularity in nature, to which the idea of nature proceeding by sudden leaps was quite alien.

Yet a further step has been taken in the conception of regularity in nature:

the ancient idea that the course of nature is strictly determined, that is, the idea of determinism, has been shaken. The road to this revelation was certainly not made easy for the physicists. The point of departure was provided by the discovery of a peculiar duality between waves and corpuscles (by de Broglie and Schrödinger). In the back and forth of mathematical approaches and experimental observations it became known that both radiated and solid matter displays such a duality. Light must be viewed partly according to the wave model, partly according to the model of discrete, minute particles. Moreover, the elementary mass-points themselves, electrons and protons, the basic constituents of the atom, cannot be completely grasped through the concept of a particle, but require for their complete description wave imagery as well. Only after this was recognized did the attempt at interpreting this information (as in Born and Heisenberg) lead to the idea that events at the micro-level can no longer be regarded as strictly determined according to laws, that future events can now be predicted only with probability. If we nonetheless observe in macroscopic processes that great certainty of functioning from which we have culled the idea of a strictly causal course of natural phenomena, this is to be explained by the common action of an extraordinary number of elementary particles, which in accordance with the law of large numbers produce phenomena of extreme regularity at the macro-level, even though only very limited regularity is found at the micro-level.

It is not easy to sketch a more or less comprehensible picture of these new conceptions about the nature of matter, but it is this very situation that characterises the present state of the exact sciences. We find ourselves today in a period of revision of all conceptual instruments of the production of new intellectual tools, and this may be the reason why querying looks are directed at the work of physicists from so many quarters. For the discoveries that are made in the effort to ferret out the internal structure of the atom are at the same time discoveries of a philosophical nature. Reaching far beyond scientific circles, they involve a new approach to the significance and possibility of theoretical knowledge in general. With the steep rise in the curve of its development, physics has merged today with a philosophical epoch.

29. ONE HUNDRED AGAINST EINSTEIN

[1931a]

It is not generally the custom for those who do not understand a scientific theory to publish a joint presentation of their misunderstandings. The *One Hundred Authors Against Einstein* (edited by Dr. Hans Israel, Dr. Erich Ruckhaber, and Dr. Rudolf Weinmann and published by R. Voigtländer, Leipzig) do not see eye to eye with this custom and have enriched the literature with a magnificent collection of naive errors.

The allegations made in this volume are truly astonishing. Einstein is said to have made mathematical errors – unremarked, apparently, by the hundreds of mathematicians who have been checking over the calculations in his theory for the past fifteen years. The theory of relativity is supposed to contain contradictions. One author, who is evidently ignorant of the elementary facts of optics, explains that the speed of light in a vacuum must be infinite. Another asserts that Einstein postulates the speed of light to be infinite. Yet a third author wishes to introduce the theory of light proposed by the philosopher Marcus, who has finally (it is claimed) refuted Einstein. Marcus assumes that light is not only emitted from a source, but also sucked in, as it were, by the receiver: "Truly, we have here a novel and wonderful picture of the world of light." At least we know now how to go about doing anti-relativistic physics: just let a philosopher figure out some way the universe might work, and that's all there is to it.

Supporters may appeal to empirical confirmations of relativity theory, but, as one of the one hundred authors informs us, "It is the height of folly for relativity theory to seek confirmation in experience." Another of this company accuses relativity theorists of downright deception: "The photographs were carefully chosen so that they would tend to confirm Einstein's hypothesis [concerning the deflection of light]." Furthermore, "Einstein found himself compelled to amputate a portion of his theories in order to reveal as if by magic Gerber's Newtonian formula, which he uses, unacknowledged, in his calculations." In the eyes of most of these authors, the whole sorry affair is to be laid at the door of mathematics, for there we can explore all sorts of things which have nothing to do with reality: "Relativity theory is a mathematical masquerade", and the square root of a negative number is not [so they claim here] in the least imaginary, contrary to the opinions of mathematicians. My own basing of relativity theory upon the arbitrary nature of co-ordinative definitions leads, in the view of one of the authors, to horrifying consequences for the administration of justice: "Is it to be tolerated that defendents have been and are still being sentenced to death on the basis of arbitrary definitions?!"

These are just a few tasty tidbits out of the hundreds of asininities served up to the world by the one hundred authors. The so-called objections have already been answered a hundred times over, yet, despite this, the editors declare in the foreword "that neither Einstein himself nor his commentators have so much as begun to make an effort to reply to the increasing list of arguments posed by their opponents". How an otherwise serious publishing house could be willing to take on the publication of such a volume must remain incomprehensible. Still, it has its compensations: the anti-relativists have over the years presented us with such an abundance of inadvertently amusing literature that I have set aside a section of my bookshelf under the label 'Labyrinths', which will now be enlarged by a new and superb example of the species.

30. IS THE HUMAN MIND CAPABLE OF CHANGE?

[AN INTERVIEW]

[1932a]

Professor Hans Reichenbach, who teaches at the University of Berlin, is a proponent of the new scientific trend in philosophy. It is the aim of this new movement to construct a precise theory concerning our knowledge of nature and the relation between thought and existence by means of a thorough investigation of the natural sciences, including mathematics; for this reason, it is more closely bound up with the empirical sciences than any other philosophical movement. The teachings of this new philosophy of science are still a matter of controversy today, facing particularly strong opposition among humanistic and cultural philosophers. This may enhance the interest of our readers in the following interview, in which this leading proponent of the movement shares with us his basic ideas.

"Professor Reichenbach, we take it to be the basic theme of your writings that, as human knowledge has accumulated over thousands of years, reaching massive proportions, the human mind has found itself undergoing constant change at the same time. What seemed to the ancient Greeks to be unshakeable scientific facts have been thrown overboard by contemporary science; think, for example, of Euclid's geometric theory and Ptolemy's theory concerning the structure of a universe revolving around a stationary earth. Now, is it really true that these changes can only be grasped by our now transfigured minds? Is our brain today different from that of the ancients?"

"I am unable to reply to the whole of your question, for the question of the structure of our brain is much more far-reaching than that as to the structure of our mind. Certainly, it can be generally assumed that there is a univocal relation between the mind and the brain, that is to say, that every train of thought and every change in our thinking corresponds to certain modifications in the brain. Detailed studies of the brain, such as those now being carried out in institutes of brain anatomy, in which slides have been prepared from cross-sections of the brains of thousands of distinguished or mentally abnormal people, have shown that human brains are, in fact, extraordinarily diverse. For instance, certain cerebral convolutions are highly developed in people of strong social concerns; these same convolutions are under-developed in anti-social criminals. The same is true for other mental characteristics. But such precise brain research has only just come into being, and we have no means of comparing our results with corresponding results in times past. That the human brain has undergone development since prehistoric times can be pretty clearly established from modifications in the structure of the skull. But whether the tremendous intellectual transformation that is expressed in our present way of thinking *vis-à-vis* that of the ancients corresponds to a demonstrable modification of the brain is a matter about which no pronouncements can be made at present."

"You think, then, that the mind has undergone change?"

"Again, I have to formulate your question more precisely before I can answer it. Insofar as it is possible to draw conclusions as to the mind of the ancients on the basis of the scientific and artistic production of a people, that is, from such manuscripts, statues, and the like as have survived – and, after all, we have nothing else to go on – a definite transformation has taken place in modes of thinking. But it is difficult to give a precise description of this change, for in spite of the many common features that characterise a people the range of intellectual abilities is very great. That is why, despite the gradual raising of the intellectual level, we nevertheless find great diversity unless, in drawing comparisons, we consider the corresponding social classes. The Greeks included in their numbers outstanding mathematicians, for example, whose mathematical knowledge and ability far excelled that of a present-day Bavarian farmer. You understand what I'm saying, of course. I have nothing against the Bavarian farmer, whose mountains I love above all others and whose view of life is certainly more akin to that of us scholars than is that of the ancient people – it's just that he isn't a mathematician.

"But this idea leads us to a more precise formulation of your question. We might put it this way: would Euclid be able to grasp modern non-Euclidean geometry, or would Ptolemy acknowledge the Copernican vision of the universe or, to go a step further, believe Einstein's theory of the universal relativity of motion to be correct? I'd give you the following answer: If it were possible for any Greek mathematician whatsoever to comprehend these modern theories, it would be a profound thinker like Euclid or an intellectual giant like Ptolemy. But I really doubt whether even these gifted men would be able to do it. In their day they were pioneering innovators, but it is not for nothing that our latest theories had to wait two thousand years to be discovered. Every single one of us scientists experiences, in the course of our individual existence, profound changes in our mental abilities. I myself can remember how, at one point, when I really worked through them thoroughly for the first time, Einstein's theories seized hold of my mind and transformed my way of thinking. At that time I had to give up as untenable a whole series of views that had previously appeared to me to be unshakeable truths. That was about sixteen years ago: in the meantime I have been able to give this new image of the universe a philosophical interpretation and have subsequently become so at home in it that I now find it difficult to visualize the older conception of space and time at all. My generation finds itself in a particularly interesting situation in this respect. As the younger generation, we forged our way into modern science and still have ahead of us today a class of older scientists who will not yet go along with many of our views. But a new generation is already coming up behind us who adopt with ease the ideas which we won so hard. By the time they come to these ideas, they have been so carefully thought through and refined that these younger scientists absorb them without difficulty, which leaves them the time and energy to make new strides. Modern quantum theory offers a typical example. Here, the decisive ideas have come partly from young scientists, who haven't yet reached thirty. In the meantime the men who were the great innovators twenty or thirty years ago have already moved up into the older generation, so that my own generation is able to observe this transition from a younger class of innovators to an older class of conservers from an intermediate perspective. It appears, if I may so put it, that the plasticity of the human mind is restricted. Each of us is able to go just so far beyond the ideas he set out with, but the younger among us starts out one step higher and can therefore advance one step farther."

"You think, then, that the boundaries the individual brain comes up against when it fails to understand certain assertions, such as that the space of the universe is closed and finite, are after all of a biological nature?"

"Perhaps, although in this particular instance it is quite possible to bring about intuitive visualization through training. Modern geometricians have learned to visualize intuitively spaces having a non-Euclidean structure. It is really just a matter of excluding certain covert presuppositions that we all carry along with us from daily experience and which restrict our mind without our noticing it at first. Certainly, it requires a lot of training and serious concentration to discern these connections; more than anything, it takes good faith to really grasp the new forms of thought."

"But if it is so difficult, how did it ever become possible for science to achieve such new forms at all?"

"It became possible because it was necessary. Through an extraordinary refinement of experimental techniques, contemporary science accumulated such vast stores of factual knowledge that scientists became unable to fit this staggering mass of factual material into the framework of traditional concepts. And the construction of a new conceptual framework proved possible because intellectual techniques had become extraordinarily refined along with experimental techniques. Scientists have learned, for instance, to construct new geometric relations, such as those of a closed universal space, in abstract mathematical formulas with which they can make calculations and draw conclusions without visualizing anything spatial at all. It was only later that the knowledge gained through this abstract process was retrospectively supplied with intuitive insights. Thus mathematical calculation, that is, expression of ideas in a technical form, leads to forms of thought previously unknown."

"But don't you think that, along with all these changes in our way of thinking, there are unchangeable forms of thought, solid buttresses, as it were, upon which the whole of our thinking rests: those forms of thought that have been described as *a priori*?"

"I'm glad you've asked me that, for that is exactly what I don't believe. Although many philosophers today still hold with the theory that there are *a priori* laws of thought, it seems to me quite untenable in view of the development of scientific thought. Too much that was described by earlier philosophers as *a priori* has fallen by the wayside as science progresses. In addition to the modern theories of space and time, there comes to mind our current conception of causality, of the cause–effect relation, which in contemporary science has taken on the character of a mere conditional connection and has coalesced with the concept of probability. Another example is afforded by modern atomic theory. While the basic idea of the atom comes from the Greeks, our present-day theory of electrons and protons is fundamentally new. Granted, the idea of atomism has achieved its highest potential in our time, for the manifold variety of atoms has been reduced to a duality of basic constituents that are identical with the atoms of positive and negative electricity. But just look at the surprising results quantum mechanics has achieved in our time! According to current theory, the same particle is now a particle, now a wave, taking the one form or the other according to the apparatus through which it is observed. Today's physicists are still scarcely able to grasp this; after all, it contradicts the ancient notions of a firm substance, developed by natural science and subsequently made a principle of the *a priori* philosophy. This new mental world no longer conforms to any *a priori* schema."

"But isn't at least logic, the science of rational inference, binding for all time? Is it conceivable that another generation may doubt the foundations of logical inference?"

"Not only is it conceivable; it has actually happened. Aristotle would find our logic just as alien as Euclid would find our geometry. Just take a proposition such as that the whole is greater than any of its parts – this proposition of Aristotle's is not true in modern set theory. For we have learned that we can talk about quantities being the same even with infinite sets and can prove that in the appropriate sense there are, for instance, as many even number as there are whole numbers. Nonetheless, the even numbers are, of course, a part of the set of whole numbers. And at that, this is a relatively mild innovation. It is much more disturbing that the mathematics of the century just past has now been shaken at its roots, as a result of certain modes of argument formerly applied without second thought now being considered untenable. This discovery was occasioned by certain logical contradictions, in comparison to which the contradictions cited by ancient philosophers appear quite harmless; these led to a crisis in the foundations of all thought, in logic."

"You believe, then, that there is no absolute truth, that every era possesses its own truth?"

"I certainly didn't say that. That thinking changes does not mean that past knowledge is cast aside as totally false. On the contrary, we could not know what we do today without the preceding levels of knowledge. If I may use a mathematical expression, there seems to be a convergence process underway here. Even if the absolute truth is not granted to us, at least we are allowed this continuous progress in our knowledge. We must only bear in mind that this increasing comprehension is meaningful in itself and refrain from burdening it with the yearning for an unattainable goal. You may call this a renunciation – I myself prefer to describe it as a very strong faith in the power of the human mind, a faith that it has within it the ability always to move beyond the state is has presently achieved to broader and broader insights."

PART III

GENERAL SCIENTIFIC ARTICLES

31. METAPHYSICS AND NATURAL SCIENCE*

[1925a]

Historically, the term 'metaphysics' has been synonymous with the demand for a system, for all great metaphysicians of the past have attempted to solve the problem of the transcendental through the construction of a system. But I do not believe that this approach is possible or meaningful today.

It has been so very common to build systems in philosophy that the route to *the one* system has been thoroughly obstructed. This is the reason that in philosophy, unlike any other field of knowledge, we do not possess a store of recognized results. We stand today before the ashes of historical systems, and when we endeavour to retrieve from them at least a few apparently undamaged pieces, we rapidly discover that the details are very much less useful than was the whole. So grand an edifice as the system of a Spinoza or Kant or Hegel at least possessed in its overall character a certain power of persuasion, exuded a productive spirit, and was capable of winning disciples and inspiring original work. But if we scrutinize the details under the microscope of exact criticism, the lovely image dissolves into nothing: we are left with nothing but a handful of broken pieces that possess no value in themselves, but were only meaningful as members of the whole.

It would be sheer sentimentality to conclude from this that it would be better to abandon exact criticism. A clearer insight, once gained, is not to be suppressed, not even for the benefit of a lost emotional value. On the contrary, it follows that we should dismiss this heap of ashes from our minds with all the abandon of youth. Nothing is more dangerous for the present than remaining in the grip of a past that is dead and gone, no matter how great it may have been; nothing can so distort the historical task of the present as a gaze focussed on the past that tries to *construct* historical continuity, rather than to instinctively *produce* it from an experience of the present. We must apply our own powers to those endeavours that have acquired new aspects with the new age.

However, this does not mean endeavoring to produce out of our own heads

* This topic was set for discussion at the conference of the Kant Society in Halle, June 5, 1925. The present article is an expanded version of the lecture given by the author on that occasion.

that philosophical system that has overcome the shortcomings of the historical systems. No matter to what extent a comprehensive system may be the ultimate aim of philosophy, this cannot be our task today. There are too many unresolved details; almost more importantly, there are too many that are *acknowledged* to be unresolved. There are still too many stones lying on the ground to start laying out the whole garden. This much, at least, philosophy ought to have learned from science: the way to find the system is not to construct it before the individual problems are solved. The exact sciences today are in possession of a system of a magnificence and a certainty that philosophy might well envy. They are indebted for this system, however, not to the speculative efforts of the so-called natural philosophers of all ages, but to the modest investigators of nature, each of whom, singly, has concentrated on an individual problem. This is why a Galileo, a Kepler, a Newton have become the founders of the most comprehensive system of knowledge that has ever been; they did not search for this system, but, instead, selected a few unsolved problems and worked them out with precision. In so doing, they not only solved the individual problems but created at the same time the *method* that turned out to be so fruitful. Mankind evidently has a greater ability to develop the method through solving concrete problems than to figure it out by means of abstract hypotheses. Philosophers ought to have far more regard for this fact. The present situation in philosophy is roughly comparable to that of natural science at the time of Galileo, and we will not extricate ourselves from this situation until we set aside for a time the idea of an all-embracing philosophy and apply ourselves to the solutions of the individual problems with the exactitude that we have learned in the natural sciences.

Thus it seems to me that the first methodological point to be made about metaphysics and natural science is a more or less pedagogical observation: metaphysics can learn a great deal from natural science about working methods and the degrees of possible precision and possible discretion. I certainly do not want to express the opinion that the methodology of the natural sciences can be adopted without modification for use in metaphysics. But once metaphysics has attained the plane of scientific method, it has taken the biggest step forward.

Nonetheless, I do not wish to fall into the error of attempting to present a philosophical method. Insight into the method is, so to speak, a byproduct that appears of its own accord along the way when concrete problems are tackled. Thus I know no better way of presenting to you the methodology of a philosophy versed in the exact sciences than presenting the problems themselves as they appear in general outline when viewed from this standpoint.

Here I am restricted by the choice of topic for today's discussion. We are to discuss metaphysics and science, not philosophy and science. This excludes a whole row of methodological and logical questions. Metaphysical inquiry wishes to proceed beyond what is known in science, asking for the transcendental significance of scientific results and wanting to learn something about the relation of the knower to things in themselves. This leads, basically, to three groups of problems. The nub of the first is the so-called question of the reality of the external world, or the problem of existence. How do I come to assert that things exist? What can I find out about them? These questions concern the passive perceiving behaviour of the subject. The second group of problems, in contrast, touches upon the active, intervening, behavior of the subject: Can I influence the world through my actions, and, if so, what does this signify? It is the problem of the freedom of the will that emerges here. Over against these two problems, which view the world from the standpoint of the subject, stands a third, that approaches the subject from the viewpoint of the world instead, posing the question of the special place of life among the phenomena of nature, the so-called question of life. Is life comprehensible by analogy to inorganic nature? If not, how is it to be scientifically classified? In these three categories of problems, which naturally enough are not independent of one another, I see the genuinely metaphysical problems of present-day knowledge of nature. In accordance with the short time allotted to me, however, I shall be able to concern myself only with the first problem.

The problem of existence was recognized very early in the history of philosophy, yet because of its peculiar complexities it has always been interpreted in a variety of ways and has been subject to mishandling, which has given rise to an unfortunate proliferation of concepts. The standpoint of the naive person who believes simply in the existence of things has been declared to be unphilosophical and, under the name 'naive realism', has been shoved aside as being, as it were, not presentable. Yet I cannot see that anyone has found any better explanations. What makes this problem so complex is the peculiar relation in which concept and reality stand to one another. There are no assertions about reality that are not formulated in concepts, that do not contain laws of thought as well as laws of reality. This is the fundamental notion that forms the starting point of investigation for Plato as for Descartes and Kant. But when it is therefore concluded that the concept is what really exists, or that the laws of thought are at the same time laws of reality, or that reality is made up of some combination of concepts and 'things-in-themselves', the inference has gone rather too far, and only the obscurities of language can mask these logical jumps. For that inseparability of thought and reality pertains only to our *knowledge* of nature; it cannot apply to nature itself – if it is at all meaningful to speak of something as having objective existence. We cannot avoid the distinction between the 'assertion' and 'the object designated by assertion'. Nor does it help to construct a concept of an object that already contains, as it were, the element of thought. We have only to take a hold firmly of such conceptual monstrosities as 'thing of appearance' to realize that they either fail to progress beyond naive realism or equate thinking with being, which no one has yet been able to carry out with a good conscience. In Kant I detect this conscience still clinging to life in the appearance of the concept of the thing-in-itself, which has caused all expositors such mental anguish and which, moreover, cannot be fitted properly into the system – but without which the Kantian system becomes a form of solipsism.

Even though Kant's answer to the problem of being cannot be upheld, it must be recognized that the problem is formulated as an analytic question. It is not a matter of establishing or denying the existence of physical objects, but of clarifying the meaning of the assertion of existence. We cannot avoid realism; it is simply a question of transforming naive realism into a well-thought-out, well-formulated realism.

Let us return to the fundamental problem. It is not possible to make any assertion about the world that does not, at the same time, contain the structure of our thinking. It is as though we observed the world through a grille; our conception of the world always bears the imprint of the structure of the grille. How can we eliminate the structure of the grille? This is the genuinely metaphysical formulation of the problem. It is no help to attribute to the complex image of the world and the grille a special existence as a world of appearance; to be honest, our greatest interest is in the thing itself.

But it is clear that there can be no direct answer, for the very reason that the grille that frames our thinking cannot be eliminated in finding this answer, either. Nonetheless, there is another approach. Instead of directing our attention to the world, we can concentrate upon the grille itself. We can discover the traces of our thinking in our image of the world and can then separate out, at least indirectly, the contribution made by reality as the non-rational remainder not included in the description of the rational, i.e., of our thought processes.

We will find the method that we must use here by analyzing the procedures leading to knowledge in natural science. We then discover – and recent developments have revealed this phenomenon – a peculiar indeterminacy in the choice of conceptual systems used for explanatory purposes. It turns out that one and the same fact may be described by different conceptual systems.

Some have viewed this as the end of unequivocal explanations of nature, but surely they are mistaken. Discoveries of such phenomena as the relativity of motion and geometry have demonstrated not the imprecision or uncertainty of a *single* description, but the equivalence of *many* descriptions, the transformability of one conceptual system into another. The objective character of knowledge is not thereby called into question; it is simply expressed in another way. What is objective is that which is described by *all* these conceptual systems: the invariant which is not touched by the transformation of the descriptions. It is not an indeterminacy in the objective content of our knowledge that finds expression in the arbitrary nature of conceptual systems, but something else; it is the imprints of our thinking upon our image of the world that have become discernible. The lack of a univocal prescription for description may be taken as a sign that we have run into the grillwork that permeates our world image, that we have exposed the subjective components of knowledge, which are arbitrary because they lack an objective meaning.

In fact, in expressing the view that space can be described in terms of both Euclidean and non-Euclidean geometry, we are not giving up on the unequivocal nature of knowledge. We are expressing, rather, the idea that geometry does not describe anything that is objective. Kant's notion of the ideality of space finds its ultimate precise formulation in this theory of the arbitrary nature of geometry. The characteristics of what is real are thereby expressed in a much more complicated manner. Whether space is Euclidean or non-Euclidean cannot be asserted. But that space can be conceived geometrically at all, that it can be described in the terms of *any* of the geometrical conceptual systems, signifies a fact lying outside the realm of reason, indicates a characteristic of the thing-in-itself. Here we recognize the peculiar nature of indirect characterization; we do not say, "The world has this or that structure," but, "The world is such that it can be described in terms of a certain class of structures."

Because it is the *class* of conceptual systems, rather than a single system, that characterizes the world, it is senseless to search for a preferable system among these systems. Not even simplicity signifies superiority; that Euclidean geometry is simpler than non-Euclidean does not make it 'truer'. Here we find differences in the degree of simplicity that are not related to differences in truth claims. Thus we speak of *descriptive simplicity*. This form of simplicity pertains solely to description and includes no assertions regarding objective reality. Here too, the scientist will certainly give preference to the simplest description, by selecting a geometry that does the least violence to the expectations of daily life, but this choice is really made from subjective motives, solely upon the basis of economy to avoid mental work. It is not connected to any particular claim to being a better or truer description of reality.

Alongside this subjective choice among conceptual systems, there is another selection procedure in science which has a completely different character. Again and again we find the phenomenon of a conceptual system being replaced by another that is *not* equivalent to the first; conceptual systems are *altered*, and the alteration is carried through with the claim that the new system describes reality better or more truly than the old. The entire history of natural science bears testimony to the truth of this assertion. It demonstrates a magnificent capacity for development on the part of human thinking; it proves that there have never been limits to thought and that, whenever it was required, the human mind discovered the more complicated conceptual system that offered a better grasp of the world. Copernicus' astronomical theory, the mechanical theory of heat, the electrical theory of matter – each of these phases in the development of natural science signifies a step forward in the progression of knowledge and the replacement of a more primitive conceptual system by a more complicated one with the claim that the new system describes the world more accurately. And we are certainly far from the end of this path. Many more tried and true concepts will have to fall by the wayside, many more old habits of mind will have to be redirected, and can be, if reality calls for it. Our conceptual armament today has taken quite a different direction than in the past. Not even the principle of causality belongs to the unshakeable pillars of the physical world view, which owes its universal validity not to the constraint of reason, but to its conformity to reality, to the things-in-themselves.

What is the criterion for this second variety of transformation? Why is one conceptual system 'truer' than the other? The natural scientist says, "Because the facts demand it." For instance, when the transition from the phlogiston theory of combustion to the oxidation theory took place, it was the facts that decided in favour of the second: bodies become heavier during combustion, whereas they would have to become lighter according to the first theory. The decision always takes place after this fashion; experiments are set up, and the facts decide. We all have the feeling that this is the proper procedure – but how can it be metaphysically justified? For it is a metaphysical question that faces us here. An assertion is being made about reality, facts are being acknowledged as the criterion of the objective state of affairs, rather than the conceptual system of the knower.

An objection has been raised against the recognition of fact as the criterion. Every so-called fact, far from being an immediate revelation of the real, is something that is read into reality and which contains the structure of thought just as much as it contains the imprints of the real world. There are no facts, proclaims the idealist, who views the whole conception of the world as a construction of reason. In a certain sense, this is true. Every experiment performed by a natural scientist presupposes a theory, in order that it may be utilized as a fact at all. This theory includes the theory of instruments, of the telescope and the galvanometer, for the numerical results that are directly read off are meaningful only when they can be *interpreted* as measurements of an increase in magnitude of an electric current. It also includes a certain theory concerning the very phenomenon to be investigated; in order to draw conclusions concerning the speed of light on the basis of observed patterns of interference, the experimentalist must presuppose the wave theory of light. This does not apply only to those experiments that are deeply rooted in theory: even the simplest facts of daily life are to some extent theory-laden. I can recognize as such a house, a wood, a person only upon the basis of a summary of past experiences, of a 'theory of daily life'. But how can facts decide between theories if they themselves presuppose theories? Instead of overthrowing the theory 'refuted' by experiment, may we not alter the theories that first made the 'thing' observed into *this* particular fact? Could we not arbitrarily establish *any* theory in just this way, by interpreting every fact accordingly? Are there any facts at all that can claim to characterize something objective?

There would be no way out of this boundless indeterminacy did we not have at our disposal a concept that offers us a choice. The concept of probability comes to our rescue. This is not to say that the philosopher now has the job of introducing the doubt-filled scientist to this concept. Quite the other way around: the scientist uses the concept, and the philosopher has the job of acknowledging it. All assertions about reality are made with a certain *degree of probability*, all facts are more or less *probable*. Thus it is possible to give a ranking of facts; some theories are of a higher order of probability than others, and the facts interpreted with their assistance can therefore be accounted certain for purposes of judging the experimental test of a new theory. Thus the results obtained from the physicists' measuring instruments are 'facts of a higher order'. It is not they that are reinterpreted, but the new facts that are structured so as not to disturb these results. Of yet a higher order are the facts of daily life, the probability of which is so great that it is usually taken for certainty.

The concept of probability becomes more apparent in the system of ranking at the boundary between the known and the unknown spheres of knowledge, where the actual working area of current research is to be found and the knowledge in our possession increased. In performing his experiments, the scientist gives no further thought to his measuring instruments. He presupposes the certainty of the higher-order facts and concerns himself with deciding between theories possessing the same order of probability. And this is precisely the point at which the method of probability becomes apparent. The scientist chooses from among the various possible explanations of a state of affairs the one that appears most probable. Here, too, we find a concern for simplicity, but in this instance a greater degree of simplicity is connected with a claim to a higher degree of probability; the simplest theory is accounted the most probable. Thus we speak of *inductive simplicity*, simplicity in this context being not solely a characteristic of description, but a mark of objectively superior evidence that can serve as the decisive factor in judging the degree of probability of a theory. Think, for instance, of the process of making a graph: a row of points, established by measurement, are marked in, and the physicist joins them together by drawing the most continuous curve possible, the 'simplest' curve. He then asserts that this curve represents the law governing the phenomenon that is confirmed by experience. Why does he select just this curve – why not one that fluctuates widely between the observed points? The choice is justified only from the standpoint of higher probability. This is the assumption that the points at which measurements were made do not represent an improbable selection. If the curve were, in truth, more complicated, it would be a 'coincidence' that the scientist just happened to hit upon a simpler sequence of points.

The objection that later measurements will demonstrate the erroneousness of the fluctuating curve inasmuch as they will not follow the oscillations, falling instead along the most continuous curve, does not hold. This is probably true, but we may never assert it until *after* we have completed our observations; we cannot predict that it will be the case for future observations. The physicist always measures a finite number of discrete points; the final conclusion always involves a leap to the continuous curve, and it is precisely this leap that is not capable of proof. While we can probably rule out certain curves, we can never prove the accuracy of the one that, in the end, we suppose to be the right one. The probability inference is the sole support of this assumption.

And yet this inference is the most remarkable hypothesis to be found in the whole of natural science. The following three assertions can be made about it:

1. The probability inference cannot be logically justified, for it goes from

observed cases to unobserved cases, about which nothing at all can be known.

2. The probability inference cannot be empirically justified. It is impossible to infer its validity from the fact that it has so often proved itself, for this conclusion is itself a probability inference.

3. The probability inference is indispensible for natural science. Without it no decision could be made, and any theory could be forceably imposed, but with it a judgment can be made about any theory, including the most general principles of knowledge.

The third characteristic is particularly important. We must get it clearly in our heads that our knowledge of nature stands or falls with the probability inference. If we refuse to use it, it would be possible to enforce, for instance, a theory of gravitation in which the distance is multiplied to the third power or a theory of electricity in which the current has no magnetic effects. The observed effects would then become exceptions arising as the result of unknown causes. Whenever a physicist wanted to control the magnetic effect of the electric current, an unknown force just happened to move the magnet – so the explanation would run. This is merely *improbable*, not *impossible*. That is why only the probability inference can help us decide between two different theories. With this inference, on the other hand, we can make a judgment regarding any theory. We cannot give a more detailed account of this point here, but it should be mentioned that even the validity of the principle of causality can be investigated once the probability inference has been admitted.

While the first two characteristics were precisely formulated for the first time by Hume, the third could be formulated in terms of a Kantian theory of knowledge. If the probability inference is a necessary condition of all knowledge, then it has a place in the Kantian theory of knowledge, according to which such necessary presuppositions preceed all knowledge in the form of synthetic *a priori* judgments. The probability principle is indeed a synthetic *a priori* judgment – if there is any such thing. There is no other principle that could be more justly so called – but it, too, is not a synthetic *a priori* judgment.

For it cannot possibly originate with the knowing subject. It contains an assertion about perceptions, and it is an acknowledged principle of all philosophical systems that the contents of perceptions are independent of the knowing subject. The probability axiom asserts that a certain statistical regularity is to be found in the contents of perceptions. Ultimately, it is an assertion of the form, "If I have seen green with blue seven times, then, on the eighth occasion, I will also see green with blue." But the knowing mind has not the least influence upon whether this will or will not happen. We have here a metaphysical axiom, a belief in the uniformity of the world that cannot be proven and that nonetheless makes a positive assertion about the world. We formulate, in this principle, the most universal feature of the real. It cannot be established upon the basis of the knowing mind; rather, it asserts something about things-in-themselves.

From thence comes the vexing character of this axiom, which Hume was already puzzling over in his day: we must be forever doubting its validity, and yet we cannot escape believing in it. It is the same as with our belief in the existence of physical objects. We possess no logical refutation of solipsism, and yet we cannot believe it; we cannot escape the evidence of reality. The analogy is not a merely superficial one. Both problems contain the same metaphysical axiom, for the assumption of the existence of physical objects can be traced back to the assumption of the probability inference. We shall now present this connection more precisely and show what a peculiar solution it entails for the problem of existence.

Let us revert to the relation between thing and perception. Our primary experience is solely of perceptions, from which we then infer the existence of things. This is, at any rate, the way in which the relation is usually presented. What is the nature of this inference?

Surely it is true that our primary experience is only of perceptions. Thus it is possible to refrain from making any existential assertions and to confine ourselves to those regarding perceptions. To every assertion a_i about real things there must correspond a combination a'_i of assertions about perceptions in which the concept of existence does not occur. For instance, we could substitute for the existential proposition, "The light of the sun provides warmth," the perceptual proposition, "Whenever I notice certain sensations of light I also notice sensations of warmth." If we wish also to account for the circumstance that we sometimes fail to have a direct sensation of warmth, for instance, because we are clothed, we must add another perceptual proposition, such as, "Whenever I note certain sensations of light (the light of the sun) and, at the same time, certain sensations of touch (my clothes), I have a sensation of warmth only after certain other sensations of touch have been added (feeling the outside of my clothes with my hand)." Thus the existential proposition a_i is coordinated with a combination of perceptual propositions that are summarized in a'_i. This procedure must always be possible, even when the connections are more complicated, for the statement of even the most remote scientific conclusions are ultimately to be traced back to perceptions. Even the existential assertion, "Matter consists of atoms," could be replaced by a combination of a large number of perceptual propositions reporting the direct perceptual images of the chemist when he checks the weight of chemical compounds and of the physicist when he observes gas-kinetic, crystallographic, and other kinds of experiments. Ultimately, we can substitute for the entire system a of existential assertions a_i, which embraces the contents of the experiences of daily life and of the whole of science, the system a' of assertions a'_i, which reports only sensations.

These ideas are certainly very old. They appear in the history of philosophy wherever the concept of being has been traced back to the concept of *possible perception.* They were brought to the fore by Mach, who proposed the solution of attributing reality only to the elements of the system a' but regarded the system a simply as offering an abbreviated way of expressing the same state of affairs. The precise formulation employing the concept of coordination must be attributed to Russell, who altered Mach's idea very slightly.[1]

Yet the systems a and a' are still not totally equivalent to one another. It is not enough to construct the system a' as a report of experienced sensations; it is necessary to add the assertion that the law of probability applies to sensations, that future sensations will display the same regularity as those already experienced. Equivalence is not achieved until this transcendental assumption P has been added: once it has, we can write the formula for the equivalence:

$$a \equiv a' + P \qquad (1)$$

If we assign the name 'scientific solipsism' to the notion that only the system a' possesses reality, we can make the following assertion: Scientific solipsism cannot completely take the place of the concept of existence unless the probability axiom is added to it; but when this transcendental assumption is added, it loses its solipsistic nature, for the probability axiom is an assertion that cannot be derived from experienced sensations; it is a metaphysical proposition that contains right within itself the whole problem of a transcendental reality.

The reason this result is so important is that it offers a completely original justification for realism. The reduction of the system a to the system a' has often been regarded as a refutation of the realistic conception of the external world: the 'thing', it is said, is nothing but a conceptual construct, an abbreviation for the combinations of experiences of sensations contained in a', which presents the genuinely real. It is particularly in Mach that such ideas are to be found. But they are incorrect in two respects. First, they overlook the fact that experiences of sensations alone cannot replace our conception of the external world; the equivalence is achieved only when the transcendental probability axiom – the P in our formula – has been added. Thus not even the positivist can avoid making a metaphysical assumption. Second, it is not admissible to transfer the existential character of the elements of the system a to the elements of the system a'. The elements of a' are 'given', but we may not regard this 'giveness' as existence in the objective sense. Things – that is, precisely the elements of a – are what exist. We always connect this interpretation with the existential assertion, and it cannot be a question of correcting this conception, but only of formulating it correctly. This is precisely what is done in our equivalence formula (1). The existential character remains with the elements of a, yet the possibility is created of connecting the existential concept with the given means of a definition. We must only avoid assigning it the simple designation, "Existence is equivalent to giveness;" we require a more complicated logical function. To be sure, we must also not be under the impression that we can now define thing and existence separately by using the equivalence formula (1). But this is not necessary, either. Our definition runs, " 'Things exist' is logically equivalent to 'I have sensations, and the probability axiom applies to them'."[2] We can, then, define 'existence' and 'thing' implicitly. This does not involve a compromise of precision; quite the contrary, this definition gives expression to the logical fact that we never use the concepts existence and thing apart from one another, that it is only their combination in the proposition, "Things exist," that is meaningful to us. The real puzzle of the transcendental is not, however, removed by means of this definition, for it appears on the right-hand side of the equivalence formula (1) in the form of the probability axiom. Nonetheless, the logical progress consists in the fact that the metaphysical element in the problem of existence is shown to be the same one that is contained in the problem of probability.

We have now gained a deep insight into the problem of existence. The obvious answer is to try to *define* the existence of things, which spares us the necessity of providing evidence for their existence. To put it simply, there is no problem as to whether or not the world exists, for we simply refer to that which we experience as the existence of the world. This idea is partially right and partially wrong. It is true that existence can be defined, and defined by being coupled with perception; the obvious procedure, in this instance, is certainly the use of implicit definition that has been worked out in mathematical logic. But it is not true that we can thereby rid the problem of its transcendental character. The totality of perception does not in itself constitute a sufficient basis for a definition of existence. We require, in addition, the probability hypothesis, for not until we have it are we able to make practical use of our perceptions – that is, to lead a practical life; and even from a purely theoretical point of view, $a' + P$ contains all that we are in fact asserting. Thus there is, indeed, a definition of existence, but a transcendental one only.

One further difficulty must be indicated. Our definition of existence has been logically constructed in such a way that the subjective side stands on the right, the objective on the left; by connecting, by definition, the objective to the subjective side, we attain the desired solution. But this definition contains another place belonging to the subjective side; the connecting equivalence sign. The afore-mentioned equivalence is, indeed, an equivalence only for a knowing subject; for none other does the proposition, "I have perceptual experiences, to which the probability axiom applies," mean the same as "Things exist." It is impossible to conceive the equivalence as an 'absolute equivalence', for this would lead us into the error of making the existence of things contingent upon the existence of a knowing consciousness – certainly an inadmissible formulation. We must never permit our solution to force us into any such consequences, for, unlike other philosophers, we are not prepared to reject a piece of primary evidence for the sake of a closed conceptual system. In such a case, we prefer to modify the conceptual system. And here that is in fact possible, for it would be senseless to demand an 'absolute equivalence'. It is impossible to assert meaningfully anything but that the two sides of the equation are 'equivalent for me'. Thus our definition, too, can only signify what existence 'means to me' or 'means to a knowing consciousness'; we cannot, in this instance, demand an 'absolute definition'. This fate is common to all definitions. We cannot, for instance, define what a circle 'means in itself', but only what a circle 'means for a mind capable of rational thought'. The same subjective element is, then, to be found in the equal sign contained in our definition of existence, and it is this insight alone that saves us from the necessity of making the existence of things contingent upon the existence of a knowing consciousness. It is only if the concept, existence of things, is to be meaningfully experienced, that a knowing consciousness must be presupposed.

We can, then, only succeed in giving the concept of existence a derivative meaning, i.e., in coupling it by definition with subjective experiences, by stipulating the restriction that *even the meaning thus given to it is merely subjective*. But nothing further can be required. We could regard this restriction as yet another transcendental element in our definition by placing a transcendent 'meaning in itself' over against the 'meaning for me', but it is questionable whether this would be meaningful at all. For inadmissible as it is to make the *thing* dependent upon the knowing subject, the *meaning*, the *sense* of an assertion must nonetheless invariably be regarded as dependent upon a thinking subject.

We cannot here explore further the question at hand. But the proposed equivalence appears to us to represent an essential result independently of this question. For it demonstrates that the transcendental element in the concept of existence is identical to the transcendental element in the probability axiom; it signals the discovery that *the belief in the existence of things is tantamount to the belief in the validity of the probability axiom*. The problem of existence is thereby given a turn that enables it to escape ever new interpretations by one system after another *ad infinitum*. The symbolizing language – from which the systems cannot escape – is replaced by the mathematical concept. The problem of existence is coupled with the problem of probability. From here on in the great mental apparatus of mathematics can tackle this problem, and it is to be hoped that, with the clarification of the concept of probability, which is becoming more and more important for natural science, new light will be shed upon the problem of existence. There are already indications that the concept of probability may have important bearings on the other cluster of problems in our knowledge of nature, which concern freedom, and it is to be hoped that a solution to this problem may be found that will do justice equally to logic and to our feelings of conviction. I simply mention this – I wish generally only to throw out suggestions. Only he who himself has tackled problems with the total dedication of the investigator, not contenting himself with merely touching upon a problem, with merely noting its existence, can comprehend how much can be achieved by means of such a method: a method that sets out upon the basis of the material of scientific knowledge and which does not regard the potent preliminary work done by mathematicians and scientific investigators as inert technical knowledge, but rather draws from it the vital philosophical powers in which it is grounded. And he will perhaps understand that in this way it is possible to succeed – not, to be sure, in building a system – but in solving one problem after another, so that the solution can be used as a building block in the system of ever-advancing knowledge.

NOTES

[1] Bertrand Russell, *Our Knowledge of the External World*, London, 1914.

[2] Note that no probability inference is drawn from the fact of sensations about the existence of things; rather, probability inference takes place entirely on the sensation side and moves from experienced sensations to those to be experienced in the future. On the contrary, we are asserting here that this process as a whole is equivalent to the metaphysical hypothesis of the existence of the external world.

32. BERTRAND RUSSELL

[1929i]

There are those who have called our era unphilosophical and held its achievements to be confined to technology and science. To the more discerning eye, however, this judgment appears unjustified, for it employs the wrong standard. To be sure, we do not find in our day philosophy as it was understood in past centuries, philosophical systems and world views – or where we do find these, they are mere pale reflections of past systems. Instead, the real philosophy of our day has developed along with the positive sciences and can only be discovered if we keep our eyes open for the special form of philosophical thinking that is an outgrowth of the construction of scientific concepts and the mental apparatus of the sciences. This philosophy has, to be sure, not yet become an objective system; for the present it is more a cast of mind than a store of factual results, and its development into a science along the lines of the other sciences is reserved for the future. This is the reason that there is so little awareness of this new philosophy. There are probably a good many philosophically minded professional scientists, but there are few philosophers. The following remarks are dedicated to one of these, the worthiest representative of such a philosophy of the positive sciences: Bertrand Russell.

Russell, the first of whose works date back as far as the 1890's, set out from the ideas of set theory, the mathematical discipline that was founded by Georg Cantor, Dedekind, and Frege and that bears so little resemblance to school mathematics that non-mathematicians would not recognise it as mathematics at all. Set theory posed the logical problem of infinity; taking its ideas as a starting point, Russell studied the question of *what a number is, as such*. This question may sound strange to the layman, who thinks that he learned in school what a number is. But, in fact, all we learn in school is to fiddle around with numbers – an ability poles apart from an insight into the meaning and significance, into the essence of a number. Russell found the surprising answer that numbers are nothing but sets – of a peculiar sort, to be sure. If we take three things – say, an inkwell, a house, and a taxi – they form a set. This is a special set; let us call it a trio. In so designating it, we have asserted something about this set, viz., that it stands in some relation to the number three, but in what does the 'three-ness' of this set consist? Russell gives the right answer: the 'three-ness' is nothing special, existing apart from the set, but is the assertion of a relation of this set to other sets, viz., to other trios. Each of these is 'just as many as' the set of persons, Brown, Jones, and Robinson, or the set of words, one, two, and three. The number three is what is common to all trios, or, as Russell reformulated it with the help of the concept of a set, is the set of all trios: a set, that is to say, the elements of which are already sets.

The significance of this theory lies in its doing away with the special logical entities with which traditional philosophical systems have struggled so often. The meaning of logical systems is reduced to assertions about real things. Everything that naive philosophical desires secretly invested in numbers is relentlessly taken away from them, and all that is left by way of meaning is just what is necessary for the use of these symbols in learning about reality, in managing daily life. Thus Russell's philosophy of numbers must be given a place in the neo-realistic trend that has become characteristic of the philosophy of the exact sciences.

The Russellian conception of number was, unbeknown to him, worked out long before Russell – as, with characteristic and admirable generosity, he later invariably pointed out – by the German mathematician Frege. But it was Russell who made it the starting point for philosophical conclusions. By this means he carried out a logical critique of mathematics which culminated in the thesis that mathematics is nothing but logic – that is to say, a particular and highly complicated nexus of basic logical concepts. In the thesis that he subsequently worked out along with the mathematician Whitehead, presented in their great work, *Principia Mathematica* (Cambridge University Press, 1910-3), he substantially transformed the logical calculus. Modern logical investigations are written in their own formal symbolism without which it would be impossible to get along.

Russell's theory gained still further significance through his handling of the logical paradoxes. He demonstrated that ordinary logic, by means of the conceptual constructs it considers admissible, becomes ensnared in contradictions compared to which the paradoxes of the ancients, such as Zeno's paradoxes, appear quite innocuous. In order to remove these paradoxes, Russell constructed his theory of types, the theory of logical levels, the most important thesis of which is that there exists in addition to true and false propositions a third variety: viz., meaningless propositions.

At first, Russell's ideas only gradually reached Germany and remained here for a time the exclusive possession of a small circle of mathematicians and logicians. More recently, Russell himself has disseminated them through more easily comprehensible writings, especially in *An Introduction to Mathematical Philosophy* (Allen & Unwin, London, 1919), and the large sales of his many works, most of which have been translated into German, show the extent to which his logic has gained a following, even in Germany.

Meanwhile, Russell has not stopped at these mathematical problems. He turned from these to epistemological questions; recently, he has summarised his ideas in the comprehensive work, *An Outline of Philosophy* (Allen & Unwin, London, 1927), in which the presentation is at once rigorous and accessible to the general public. It is, indeed, generally the mark of the truly original thinker that he is able to present his theory without employing an elaborate technical apparatus; even if the theory rests upon the results of the sciences, such a thinker is able to encompass within his view both the essence of the matter and the approaches to it and is thus able to lead the reader without effort from the simplest facts of daily life to the most profound questions of philosophy. This is the very gift that Russell possesses to such a high degree. He sets out from the policeman's response concerning the whereabouts of a street and ends up with the epistemological doubt as to the existence of beings outside the mind, or he uses information in a schedule about the departure time of a train as the starting point for an analysis of the concept of truth. The warm humor which such realistic examples impart to the presentation is thoroughly enjoyed by Russell, and we can discern a secret smile whenever he takes the opportunity afforded by the analysis of such an example to make an ironic aside aimed at academic philosophers or Oxonian legalists.

An Outline of Philosophy is principally concerned with 'man from without' and 'man from within'. To the first cluster of problems belong speech, perception, memory, the process of learning in children and in animals – in short, all those functions that link man to his environment. To the second cluster belong the questions of introspection, of imagination or ideas, of consciousness, of the will, and of other such phenomena. Between these two appears a section about the physical world, while a section on general philosophical questions forms the conclusion.

It is the problem of existence that always stands in the foreground of epistemological discussion for Russell, who attacks it with penetration and with an open mind. His logically sober methods, familiar to us already from his analysis of the concept of number, led him – just as did his realism – to positivist ideas, and he had thoughts about the concept of existence similar to those of Ernst Mach, to whose positions he came close, especially in his postwar works. According to these conceptions, the 'thing', the 'external world', does not exist 'in itself', but is a conceptual construct that must be analysed into its special elements; the elements that constitute what really exists are perceptions, the immediate data of experience. These are the neutral 'stuff' of which all things consist. This turn of mind brought Russell close to the school of the behaviorists with respect to the concept of consciousness; they dispute the existence of a 'consciousness-in-itself' in much the same way that positivism disputes the existence of 'things-in-themselves' and see the real as consisting in individual acts, while renouncing any distinction between the mental and the physical. But Russell goes beyond both schools in that he applies his mathematical and logical methods to this problem with great adroitness, presenting, for instance, the objects of the external world as sets of elements in much the same way that he demonstrated number to be a set of elements.

Here, again, we unquestionably find an essentially novel feature of his philosophy that is perhaps more important than the underlying positivistic thesis itself. Today the conviction is becoming more and more widespread that epistemological problems are only to be solved by using mathematical-logical methods. Precision of formulation is here coupled with deep insights that must remain an eternal mystery to the logic of the old school, wedded as it is to word and grammar. Russell's attack on the concept of a judgment as a proposition made up of subject and predicate, his theory of relations and of definition by means of coordination of assertions proves to be vital for every investigation of the problem of existence, and it is to be expected that the next decades will bring solutions to this ancient philosophical problem that bear the same relation to classic systems of philosophy as the quantum theory of modern physics bears to the atomism of the ancients. To be sure, the layman will have to vastly extend his philosophical knowledge in order to comprehend at all the point and the meaning of the questions, and he will notice once again that humanistic education with its standards constitutes just about the worst possible preparation for understanding the world of today.

Russell is now fifty-five years old, but his great productivity during the past decade shows that his development has not yet come to an end. During this period we have come to know him not only as a great logician and epistemologist, but also as a great social philosopher. Russell's is one of those rare expansive natures that do not find fulfillment through work in a single sphere of knowledge and tackle the questions of life, and even the shape of their own personal life, with the same philosophical earnest. Coming himself from an ancient English noble family whose history extends back as far as the time of Henry VIII and who put forward several famous leaders of the Whig party and several statesmen (such as Russell's grandfather, the Earl of Kingston-Russell), Russell did not hesitate for his own part to break with all tradition and to lead the crusade against the war at the time of the mass murder of the people. He openly advocated pacifism and had to spend some time in prison as a conscientious objector, as having refused to do military duty on grounds of conscience, thereby losing his teaching position at Trinity College, Cambridge. Undeterred by such opposition, he presented the basic principles of his social philosophy in a number of books. After outlining his ideas in *Principles of Social Reconstruction* (Allen & Unwin, London, 1916), he gave a more detailed version of his program in *Prospects of Industrial Civilization* (Allen & Unwin, London, 1923). In this same group are to be included his books about China and Russia, written on the basis of his experiences during his extensive travels, and also the inspired little pamphlet, *Icarus, or the Future of Science* (London, 1929).

It is impossible to characterise Russell's basic political philosophy by identifying it with the program of any particular party. Much as Russell advocates socialism, he is no party man. On the contrary: the great value of his writings consists precisely in the fact that he is as forthright in his criticism of the parties of the left as of those of the right. Not that this makes him a man of the center, of compromise; radicalism and consistency are his outstanding features. His politics span a broader field. He focusses upon ultimate human values; politics and the organisation of the state are not an end, but only a means of achieving a happy life for the whole of mankind. That is why the power of the state, even of the ideal state, ought not to become too great, as Russell points out in *Prospects of Industrial Civilization*. It is far more important to secure the greatest possible freedom for the individual person and to offer him opportunities for worthwhile pleasures and happiness alongside the necessity for daily work. Work does not ennoble life, at least not for the great mass of mankind; the purpose of life is idleness. To be sure, this goal is still far removed from us, and our time is a time of struggle and transition. Four great social forces are at war – capitalism and socialism as forms of industrialism, imperialism and self-determination as forms of nationalism – and the outcome of this state of tension can only be awaited with pessimism. If we are permitted to hope that uplifting tendencies are nonetheless at work, we can only base this hope on a belief in human reason, the influence of which in past epochs, too, is not to be as greatly underrated as is held by the advocates of dogmatic theories of history. "Men who genuinely believe in reason, and at the same time possess a vigorous intellect, have a power over opinion which is incalculable, because it is more lasting than any other power. . . . To save the world requires faith and courage: faith in reason, and courage to proclaim what reason shows to be true. It is not a hopeless task to save the world, but it will never be achieved by those who allow themselves to think it hopeless" [p. 221].

Russell excells in his courage as a prophet, but he also possesses the power to convince to a degree rare in a man of science. All his writings show him to be a master of presentation: his concise, lapidary speech, his dry irony, his straight-forward sentences are as aesthetically pleasing as they are convincing. Observe, for instance, how in his book on China he presents the history of China in somewhat under thirty pages; after reading this description, we really have some sense of China. Or read, in chapter sixteen of *Introduction to Mathematical Philosophy*, about the definite article 'the', of which he says, " . . . I would give the doctrine of this word if I were 'dead from the waist down' and not merely in a prison" [p. 167]. In this chapter he disputes the existence of ideal objects, asserting that "A robust sense of reality is very necessary in framing a correct analysis of propositions about unicorns, golden mountains, round squares, and other such pseudo-objects" [p. 170]. This true writer is to be compared among the ranks of philosophers with his great compatriot Hume, to whom he bears marked similarity both in his style and in the sober sense of reality displayed in his methodology. And indeed, Russell, like Hume, has the power to arouse some thinkers "from their dogmatic slumber".

Surprising as the combination of mathematical logic and a pacifistic and anti-capitalistic ethic found in Russell may appear, it becomes comprehensible when pursued to its psychological roots. He who succeeds in toppling the most ancient traditions in logic and in divesting old ideals of their aura through sober scrutiny will also take a freer approach to the ideals of bourgeois morality and will not be afraid to sacrifice values that other, more tradition-bound natures are unable to give up. That such a freedom of spirit is not tantamount to a lack of conscience is shown by Russell through the earnestness of his character, the sincerity of his conviction, the nobility of his sentiments. He belongs as much to the philosophers of action as to the philosophers of theory, and in this human achievement may lie his greatest significance for an era in which thought and action are all too widely separated.

33. THE PHILOSOPHICAL SIGNIFICANCE OF MODERN PHYSICS

[1930e]

I

Alienation between the world of science and the world of everyday life has emerged in our time with a force never known before. The extraordinary refinement of scientific methods has brought about such a transformation in the stores of knowledge, both in the physical and in the biological sciences, that the non-scientist is no longer able to relate scientific knowledge to his daily experience. He hears, no doubt with astonishment, of the results of relativity theory or biochemistry, nor does he dare to question what is presented to him with authoritative claims to learned scientific truth, yet he is unable to make anything of these reports about alien worlds. He does not see how they are supposed to relate to the objects of experience that he calls world, environment, reality, life, and, no matter how greatly he admires them, he is unable to overcome a feeling of emptiness that prevents him from developing a sympathetic concern for the fruits of science, however sincere his desire to do so. This bewilderment and alienation is the lot not so much of the educated man, who through extensive training has become all too accustomed to divorcing affairs of the mind from affairs of the heart, but of that large class of people who have continued to be excluded from higher education by social privation and who, with a sound instinct, cannot become used to leading a double life, alternating between the world of science and the workaday world.

The situation described is all too familiar, and efforts have continually been made to overcome this tension through special educational arrangements, be they extension courses, study groups, or popular literature. That, to date, these efforts have met with no extensive success must have more than superficial reasons. In fact, we do not believe that anything very decisive is to be accomplished through mere organisation, through the attempt to extend familiarity with the results of science to further classes of people by means of artificial educational efforts. For the source of tension seems to lie rather within the intellectual classes, who themselves have by no means succeeded in bringing technical science and daily life together in some form of internal unity. As long as the educated acquiesce in a double life, the uneducated, who find nothing enticing in such intellectual dishonesty, will continue to regard science as a worthless commodity.

The real root of the cultural tension is to be found in the fact that the intellectual classes have failed among themselves to combine scientific discoveries and the workaday world into a unified world view. Here philosophy bears the preponderance of guilt, for instead of digesting the results of the sciences and transforming them into a new way of viewing reality, philosophers have drawn even sharper boundaries between science and daily life. They have failed to notice that technical science simply carries out a continual modification of those basic concepts that we employ in the very same way in daily life and have founded instead a two-worlds theory. As they were not in a position to give a philosophical interpretation of scientific knowledge, they certified the naive forms of knowledge used in daily life to be necessary forms of thought, *a priori* categories. Having displayed this narrow-minded lack of vision, they were compelled to grant science a special arena of existence in which there was no need to worry about necessary forms of thought – a sort of playing field for mental gymnastics – and which, despite all the frequently expressed admiration for such mental gymnastics, is therefore isolated from truth in the philosophical sense by an unscalable barrier. By describing discoveries such as the relativistic theory of time, non-Euclidean forms of space, or the limitations imposed upon the principle of causality by quantum mechanics as working hypotheses or as conceptual fictions, philosophers have drawn this boundary. Here, then, lies the source of that unfortunate rift, and, with all the diligence in the world, the scientifically untutored will not be able to bridge it unless philosophy, on its part, shows the way to unification. Thus we view the work of present-day philosophy of science not only from the standpoint of its scholarly significance, as a clarification of basic scientific concepts, but also at the same time from the standpoint of society. Seen from this vantage-point, a clarification of basic concepts is at the same time a reinterpretation of out-moded philosophical ideas, and only the disclosure of the continuity between the workaday world and the scientific world will be capable of carrying out that incorporation of the cultural fruits of science; today, we must still suffer from a general inability to bring this incorporation about. It will not do to pronounce the physicists' space or substance or laws things existing by themselves, fictitious entities fundamentally different from those designated by the same words in ordinary life. It emerges all too clearly that for a philosophy to take this view means nothing other than a dogmatic clinging to the space, the substance, and the concept of law which were created by physics three hundred years ago and which have become untenable in the face of the wealth and depth of the knowledge possessed today. Only when this insight has become the common property of leading scholars, only when these leaders cease to construct their philosophy out of the philosophical systems of the past and draw instead upon the scientific theories of the present – only then will the artificial boundary between science and man, between the world of formulas and the world of experience, disappear, and people will come to understand that all those formulas are nothing but an interpretation of experience, an interpretation to which scientists were led by the very same path upon which every one of us treads daily.

We shall endeavor to give here a general outline of the way in which the modern physical view of the world is to be grasped and how it is connected with the concepts of everyday life, once we abandon the claims of the philosophy of the *a priori*. (We shall, however, bypass a corresponding study from the biological standpoint.)

II

To begin with, we shall not dispute the fact that there is a difference between the world-view of science and the world-view of everyday life. On the contrary, we wish to present a clear outline of this difference.

The everyday world is a world of tangible and visible objects about us, constructed of solid matter and conforming to a few simple and reliable laws. From wood, iron, or stone, we are able to construct instruments or machines with which we put these regularities [of natural laws] to use. We know that matter can take the form of a solid, a liquid, or a gas, that there is heat and cold, that there is light, color, and sound; our senses transmit to us vivid information about all these things. This world picture also includes such entities as electricity, which, while we cannot see it, we deal with it so extensively that we have completely forgotten its abstract character; we can produce it in machines, send it through wires, and light the room with it. This world-picture possesses, too, a simple cosmology. There is a sky above us through which the stars take their course, and the whole universe is set in space and time, the basic features of which are familiar and require no explanation for their certainty. For at school we all learned, under the name of geometry and mechanics, those simple laws which appear to sum up the totality of human wisdom concerning space and time.

But if we now step from this naive point of view into the world-view of science, everything is instantly transfigured beyond recognition, as if by magic. Here the uniformly solid substance of the naive world view does not exist; instead, there are nothing but tiny particles that swirl about in a great state of agitation. The still, clear water of a lake corresponds, according to the scientist's conception, to something more like a swarm of gnats in which everything is whirling about every which way; the water has no surface, but only a fuzzy boundary from which particles of water are constantly shooting up into the air, causing others, in turn, to enter the water from the air. And the iron supports of the bridge rising out of the water, considered the symbol of immobility and supportive strength, reveals itself upon closer inspection to be a quivering construction the particles of which wave back and forth like the tiny blades in a clump of grass and are not even held together by any sort of connection, but only by the mutual forces of attraction they exercise from afar. If a train goes over the bridge, we are not to suppose that its wheels touch the tracks; what really happens is that the surfaces of two systems of quivering particles come so close to one another that repelling forces drive the particles apart. In this world of quaking grass and insect swarms there is no light, no colour, no sound. Those, too, are simply phenomena of vibration of a different variety, distinguished from one another only by their frequency; what we ordinarily call light is nothing more than electricity. The distinction between energy and mass has also become questionable. They are fundamentally the same, and since mass is, in turn, basically nothing but electricity, the entire world consists of electricity. The laws of this electrical world of vibrations are completely different from those governing the crude objects of everyday life, which turn out to be laws of averages for a basically far more complicated world which takes on such a simple character only through the large number of participating individual particles. Finally, even the laws of space and time – that is, the most general laws governing this world – are substantially different from those we learned at school: space is curved, the sum of the angles of a triangle is not equal to the sum of two right angles, a straight line eventually joins up with itself, and time takes on a peculiar species of indeterminacy as soon as we attempt to measure it.

These two pictures of the world are genuinely different, and it must be admitted that expecting the non-scientist to believe all that is asking a lot. How does the scientist come to reinterpret in apparently so absurd a manner objects that present themselves in daily experience in such a clear and simple form? Is it not a mere addiction to speculation, an imaginative delight in creating aimless complexities, that drives him to dream up such mental creations? From where does science obtain the right to remove itself so far from the immediate intuitive vision of the world? Can we go on speaking of continuity between the two worlds at all when science loses itself in such heights of abstraction that the full weight of the authority of the scientific profession is required to keep the public from openly declaring it to be nonsensical, absurd, and ridiculous?

If, despite all this opposition, we wish to attempt to prove that one and the same mode of thinking has led the man of the workaday world to the first world picture and the man of science to the second, we will first have to set ourselves the question by what means the scientist actually reached his picture of the world. In replying, we shall have to enumerate four sources from which modern scientific knowledge has sprung.

The first source to be considered is the increased precision of instruments of observation as compared to the powers of simple sense perception. The microscope has secured for us glimpses into the structure at the micro-level, where the naked eye supposes there to be nothing but uniform unstructured substance. The telescope has revealed to us the forms of remote heavenly bodies that appear to the eye to be mere points. Modern technology of measurement can distinguish even one one-millionth of a millimeter or one one-thousandth of a second, and the one-trillionth portion of an electrical current flowing through the filament of a bulb is well within the scope testable by a modern galvanometer. It is hardly surprising that such intensified observations teach us something different from the crude methods of our sense organs.

The second source of our scientific knowledge is experiment. Natural science does not restrict itself to observing nature just as it is; it attempts, instead, to place nature in quite new conditions by artificial interference, i.e., by experiment. Iron rods are stretched and bent so that we may become acquainted with their strength. By sending an electric current through water we discover the elements making up this substance, which to the unpractised eye appears homogeneous. Electricity is also sent through a vacuum in glass rods in order to discover its own real nature. We set a plant into ground from which a certain nutritive salt has been removed and watch what happens to the plant. Thus our knowledge of the world is vastly increased through purposeful modification of natural conditions; we learn in this way about many phenomena we would never have an opportunity to observe if we merely passively waited around.

The third source is the quantitative description of observed connections. That a gas has to be suitably compressed in order to be introduced into a smaller amount of space is a fact familiar in daily life, but the scientist investigates the quantitative law of the relation between volume and pressure. Once he has discovered this law, he adds to it another variable temperature, and constructs with it a more general law. Astronomers, likewise, have reduced the motion of the heavenly bodies to quantitative, i.e., mathematical forms. Similarly, the amplifying valves used in radio transmission have been studied to determine the interdependence of electric currents and electric voltage. Not only is our knowledge extraordinarily heightened by such a quantitative comprehension of natural relations, but new and otherwise unsuspected relations are discovered. Recall, for example, that atomic theory was introduced into modern chemistry essentially upon the basis of the quantitative knowledge of equivalent weights.

Finally, the fourth source is the intellectual penetration into the facts which has had such extensive success in modern science. We no longer content ourselves with enumerating many individual regularities, but try to reduce the multiplicity and encompass the widest possible range of facts with the fewest possible presuppositions. This is actually the procedure we call explanation: to explain, to comprehend means essentially nothing other than to combine under a unified law. An exemplary model is the Newtonian law of gravity, which combines into a single formula the laws of such men as Copernicus, Galileo, and Kepler. It is precisely this part of the scientific process that is impossible in the absense of the process of quantitative comprehension described above. On the other hand, the process of intellectual penetration is the most interesting part of research, for it leads to that construct that we call a world-view.

Summarising this overview of the sources of scientific research, we arrive at the following formula: *the scientist wants to know a great deal more about the world, and to comprehend it with far greater precision, than the man in the street*. No wonder that he reaches different results. The first two sources showed particularly great differences between everyday knowledge and the more precise and detailed scientific findings concerning nature. But the last two sources also revealed insights into nature such as the short concentration span of daily life would never be able to discover. This discrepancy holds not only for the science of the past few centuries, although it is indeed to be found there; it has been renewed with surprising force in our generation, which is experiencing the transition from classical physics to quantum physics.

The result of this transition is that we know today that the universe looks very different at the micro-level and at the astronomical level than it does in the intermediate range directly accessible to our perception. Our knowledge of the astronomical sphere stems principally from Einstein's relativity theory, which demonstrated that the space of the universe has a non-Euclidean structure, that simultaneity as applied to cosmic dimensions rests upon an arbitrary element – that, then, terrestrial measuring rods and clocks would produce basically different spatio-temporal relations if they were used in outer space to make the same measurements common on earth. For the world at the micro-level, this result was established largely through quantum mechanics, which has shown that events at this level lose the continuity which is to be observed at the macro-level and are not even subject to the strict regular connection that we have come to know as the causal principle for the world of intermediate dimensions.

Transformation of this sort would be relatively harmless if they had no more significance than a replacement of the content of ideas such as we have often undertaken in the macro-sphere and in the micro-sphere. That astronomical space as a whole is not filled with the compact matter comprising our immediate surroundings, that matter analyzed at the micro-level consists of tiny separate particles and the solidity of macroscopic bodies is merely illusory – all this would be, relatively speaking, easy to overcome. After all, every traveler's report of the people and objects of distant lands is full of factual oddities, and we must believe them if creditable evidence is presented. But we find it much harder to believe reports of a violation of those customs and traditions that are considered in our part of the world to be inviolable principles of universal human order, and if the reports concerning foreign lands were not repeatedly confirmed by so many witnesses, there would assuredly be among us yet some preachers who would dismiss as impossible certain facts about sexual customs or property relations among primitive peoples. The discoveries of modern physics not only brought about a complete overhaul of the contents of an existing conceptual framework, but also called the conceptual framework itself into question and brought about its transformation. Even our *conceptual* world has revealed itself as suited only to intermediate dimensions. The concepts of space and time had to undergo a transformation for the astronomical world, as did the concepts of substance and regularity for the micro-sphere, that demonstrated the corresponding concepts as previously developed from the world of intermediate dimensions to be approximations applicable solely in spheres of an intermediate order of magnitude. This is a development comparable only to Copernicus, a profound discovery: the discovery that the traditional basic principles of science apply only to intermediate dimensions is the Copernican revolution of our time. It has dethroned a realm of concepts, just as Copernicus dethroned the earth. It shows these concepts to be too narrow, arising solely from the limitations of our own dimensions in the universe.

III

But how is such a transformation possible? Did not the philosophy of the *a priori* compile a series of proofs establishing the immutability of basic concepts for all time?

The ideas of the philosophy of the *a priori* have themselves undergone a number of transformations, but at present we will examine in detail only that form which has been constructed as a continuation, and also a modification, of Kantian ideas. According to this philosophy, the fundamental concepts such as space, time, causality, and substance are methodological principles invariably presupposed by scientific investigation. It must, to be sure, appear doubtful whether they can still be described as true from this viewpoint, since, as methodological principles, they assert nothing about the world of reality but are merely forms for the ordering of human thought. What is certain, however, according to this conception is that scientific knowledge would be entirely impossible without such principles. In gathering knowledge we must order the material of experience in accordance with certain principles. It is for this reason that it must appear impossible ever to refute by experience the principles used for organisation. In order to measure space, for instance, we are obliged to use instruments of measurement the designing and mechanical production of which presupposes Euclidean geometry; hence the attempt to establish the validity of non-Euclidean geometry by measurement always appears self-contradictory. The same holds for the categories, or basic concepts. If, for instance, we wanted to test the principle of causality, we would have to perform experiments that themselves presuppose the causal principle. Thus a similar sort of contradiction seems to appear if we try to infer irregularity in nature. This proof of the immutability of certain fundamental principles of knowledge, which stems from Kant, has played a major rôle in philosophy and is still upheld today by the majority of philosophers. However, modern physics has in the course of its development found the means of refuting this conception.

When we do experiments, our observations always take place in the world of intermediate dimensions. This applies equally to investigations of the expanse of the heavens, in which our data stem ultimately from telescopic pictures or – a yet clearer example – from photographs, and to the world of atomic dimensions, about which we draw inferences on the basis of measurements of pressure and temperature, i.e., of data gained by means of instruments of the intermediate order of magnitude, and also through photographs of specks and streaks, as with crystal scintillations and Wilson cloud chamber tracks. Therefore, it is permissible to employ the conceptual world of the intermediate dimensions in making intellectual use of these means of observation. We might nonetheless infer a quite differently structured world in other dimensions if this world is structured in such a way that the structure in the intermediate dimensions converges with high approximation toward that conjectured by the old hypothesis. To put it more clearly, we know that within small areas non-Euclidean geometry does not differ appreciably from Euclidean, and we therefore are able to continue making calculations for the order of magnitude of astronomical instruments on the basis of Euclidean principles without erring conspicuously even if a form of non-Euclidean geometry applies to the universe as a whole. There is therefore no contradiction involved in inferring the applicability of a non-Euclidean geometry at the astronomical level from the presupposition that Euclidean geometry applies at the intermediate level. The same is true of inferences about the world at the micro-level. We know that laws possessing almost strict certainty result from great numbers of occurrences of elementary processes, even though the elementary event itself is subject to the uncertainty of probability. But our observations of quantum events are themselves carried out with the assistance of instruments of intermediate dimensions, i.e., with the help of processes in which a large number of elementary events participate, and we are therefore allowed to presuppose the concept of strict causality in the theory of these instruments while nonetheless inferring deviations from strict causality at the micro-level without incurring a contradiction. For instance, when we observe interference phenomena with a telescope, we may calculate the path of the ray within the telescope in accordance with the strict laws of geometrical optics without taking account of the essentially quantum-like process of the propagation of light, with its separate collisions and irregularities, and no contradiction is involved in our inferring the probable nature of the elementary quantum event upon the basis of such observations. Even when we infer the statistical nature of events in the kinetic theory of heat, we make use of the thermometer and the manometer, changes in which are, viewed strictly, only results of the averages of elementary processes which we treat as thermal and mechanical processes in accordance with macroscopic physics.

Thus the procedure that physics has introduced for the modification of its presuppositions may be described as *a process of continuous extension*. It is possible to describe the universe as having essentially different structures at the astronomical level and at the micro-level as long as the described structures are such as to converge with high approximation with the world of intermediate dimensions. This convergence is realised for the large-scale world by adoption of the previous structure as an infinitesimal principle, i.e., as applicable to the small-scale world. Conversely, the previous structure must be accounted an integral principle for the small-scale world, i.e., as applicable to the large scale. Typical applications of this idea are to be seen in the Riemannian construction of non-Euclidean geometry from the postulate of Euclidean geometry taken at the infinitesimal level, and in Bohr's complementarity principle, which asserts that an increase in the size of quantum numbers is accompanied by a gradual transition from the quantum model of the atom to the classical model, i.e., the model patterned after the intermediate dimension. Many other examples could be named. In carrying out the process of continuous extension, physicists have discovered a powerful instrument of conceptual generalisation of which they make constant use in their new theories and to which they are indebted for the startlingly novel features of their world picture.

IV

This clarification of the possibility of introducing new basic concepts into science turns out at the same time to have substantial bearing on the explanation of the psychological reasons for the unconditional claims made for the traditional basic concepts.

For the considerations presented here demonstrate clearly that the basic concepts of the past were only able to arise in the minds of men because the world surrounding them displays certain general types of relations that suggest application of these concepts. For instance, it is a characteristic of rigid bodies that they conform, when measured in areas of intermediate dimensions, to certain laws consequently formulated by Euclid as the axioms of geometry; the intuition of Euclidean space, then, arose from a preoccupation with the objects of the immediate world, which display corresponding laws. The concept of strict causality, too, was urged upon the human mind by the great regularity seen in the spatio-temporal changes of macroscopic objects, or mechanics. If, over and above this, we were to postulate these concepts to apply to the astronomical world and the micro-world, we would be making an erroneous extrapolation. But if we go on to conclude that these worlds are inconceivable because they do not conform to the old concepts, we fail to recognise the origin of these concepts in the physical world and become guilty of erroneously interpreting physical fact as conceptual necessity. Our so-called basic concepts are determined by our surroundings; their repeated daily use rendered them familiar, and habituation was taken for conceptual necessity.

Once this has been acknowledged, there is nothing to prevent the development of a corresponding habituation to concepts of another sort. If we lived in an environment in which rigid bodies conformed to the spatial relations of non-Euclidean geometry, non-Euclidean geometry would appear to us just as intuitive as Euclidean geometry. We would have learned to project congruence into space differently from the way we do now and would therefore be able to read off Riemannian geometrical axioms from our intuition of space, just as we are able to read off Euclidean geometry from our present spatial intuition. And if the bodies about us did not realise the strict regularity of mechanics but followed instead the random laws of dice-throwing whenever we attempted to move them by mechanical force, we would soon accustom ourselves to the idea that we cannot, by the production of certain causes, uniformly determine future events, but can only hope for a probable outcome.[1] There is, then, no fundamental difficulty involved in infusing the more general conceptual structures of present-day physics with an intuitive content in just the same way it can be done for the basic concepts of classical physics. Nothing more is required than the filling in of the corresponding events in our environment, for it turns out that for the modification of our viewpoint such events need not even be realised; a mere intuitive imagination of corresponding experiences is sufficient to induce the mental adjustment. Or at least this intuitive rendition of such unfamiliar experiences is sufficient for those whose intuitive capacity has been sufficiently schooled in abstract thinking and whose will is firmly committed to making the change.

The epistemological critique of the so-called *a priori* concepts shows not only the possibility of logically overcoming them but also the path to the intuitive realisation of new concepts. This idea constitutes the very nucleus from which the world picture of modern physics must be constructed. It is not true that physics, as constituted by its latest results, is not intuitive; we succeed in making physics intuitive at both the astronomical level and the micro-level if we simply imagine events within the world of the intermediate dimension in ways corresponding to the behavior of bodies taken in these extreme dimensions.

Although these ideas stem, indeed, from the sciences, they are nonetheless by no means universally acknowledged among professional scientists. Instead, there has grown up among scientists the unfortunate notion that the characteristic of being intuitive is one to be rejected in the interests of exactitude, one representing, as it were, a requirement of the layman from which the scientist must free himself. Thus physicists have erected numerous warnings along their path into the new conceptual world, forbidding intuitive visualisation at critical junctures and promoting the belief that true physics now consists entirely of mathematical formulas. This system of prohibitions has doubtless proved successful in practice, for it has prevented obsolete intuitions from inserting themselves in the new conceptual world with unjustified claims to validity. The construction of a new form of intuition can generally succeed only once the conceptual structure is firmly established; there are ways to circumvent traditional intuitions by means of concepts before pressing ahead to new intuitions. The development of modern physics is convincing evidence for this psychological fact. Nonetheless, the system of prohibitions is of merely temporary significance; it is a working principle, not a destination, and ultimately we must try to reach a situation in which everything can be rendered intuitive just as it can be formulated conceptually. We ought not to content ourselves with the thought that the deviations in the world at the astronomical and micro-levels have so little influence upon the sphere of intermediate dimensions that the alteration of intuitive images is superfluous. If we do, we will not achieve genuine intuitions, but will instead fail to grasp intuitively those differences that become noticeable only in the extreme dimensions. If only for this reason, we must maintain a more extensive possibility of intuitive visualisation, for it does not seem to be impossible in principle that such modifications will some day have a noticeable influence upon the world of intermediate dimensions as well. It is, then, in principle possible to depict the corresponding modes of behavior for the world of the middle dimensions, but at the same time that is all that can be required of an intuitive visualisation, all that can meaningfully be called intuition.

At this point, the philosopher's train of thought coincides with the tendency of the layman to transform science from an abstract system of mathematical formulas to a concrete system of events that can be intuitively grasped. It is not this demand of the untutored that divides him from science, for the demand is justified: it is just that the results of carrying it out look fundamentally different from what the non-professional would have liked to believe. And yet there is something so intuitive, so convicing, about the depiction of possible experience that it can win over even those who are separated from the exact sciences by their ignorance of mathematical formulas. The alteration of our way of thinking really requires no more than a readiness to recognise every-day experiences as only relative to prevailing conditions – a readiness for which the all too rapid development of modern technology has already successfully paved the way. Not physics, but the man who denies that modern physics can be intuitively visualised, be he physicist or philosopher, bears the guilt if the layman knocks in vain at the gates of modern science. It is, then, in the solution of the problem of intuitive visualisation as it is carried out in modern philosophy of science that we find the sole source from which a unification of science and the workaday world can spring.

V

There still remains to be refuted one last significant objection on the part of the philosophy of the *a priori*.

It may be – so runs the argument – that physics possesses the capacity to overcome *a priori* conceptions, but has it the right to make use of this ability? Would it not thereby call into question the value of all its results as knowledge? Knowledge, it is said, does not mean the blind construction of a system of formulas, but the ordering of the material of knowledge into pre-established, definite forms, the *a priori* categories. If physics ignores these categories, its results will at best be a fiction possessing practical usefulness, but they will never be the answer to the question as to the essence of the universe, will never constitute genuine knowledge.

This idea is characterised by its presentation of certain criteria of truth, unified into a system of categories, the collective fulfillment of which alone can bestow upon the physical picture of the world the character of truth. But it rests upon an error, as can easily be demonstrated.

It is impossible to predict *a priori* whether or not the system of categories, during the course of its continued application to the objects of our experience, will ultimately lead to contradictions. To be sure, there will always be objects of experience that are compatible with the system; that is clear simply from its historical origin as the mode of thinking of physicists of past centuries. But what remains to be proven is a far more sweeping requirement: namely, that the system be compatible with all *conceivable* objects of perception. While the execution of such a proof in a fully universal form appears scarcely possible, it is, in contrast, extremely easy to find individual cases that refute the claim. We need only imagine a perceptual world in which contradictions occur whenever we attempt to maintain the totality of the *a priori* concepts. Conceivable types of experience of this nature have been constructed; most notably, it is demonstrable that for such imagined empirical states the categories of normal causality and the categories of Euclidean geometry come into contradiction. Thus the idea propounded by the philosophy of the *a priori* that knowledge is essentially bound to a system of categories is untenable, for it leads to contradictions.[2]

But how is it possible, even in artificially constructed worlds, for experience to offer such sweeping decisions? Experience, after all, presents us solely with individual items of observational data. Is there no possibility of rescuing the scheme of categories by introducing latent causes or hypothetical forces? This possibility would indeed exist, were it not that natural science has at its disposal the principle that such artificial assumptions may be rejected – that is, the principle of induction.[3]

Admittedly, this principle must be presupposed in order for our proposed proof of the possibility of contradiction within the system of categories to be conclusive. But induction is absolutely unavoidable, for it signifies the real means whereby truth is judged in science. If we were to forfeit this principle, science would become arbitrary and any assertion whatsoever concerning physical nature would be compatible with a given set of observations.

For it is only the principle of induction that permits the inference of the general law from the facts. We observe, for instance, that a current sent through a wire loop deflects a magnet and conclude that this will always occur, or we notice that energy has not increased or decreased in any of the phenomena known to us to date and conclude that energy remains constrant in absolutely every case. This inference may also occur in more complicated conceptual relations if the store of observations is embedded in a larger theoretical framework and thereby confirms indirectly presuppositions of quite another variety. For example, we regard observations of the deflection of light on the sun and the perihelial displacement of Mercury as constituting indirect confirmation of Einstein's assertion that non-Euclidean geometry applies to outer space. This, too, is an inductive inference, as a more general law is inferred from a store of observations, yet here the law is not simply the generalising of what has been observed to extend to other similar instances, but also the transition to a more fundamental law.

Without the principle of induction, it would certainly not be possible to infer general laws from observations. For the individual case observed could always be the exception; for instance, that the magnetic needle is deflected every time an electric current is sent through the wire coil could be a merely accidental coincidence of events not representing any law. It could also be a mere coincidence that the perihelial displacement of Mercury predicted by the Einsteinian theory of gravity equals precisely 43 degrees, the exact amount found by astronomers independently of Einstein's theory, or that the spectral lines of incandescent materials, insofar as they have been observed, conform to the very laws expressed by Bohr's formula. But the physicist does not believe this for a moment. Indeed, this possibility seems so absurd that no one would entertain it seriously. We simply do not believe in such curious coincidences, especially when the agreement is observed not just once, but in numerous instances, as is the case with, e.g., the spectral lines. The ground of this inference is none other than the principle of induction.

The objection has been raised that in numerous cases, and in precisely those that are significant, physics passes judgment on conflicting theories by means of a single experiment, without waiting until a quantity of observations have been made. According to this view, for instance, only a single demonstration of the interference of light waves would be required to refute the Newtonian theory of emission. But more thorough investigation shows that this constitutes a merely apparent exclusion of the principle of induction. For in attributing to the arrangement of the experiment that significance which is required for the result – in calculating the passage of light through the telescope in accordance with the laws of geometrical optics or in interpreting the results registered by the amperemeter as a measure of the strength of the electric current, for instance – we have already presupposed the principle of induction for past objects of experience. There are many primary objects of experience that have acquired, upon the basis of induction, such a high degree of probability for a certain interpretation attributed to them that we need realise them only once in order to assign a high probability to the assertion of this interpretation. In such cases a single realisation is sufficient for the crucial experiment because its interpretation can be supported by a series of primary interpretations of its components, which, in turn, contain the principle of induction. If we look still more closely, we find that the principle of induction occurs yet once again as soon as it becomes a matter of attaining quantitative results from experiments. For a great many measurements are required in order to eliminate errors of observation, as this can only be done by means of statistical methods. Thus a modern physical experiment looks substantially different from the way it is pictured by the layman. Michelson's famous experiment, an exemplary model of a crucial experiment, rests, in its most recent version, upon five thousand separate observations. This 'one' experiment then, is really five thousand experiments. In applying the theory of error, we again come upon the principle of induction; here it asserts itself only in the narrow sphere of the regulation of precision, since the greater sphere of more extensive possibilities has already been excluded through the primary inductions in the primary components of the experiment.

At this point we will simply indicate the significance of the principle of induction, rather than embarking upon an investigation of the source from which science derives the right to use inductive inference. For this is one of the thorniest problems of modern philosophy of science. (The present author's thoughts on the subject will appear in the next issue of this journal [1930g].) For the present, let us merely emphasise that the principle of induction is whole-heartedly embraced by the whole of the scientific world and that no man exists who seriously disputes this principle, in daily life or in science.

We have referred to the principle of induction as the step from facts to general law. But wherein consists the claim to universal validity which is associated with laws of nature? Natural science is far from interpreting universal validity as some sort of mysterious claim to metaphysical necessity. Declaring a law to be true is intended to assert only one point: that the law permits inferences concerning future perceptions. This inference may be of a very simple form; if the deflection of the magnetic needle by the electric current is referred to as a law, for instance, what is asserted is that the coincidence of the two events, 'production of the electric current' and 'deflection of the magnetic needle', will be found not only in past observations, but in all future observations. If we said earlier that such a coincidence could not be a mere accident, we now have the more precise determination of this conception; to say that we have here not an accident, but a law, is to say nothing other than that the coincidence will invariably be repeated. The situation is analogous for more complicated connections. If we assert the validity of non-Euclidean geometry in the astronomical sphere, we declare that not only the observed phenomena of light deflection and the displacement of Mercury's perihelion are conclusive, but also a number of further phenomena that can be calculated by drawing on the concepts of non-Euclidean geometry, yet were not directly testable up to now. The claim to validity made by scientific laws consists, then, in a prediction, and the principle of induction, upon which it rests, signifies nothing more than the inference from observed cases to unobserved cases.

VI

Nowhere is the anti-metaphysical orientation of modern science expressed so visibly as in this conception of the problem of validity. Scientific investigation is not concerned with fitting experiences into a fixed schema prescribed by reason, as Kant believed, nor is it concerned with arranging observations in space and time or with categorising them under certain compelling concepts, such as substance or lawful necessity. Its sole concern is the prediction of future experiences on the basis of past experiences. Since modern science wishes to do no more than this, it is in a position to reject the ballast of the traditional conceptions in which the historical development of thought has abounded. The world-picture of daily life still lags behind that of science by the length of this decisive step. For it is filled with graphic descriptions construing nature according to the pattern of human nature: the force of a tensed spring is compared to the straining of a man lifting a heavy burden; the course of the falling stone as described by Galileo's law of falling bodies is likened to the conformity of citizens to police laws; light is viewed as a delicate colored material, just as it appears to our eyes. Such comparisons endow the laws of nature with intuitive contents that are erronously transferred from other realms of experience to the physical world. The primary demand that must be put to anyone who wishes to comprehend the physical view of the world is that he refrain from making such unjustified extrapolations. We may regret this dehumanisation of nature, we may regard it as depriving physical nature of its soul, thus divorcing it from life and rendering it uninteresting. But the physicist does not deal in such concepts, for they measure scientific investigation by standards derived from the world of the poet and the artist, which therefore are meaningful only in another sphere. We must, rather, recognise the inapplicability of such standards to scientific research as a matter of intellectual purity. Precisely those possessing artistic sensibilities should be able to muster the strength for carrying out such a purification; after all, artistic creation is imbued with a moral power of purification of a similar sort. The rejection of emotional viewpoints in science does not mean that we wish to deny the value of the world of the artist; it simply means that we reject the transference of the artist's concepts into a sphere in which they do not belong. The artistic conception of nature possesses its own value. But by the same token, it is unable to achieve precisely what science can achieve: the prediction of future phenomena.

We have called this basic feature anti-metaphysical; we might also describe it as stripping nature of all supernatural qualities. Older peoples imbued nature with life by the imagined presence of gods and demons, for they were unable to conceive of the occurrence of natural phenomena except through the model of men-like creatures who regulate events and the regular decrees of whose wills appear to us to be laws of nature. Though this polytheistic world-picture of primitive peoples may long since have been superseded, filling natural science with conceptual constructs of a metaphysical nature is in substance no different. The space–time concept, the concept of substance, the metaphysical concept of power and law, all unmistakeably anthropomorphic in their origin, are today nothing but graphic embellishments bearing no relation to the experiences that actually form the basis of physical knowledge. Modern science consists solely of these experiences and their order in accordance with a predictive theory. There has perhaps never been a greater revolution in the history of mankind than this gradual transition from the divine nature of the primitives through the metaphysical nature of the philosophers to the more sober nature of present-day physics, which contains nothing but facts and conceptual relations between facts.

To be sure, this picture of the world looks different from that of the workaday world, but it is only in degree, not in kind. For the ultimate criterion of scientific truth, success in the prediction of future phenomena, is equally the ultimate root of all claims to knowledge in daily life; no assertion, no belief, would be maintained in daily life were it not substantiated by the occurrence of experiences confirming it. There are, then, not two worlds containing different kinds of knowledge. There is, rather, *a single* truth, the demands of which determine equally the attitude of the scholar and the layman. While recognition of this criterion of truth compels science to make radically altered assertions about the universe, it combines with the belief in the new views a conviction that, theoretically, anyone attempting to comprehend the events of our experience with the same thoroughness, the same earnestness, and the same openness of mind would attain the same results.

And if the path of science has been a sobering one, the remoteness of which from the traditional worlds of fantasy could not be borne without moral strength, this constitutes a feature which appears in the very same form in the evolution that the world picture of the ordinary man has of sociological necessity been undergoing for some time. For divesting the world of anthropomorphism and magic is a fundamental feature not only of scientific investigation but also of our daily existence; it is the aspect under which we must view the present. Here again, we may regret the loss of emotional values with which past centuries imbued the world, naive and not yet technologically rationalised, but regret will not win them back, any more than magic will entice back our childhood in our later years. But he who feels within himself sufficient vital strength to affirm the present will not direct his gaze to the past when faced with a sobered world. As far as our knowledge of the world goes, nothing counts but the truth; we cannot be bound by a nostalgia for obsolete feelings. And as far as our emotions go, there is still quite enough at which we can direct them and through which we can give them expression. Let our emotions make their impression not upon the theoretical world-picture, but upon life itself.

It is here that the moral possibility, as it were, appears to open up that the layman, too, will make the decisive step in his view of the world that will close the gap which heretofore has isolated him from the world of modern physics. The collapse of traditional emotional values constitutes today the problem for the life of every one of us, for every-day life, and although the revolution in physical science may have had its source in logical critique, its result is simply that science has taken its place in a sociological trend of our times. The caving in of the system of *a priori* categories and its replacement through the sober principle of induction, through the postulate of predictability, mirrors with the utmost clarity the experiential situations of daily life; the same battles between outdated doctrine and new experiences are waged into both arenas. Once we grasp this, we have triumphed over the theory of two worlds. It is just not true that technical science moves in an alien world, an impersonal world with no basis in experience. He who is acquainted with physics only from afar, who sees it as filled with nothing but strange names and mathematical formulas, will, to be sure, come to the conclusion that it is the affair of scholars alone, constructed cleverly and with discernment but without significance for circles of society having other interests and other concerns. Yet there can be no greater misconstruction of physics than recoiling from the tough layer of technical terminology with which it has surrounded itself. Whoever succeeds in penetrating this wall will find behind it a science full of lively investigation, full of inner vitality and the great excitement of finding answers to the questions posed by the mind thirsting after knowledge. If the answers look completely different from what was originally expected, they are not thereby rendered any less valuable. Quite the contrary; it is one of the very greatest achievements that, along with the contents of natural science, new forms are created for the intellectual system within the framework of which nature is confined by the process of knowledge. And it is, perhaps, to be regarded as the most significant result of modern scientific knowledge that the picture of the world to which it has led has at the same time brought to light a new vision of man as a thinking mind, for science has shown us that reason is no rigid scaffolding of logical pigeonholes, that thinking does not consist of the endless repetition of outmoded norms. She has taught us instead that man grows with his knowledge, that he carries within him the capacity for forms of thought of which he could not so much as conceive at earlier stages of his existence.

NOTES

[1] It is not at all difficult to imagine such situations. Suppose that a chair that we shove in a certain direction reaches the intended goal only occasionally, sometimes ending up in completely different places, or that a vehicle propelled by a machine cannot be uniformly steered, but instead describes zigzags, like a Brownian particle under a microscope. In such a world the concept of strict causality would not have evolved; on the other hand, we would have learned to construct mechanisms which on the average, at least, would be somewhat likely to attain their goal. Yet the difference between such a world and the one we know is simply a matter of degree. Our vehicles, too, sometimes fail to reach their goal. We must rest content with probable prospects, and even if the degree of probability is relatively high, we must always contend with a certain number of failures of the sort that would merely occur more often in the world we have been imagining.

[2] Cf. Reichenbach, *Relativitätstheorie und Erkenntnis a priori*, [1920f] and *Philosophie der Raum-Zeit-Lehre*, [1928h].

[3] This form of inference is to be described more precisely as incomplete induction, in contrast to complete induction, which occurs in mathematics.

34. THE KÖNIGSBERG CONFERENCE ON THE EPISTEMOLOGY OF THE EXACT SCIENCES

[1930f]

Last year a conference on the epistemology of the exact sciences was held in Prague in connection with the conference of mathematicians and physicists. This year, the congress of physicists, mathematicians, and natural scientists held in September in Königsberg was made the occasion for a second conference on this subject.

The presentations were grouped around two central themes. The first was the foundations of mathematics. Each school of thought in this area was represented by one speaker. Mathematical logic (Russell) was represented by R. Carnap, of Vienna; intuitionism (Brouwer), by A. Heyting, of Enschede, Holland; formalism (Hilbert), by J. von Neumann, of Berlin; and linguistic critique (Wittgenstein), by F. Waismann, of Vienna. These problems deal with the security of the position of mathematics in the face of the enormous difficulties that have arisen with the increase in the precision of thought brought about by symbolic logic. Against the attacks of intuitionism, which does not extend unconditional recognition to certain important logical and mathematical principles, most notably the law of the excluded middle, Hilbert has defended himself through a new justification of mathematics in which ordinary mathematics is translated into the symbolic language of a calculus which then becomes the object of a meta-mathematical investigation supposed to establish the consistency of this calculus and thereby to justify ordinary mathematics. This transition from material thought concerning content to a purely formal calculus was presented most convincingly by von Neumann. Heyting, on the other hand, posed himself the task of presenting Brouwer's ideas in the most precise form possible, freeing them from the author's original mode of presentation, which was very difficult to comprehend. The Russellian view, according to which mathematics can be completely reduced to logic, was given a clear presentation by Carnap, as was also Russell's theory of types, according to which certain apparently meaningful propositions must be designated as meaningless, as having no sense, if contradictions are to be avoided. Wittgenstein's peculiar notions, according to which logical difficulties spring essentially from a misunderstanding of language, were explained far more lucidly by Waismann than they are in Wittgenstein's publications. As a result of this discussion, it can be established that the difference between the various viewpoints are no longer as sharp as they once appeared to be. Today, for instance, the intuitionists acknowledge the consistency of Hilbertian mathematics, while the formalists on their part admit that their metamathematics essentially follows trends that accord with those of Brouwer. Mathematical logic, on the other hand, unites in itself elements of the two opposing tendencies. A basically compromising mood also prevailed in the free-ranging discussion.

The second theme of the conference consisted of the philosophical questions arising from quantum mechanics. W. Heisenberg, of Leipzig, gave a paper on causality and quantum mechanics. This was preceded by a talk by H. Reichenbach, of Berlin, on the physical concept of truth, which maintained on the basis of a philosophical critique of earlier physical theory that the appearance of the concept of probability in physics actually led some time back to a revision of the physical concept of truth: that two-valued logic, formerly the only logic recognised, was replaced by probability logic, in which a proposition may possess any degree of probability on a continuous scale reaching from 0 to 1. These ideas led smoothly into Heisenberg's talk, according to which strict assertions concerning natural phenomena at the micro-level can no longer be made and must therefore be designated meaningless. These presentations, too, which revealed a remarkable agreement between the results of research in physics and in the philosophy of science, were followed by a lively discussion, in which numerous additional points were clarified.

There were also a few other sessions on other topics; mention will be made here only of the lecture by O. Neugebauer, of Göttingen, on the history of mathematics before the Greeks.

The wide-spread interest aroused among the professionals in various fields by the strictly scientific treatment of philosophical questions presented publicly on a large scale for the first time at these conferences has resulted in plans to continue the conference on epistemology. The complete report of the conference will appear in *Erkenntnis* (Meiner, Leipzig), in which the report of the first (Prague) conference has already been published.

35. THE PROBLEM OF CAUSALITY IN PHYSICS

[1931i]

The transition occurring in present-day physics from the idea of a strict and regular connection between individual events to the statistical lawfulness of mass phenomena is regarded by many as a violation of a fundamental requirement of traditional science, as the beginning of a revolutionary epoch in physics, so to speak. The word has got around that science is undergoing a crisis, and there are even those who believe that the break with the classical tradition of physical investigation which they hold to be taking place is tantamount to a renunciation of science in the true sense, to a surrender of precise investigation conforming to laws to imprecise methods of chance connections between natural phenomena.

That physics is currently experiencing a transformation in the concept of a law corresponding to a transition from strict certainty to probability is certainly true, particularly in its latest body of theory, quantum mechanics. Yet it is wrong to see in this development a break with past scientific methods, a discontinuity in the historical evolution of physical investigation. For closer observation reveals that classical physics itself contains as a vital constituent that basic concept which has emerged as the basic concept of regularity in quantum mechanics, the *concept of probability*, and that the transition to the probabilistic regularity of quantum mechanics simply represents a consistent extension of concepts that are to be found in classical physics but have been obscured through inaccurate interpretation. Knowledge of these relations indeed presupposes a more exact scientific investigation of physical methods than has ever been developed in the academic philosophy that has held sway up to the present time. But as we have some time since come into possession of a new philosophy of science, developing from a branch of mathematical and physical research, we are in a position to carry out just such an investigation. Its execution is very greatly facilitated by a singularly fortunate coincidence. For it turns out that the investigations of causality in this field have from the first employed that very form of the classical physical concept of causality which quantum mechanics uses now; specifically, the universalisation in quantum mechanics of the concept of regularity was foreseen some time ago as a possible direction of expansion. Not long ago, Schlick expressed in these pages[1] the view that empirical research into the concept of causality has given rise to substantially broader forms of knowledge than philosophical critique ever dreamed up. I readily concur with this comment insofar as it touches academic philosophy. However, the very different situation that exists with respect to the work being done by the new scientific philosophers appears to bode so well for cooperation between science and philosophy that I should like to call attention to it right from the beginning of my presentation. This claim will be substantiated by the following discussion (cf. especially Section III).

I

The concept of probability first occurred in physics one hundred years ago. It appeared at that time in two entirely different contexts: on the one hand in the theory of error, which at first was applied to astronomical and geodetic measurements but later extended to physical measurements of other kinds, and on the other in the kinetic theory of gases and liquids, where it was most notably used in the statistical interpretation of the second law of thermodynamics. That, at the time, scientists did not recognise the fundamental significance of this development or realise that it was the same extension of the causal concept that appeared in these two places stems from their clinging with excessive tenacity to an ideal model of physical science they were loath to abandon – that model, in fact, which today has finally been firmly rejected by quantum mechanics. To be sure, the situation at the time did not make discernment of these connections easy for the physicists, especially as the philosophers offered them no assistance whatsoever.

As far as concerns the initial appearance of the probability concept, it appeared relatively easy to defend the ideal model of strict regularity against the claims of probability. The theory of error was interpreted from the first simply as concerning the precision of observation, as required because of the minor variations in the apparatus employed and the limited character of human sense perception. The sphere of the concept of probability appeared, then, to be restricted to the narrow interval within which precision is to be regulated; outside this interval, it seemed, strict causality in the classical sense was applicable. Moreover, even with this narrow interval the concept of probability appeared to hold sway solely because of the limited human capacity for observation. An ideal observer would, it was thought, be able to dispense with this concept within this interval as well. Laplace, who, with Gauss, was the most prominent founder of the theory of error, clothed this idea in its most familiar form in his image of an ideal scientist able to predict future events with absolute certainty without using the concept of probability. He expressly chose to present his idea in his *Philosophical Essay on Probabilities,*[2] intending thereby to emphasise the provisional character of the concept of probability, which he confined to techniques of observation.

With the theory of gases, matters at once became more complex. The concept of probability was no longer restricted to the utilisation of observations. Instead, it assumed a role in the very formulation of physical law: the second law of thermodynamics, previously regarded, along with the first, as a causal law in the strict sense, became transformed into a law of probability. No doubt some scientists expressed even then the thought that the first law, too, and perhaps all causal laws, would meet with the same fate,[3] but others, and probably the majority of physicists, could not make up their minds to recognize the concept of probability as enjoying a status equal to that of the concept of causality and attempted to assign to the probability concept a merely provisional role in the theory of gases as well. Indeed, it appeared possible to offer for the appearance of this concept in the theory of gases an explanation analogous to that offered ever since Laplace for the theory of error. The precise determination of the initial positions of the gas molecules in a system and, likewise, the calculation of their innumerable paths is a task that can never be achieved by human powers; here, too, then, the statistical law could be regarded as an aid called for by human imperfection, and the fiction could still be maintained that the events occurring within a gas 'as such' unfold in accordance with strict causal laws, that here, once again, the ideal scientist could dispense with statistics and also avoid using, e.g., Boltzmann's interpretation of the second law. Boltzmann himself was evidently satisfied with this conception, as is shown by his manifold attempts to establish his statistical principle as a consequence of the basic laws of mechanics; to be sure, his and all successive attempts in this direction failed.

Despite these efforts to rescue the causal picture of the world, there were always a few voices here and there expressing dissatisfaction with the interpretation of the concept of probability as an interim device. The decisive objection that can be brought against any subjective theory of probability representing the concept of probability as a function of human imperfection is that the laws of probability hold good in the objective world as laws of frequency. If the probabilistic account of phenomena arises solely from human limitations, it is impossible to understand why nature should conform to its demands. If we knew *nothing* about the initial positions of the molecules and their paths, then we could make *no* predictions as to the future of the gas system; we would be completely unjustified even in making assertions as to probabilities, for we can hardly expect nature to take such great heed of human ignorance that it will produce predictable events, even though we do not know the initial conditions of the phenomena. If the fact of the validity of statistical laws, which are, after all, precise in a very high degree, is to be comprehensible, it must be that these laws reflect an objective basic feature of natural phenomena, just as much as do the laws of causality. Back in the 1840s, the French mathematician Cournot expressed this idea and responded to the Laplacean concept of the ideal scientist by pointing out that this ideal man, far from avoiding the use of statistical laws, would be a superb probability mathematician as well as an expert in the theory of mechanics and would be as superior to us in his deployment of the mathematical probability calculus as in his use of the differential equations governing motion.[4] In fact, this point reduces every subjective theory of probability to an absurdity, and today it seems scarcely comprehensible that the subjective theory of probability held sway for so long. The objective theory has been increasingly in evidence in more recent advocates of probability theory; I cite here only the works of von Mises and Zilsel, along with my own writings [1929g].

II

The development described so far brought the concept of probability into a position parallel to that of the concept of causality. In addition to the principle of causality, a principle of distribution had been recognised to hold for nature. This principle asserts the existence of regularity in mass events, just as the principle of causality asserts the regularity of individual events. The principle of causality is applicable to such phenomena as planetary motion or electrical events, the principle of distribution to phenomena in which a large number of individual events of a similar kind are combined, as in the theory of gases or the theory of observational error. The next step in this development seems to me to be taken with the realization that these are not separate forms of regularity, but that the two are closely related – that, indeed, the causal principle cannot be formulated at all without making use of the principle of distribution. As I presented this idea eleven years ago in this journal [1920e, Section I], I can confine myself here to a brief summary.

Assertions of causality are distinctive in that they must always take the form of an implication: *If* condition A occurs, then condition B will follow. This is all that is asserted; nothing, of course, is said as to whether condition A in fact exists. Ordinarily, no attention is paid to the consequences entailed by this state of affairs. If we knew nothing more than the implication, it would be totally impossible to employ the principle of causality in even one single case. This stems from the fact that the precondition *A* is never fulfilled in nature. All we can ever do is grasp by observation individual parameters of natural events, while knowing very well that other unknown parameters, 'remainders', also participate in determining the event. If we are nonetheless able to apply the principle of causality to natural events, it can only signify that the implication between *A* and *B* is applicable even when a condition somewhat different from *A* is present. That this is possible – that we can, under certain conditions, apply the implication even when *A* is *not* present – points to an underlying assumption extending far beyond the causal principle.

From time to time we employ the supplementary hypothesis that the connection between *A* and *B* must be continuous. In such cases it is to be expected that the existence of a state only slightly different from *A* will result in a state only slightly different from *B*. But it is easy to see that this supplementary hypothesis is of no use, for we do not even know that a state slightly different from *A* is present. What we know about *A* can never be expressed other than in the form of a proposition about probability: We know with a high degree of probability that a condition slightly different from *A* exists. This is expressed most clearly in the mathematical form of the Gaussian law of error, the curve for which never reaches the abscissa axis, approaching it asymptotically instead. If we predict the consequence *B* despite the fact that we can only make a probabilistic report about *A*, it signifies our recognition of the principle of probability. We predict *B* only with probability, never with certainty. Every causal assertion as applied to the prediction of a natural phenomenon takes the form of a probability assertion. Thus, in every prediction that rests upon a causal law, we proclaim our acknowledgment of the probability principle. We should be glad that at least this means of prediction is open to us; otherwise causal assertions would be devoid of content, as their presuppositions are never strictly fulfilled.

Only by means of this conception is it possible, I believe, to render comprehensible the phenomenon of regularity in nature. It has been objected[5] that the concept of probability used in statistics has nothing to do with that used in the application of causal propositions or – which amounts to the same thing at a more general level – in the assertion of the applicability of physical theories to reality, but this position seems to me to be quite untenable. Indeed, it cuts off every possible route to achieving an understanding of the phenomenon of regularity in nature. It is, by the way, easy to discover the psychological roots of this erroneous conception. It rests upon the schematisation in which the course of events in nature is pictured after the model of a strict sequence, all deviations from which are to be regarded as a reflexion of human limitations. But it is totally inadmissible to describe nature by means of a model such that it is impossible to resolve the idealisation in terms of a convergence procedure. Propositions about the model may not have more or less import than propositions about the convergence procedure. Erroneous ideas about strict causality all stem from the fact that this equivalence has been forgotten and characteristics have been built into the model that have no comprehensible significance within the convergence proposition.

If we apply this fundamental precept to the principle of causality, it takes on the following form: If an event is described by means of a finite number of parameters, the future course of this event may be predicted with probability, the probability converging towards 1 (unity) as the number of parameters taken into account increases.

The concept of probability employed here is, of course, identical to that used in statistics. This fact is usually forgotten because the prediction is limited to certain intervals of precision, enabling us to predict with a high degree of probability the occurrence of the result within this interval without worrying our heads over the magnitude of the degree of probability. Yet it is just as inadmissible to distinguish such approximate estimates of probabilities of a known magnitude as it is to regard the rough comparisons of geometrical quantities often used in daily life as different in principle from geometrical measurement. That the statistical concept of probability is invariably present in physical propositions, which can, therefore, be translated into frequency propositions, becomes clearest if we consider the application of the theory of error. Quantitative probability claims have retreated into a very narrow margin as regards physical propositions, since the probability is very high outside this margin. Within it, however, all the characteristics of probability familiar from statistics are unmistakably in evidence.

It is, therefore, a mistake to see the theory of error simply as a set of rules of thumb regarding precision. In it is displayed, rather, a fundamentally necessary use of the principle of probability. Certainly, if we go on to assert that the concept of probability found in the theory of error is identical with that used in statistics, we are called upon to offer some sort of evidence. The claim can, in fact, be substantiated. For, by employing an idea stemming from Poincaré, we can demonstrate that probabilities pertaining to statistical occurrences such as games of chance can be reduced to the assumption that a certain probability function exists. For instance, if we conceive of the angle of rotation of the roulette pointer, reckoned in multiples of 2π, as a physical factor repeatedly reproduced, this factor will conform, in the frequency of its repetitions, to a continuous probability function. The division into red and black sectors represents a subdivision of the abscissa into equal intervals, and it is easily shown that the equiprobability of red and black can be inferred from this fact. I refer the reader to my earlier article for a more detailed argument for this thesis [1920c]. Let me simply stress once again the fact that by means of it the identity of the presupposition for statistics and for the theory of error is proven. Thus the two historical sources of the concept of probability in physics are merged, and the concept of probability shows itself to be not a disruptive outsider horning in, but a necessary constituent of every proposition about reality, through which, too, the principle of causality finally takes on a comprehensible meaning.

III

Formulating an idea in a conceptually rigorous manner has the advantage of revealing at the same time the conditions under which the idea is applicable. Thus, with the resolution of the causal proposition into the assertion of a convergence phenomenon in which the probabilities converge towards 1, the possibility of generalisation became immediately recognizable. As I already expressed this generalization in a previous article, published in 1925 [1925d, note 2] the reader will permit me to formulate it as I did there.

Whether the probability can in every instance actually approach 1 without limit or whether, instead, limits appear in certain places before 1 is reached is a matter open to dispute. These limits might also turn out to be, in fact, unattainable, so that the proposition that for every level of precision achieved a higher level exists would remain true. Justifiable as such an assumption – which would be confirmed if quantum theory abandoned all attempts at causal explanation and contented itself with the probability jumps of electrons – may appear, we shall not discuss it here, and everything that follows is also compatible with a probability capable of approaching 1 without limit.

A philosophical theory can, of course, never undertake the task of making specific factual predictions as to the future development of scientific fields; it can only point out the *possibility* of more general forms. For this reason I made the more general discussion independent of the existence of the particular situation treated in the afore-mentioned article. Whether a particular situation exists or not can only be decided by means of physical research. In the meantime, however, this decision has, in fact, been made. The existence in principle of a limit for the probability of prediction asserted by Heisenberg's indeterminacy principle (1927) is tantamount to the very idea formulated above.

To be sure, Heisenberg appears to establish the indeterminacy principle upon other kinds of considerations, grounding it upon the reciprocal effects between the instrument of observation and the object of observation. Yet closer inspection reveals that the arguments used by Heisenberg are of the same sort used here. For Heisenberg's indeterminacy principle cannot be established *solely* upon the basis of the fact that, in the course of observations on the micro-level, the instrument of observation affects the object. Disturbances of this nature are also known in macroscopic physics, where physicists respond by including the instrument of observation in the theory and constructing a comprehensive theory of phenomena in which the instrument and the object of observation occupy analogous positions as a matter of principle. Think, for instance, of the correction factors that become necessary with the introduction of a thermometer in the measurement of temperatures. It is no use at all asserting that with respect to the macroscopic level the influence of the instrument of observation is, at least in principle, eliminable. In the first place, this is not even true; in the second place, it is a quite unnecessary dodge, precisely because it is possible to construct a comprehensive theory in which all contributory factors, including those resulting from the instrument of observation, enter in as unknowns from the very first. It is not necessary, for instance, to characterise the electrical field strength by means of a transition to a limit in which the attracted body being tested becomes infinitely small. It is possible to use instead the method of 'integral definition', in which the entire field affected by the body is conceived as a function of the factors to be defined, in such a way that, conversely, these factors can be calculated from the integral effect. The division into instrument of observation and object of observation certainly simplifies physical experimentation, but it is not conceptually necessary for rigorous determinations.

It is, instead, a second vital idea that leads to Heisenberg's indeterminacy principle. This principle asserts it to be impossible to draw a univocal inference from the observed integral effects to the factors present during the observation. This assertion is essentially confirmed by the knowledge that the waves determining an occurrence establish in a peculiar two-fold manner the position and velocity of the particle; the position of the particle is determined by the length of the wave, the velocity by its frequency, and as the length and the frequency of the wave are bound together, for mathematical reasons, by a particular sort of cross-section law, it emerges that only one of these factors can be determined with any degree of precision at any one time. It is, then, the wave principle which, together with the concept of integral definition, leads to the indeterminacy principle.

This principle also includes an interesting variant of the idea that there is, in principle, a limit to predictability. As is well known, Heisenberg showed that any one of the factors under consideration could be determined with the desired precision only at the expense of the other and, further, that even the prediction of future values for any given factor can, through suitable experiments, approach towards certainty without limit, but only at the expense of the other. The restriction of the predictability to a probability limit less than 1 applies only to the combination of all contributing parameters, which is given by the product of the individual probabilities. This does not, of course, involve any fundamental deviation from the idea formulated at the beginning of this section, for the total event is determined only by the totality of all the parameters, and it is precisely this that cannot be predicted with an unlimited degree of probability.

This is the point at which to return to our original comment that the development of quantum mechanics is not to be regarded as a crisis of causality but as an unbroken extension of a line of development reaching quite far back into the past. Our presentation established that the probability concept occupies in principle the same position in classical physics as in quantum physics – that is to say, that the claim of classical physics concerning its applicability to reality cannot even be formulated without using the concept of probability. The difference consists solely in the fact that in classical physics the probability of the prediction of individual events can be increased without limit towards 1, which is impossible in quantum mechanics. This is a difference analogous to that involved in the transition from Euclidean space to Riemannian space. Just as this transition was felt to be a crisis only by those who believed in the *a priori* validity of Euclidean geometry and were confused as to the fundamentally empirical and therefore invariably generalisable character of the principles of knowledge, the present expansion of the concept of regularity in classical physics to include that of quantum physics can only appear to be a crisis to those who have permitted *a priori* dogmas to blind them to epistemological insights. The unbiassed epistemologist will recognise here a natural and continuous line of development.

IV

I will now turn to the objections recently raised by Moritz Schlick[6] against my conception of causality. As I am in fundamental agreement with Schlick with respect to the basic conception of philosophical work, a concurrence expressed most particularly in our high regard for the mathematical sciences as a tool for philosophical research, the explication of our differences is a matter of great importance to me.

First of all, I am happy to see that Schlick, in his presentation, likewise espouses the view according to which scientific propositions are to be seen as predictions and the confirmation of predictions regarded as the sole criterion of their truth (Schlick, p. 150). In pursuing this conception, Schlick has adopted my distinction between descriptive and inductive simplicity, although not making use of the terms I introduced [1924h and 1929g, Section II]. In making this distinction, I intended to encapsulate the variations that are found in physics in the usage of the word 'simplicity'. Descriptive simplicity merely concerns economy of description; thus the metric system is simpler than a non-decimal system of measurement. Inductive simplicity, on the other hand, makes a claim about predictions; if, for instance, we draw the 'simplest' curve through a series of measured points, we are thereby asserting that any further measured points would lie approximately along this curve. I have therefore called inductive simplicity a probability principle.

Here Schlick voices an objection; he asserts that the term *probability*, when used with respect to predictions, means something completely different from the concept that is used in probability calculus and which occurs in statistical physics. As I have always opposed this view, having shown yet once again in Section II of this article that it is untenable, I can now be brief. The reducibility of statistical mechanics to the existence of probability functions demonstrates that the same concept of probability is employed in statistics as in any physical assertion. Any precise astronomical measurement can be transformed at a single stroke into a game of roulette by dividing the error function into intervals and numbering off the outcomes analogously to the divided sectors. The assertion of regularity in the statistical repeated series is nothing other than the purely logical transformation of an assertion as to the presence of certain physical outcomes.

Schlick comes to this conception through his advocacy of the so-called *Spielraumtheorie* of probability. This conception, which hails back to more traditional probability theory (as that of Kries and Stumpf) and which has been recently espoused by Waismann, is nonetheless untenable, as the determination of the degree of probability must, according to it, always remain arbitrary. For a more detailed demonstration, I refer the reader to my own work [1924g, Section 22] and to that of von Mises,[7] who rightly offers a detailed refutation of this view.

Schlick proceeds to object to the principle of distribution, also set forth above, which expresses an assumption regarding the influence of the 'remainders' upon physical events. Schlick takes the view that it is not an assumption that appears here, but a *definition*, viz., the definition of causal independence. In making this remark, Schlick is evidently suggesting that an event is to be called independent of certain physical factors if these factors influence it only within the margin of observational error. No doubt such a definition is in use, but Schlick completely overlooks the fact that the definition does not eliminate the pertinent assertion as to the state of affairs. For it is possible, by singling out a finite number of parameters, to characterize an event in such a way that the remainders influence the event solely in accordance with the laws of probability: this is a *fact*, and its factual nature is not eliminated by the introduction of the designation 'independent' for the relation between the event and the remainders in the case under discussion.

Schlick's further objections are directed against my attempts to utilise elsewhere [1925d] my critique of determinism, and the probabilistic connection between events in the universe offered as a substitute for the distinction between the past and the future. Schlick believes that the characterisation of temporal direction I offer will not hold up, that, on the contrary, only the Boltzmann concept of entropy is able to achieve the required distinction. For my refutation of this view, I refer the reader to another article [1924g, Section 21] in which I show that, if determinism holds, it is impossible to define an asymmetry in temporal direction on the Boltzmannian concept. Schlick goes on (Schlick, p. 161) to cite a remark of mine indicating that we draw a distinction in our actions between the past and the future in that we plan an action for the coming day but not for the past day – a distinction that is meaningless for determinism, according to which the past and the future are determined symmetrically from the standpoint of the present. He claims that I have hereby confused determinism and fatalism. What he means by this remark is beyond my comprehension, for my comments have nothing to do with fatalism, but simply point out the logical fact that the asymmetry of action is coordinated through determinism with a symmetry of the determinacy relation, giving rise to a contradiction – a fact that has been neglected heretofore and with which no deterministic solution of the problem of freedom can be reconciled unless it emphatically denies all meaning to what is obviously a basic tenet of experience. Schlick explains further that "our actions and intentions clearly have significance *only* insofar as the future is determined by them." This remark does not constitute a refutation of the argument I present, for it is irrelevant to it; it is, rather, an argument of an entirely different sort which is supposed to make plausible a deterministic solution of the problem of the freedom of the will. As this argument is frequently brought against indeterministic solutions to the problem of the freedom of the will, I should like to give a brief presentation of the problem from the viewpoint of probability theory.

Needless to say, the indeterminist, too, wishes to maintain the existence of a dependence of future events upon his own actions, as he certainly wishes to have an influence upon future events. But this state of affairs is preserved by probability relations, as the future is still held to be predictable with a high degree of probability. If, conversely, Schlick uses this latter fact as an argument against indeterminism on the grounds that "the remaining uncertainty is so minimal" that the freedom of the will has too little room, as it were, in which to operate, he thereby displays a misunderstanding of the whole problem. First, there is a fundamental difference between high probability and certainty; it is, to be sure, not essential for the practical employment of predictions, but it plays a decisive role in the theoretical interpretation of actions. Second, an indeterminist solution to the free will question naturally cannot directly interpret the indeterminacy of chance events as the freedom of the will. Such naive interpretations of probability relations may indeed have been attempted, but it goes without saying that any theory that is to be taken seriously will define the concept of the freedom of the will in a completely different and far more complicated way. What such a theory will look like I do not know, as we do not as yet possess a serviceable model. Only one thing appears to me to be clear: All earlier solutions of the free will problem have set out from the presupposition of strict causality and have reconciled themselves to it, regardless of whether, as in Spinoza's theory, they interpret determinism so as to favour the freedom of the will or whether, as in Kant's, they believe it necessary to make a leap into metaphysics in order to escape it. Since the fundamental assumption underlying all these philosophical theories has been altered in that physics itself no longer upholds determinism, a whole new situation has arisen. It remains to be seen how philosophers will react to it, but under no circumstances should a dogmatic insistence on former solutions be permitted to interfere. I regard my new method of distinguishing past from future on the basis of the theory of probability relations as a step towards a new approach.

In this connection I would like to go briefly into the objections that H. Bergmann[8] has made to my demonstration of the asymmetry of temporal direction, particularly as Schlick sometimes appeals to these objections. Bergmann has fallen into several decided errors in construing my view. He believes (Bergmann, p. 19) that, in calling the cause–effect relation asymmetrical, I confuse partial causes with causes. In fact, I have always expressly stated that the method I have adopted pertains specifically to the connection between partial cause and total effect on the one hand and between partial effect and total cause on the other. There is not a single place in my writings in which the confusion suggested by Bergmann occurs, as can best be seen from the passage from my work that he cites (p. 21). Bergmann makes a further attempt to refute my idea by replacing the probability implication I propose by general implication and then constructing contradictions from the result. Of course, contradictions arise in this way; the reason is that Bergmann replaces probability with certainty. For probability implication there is nothing contradictory in the fact that a predicted event occasionally fails to occur; such failure is a contradiction only for general implication (p. 23). I wish to note Bergmann's error only at decisive points in his argument; I will not go into his ideas as a whole, for he works entirely within a Kantian framework, and I have set forth my differences with Kantian philosophy in a number of other writings.

V

It has now been shown that the meaning of every assertion about reality is given through a prediction, and in just the same sense that the meaning of every probability assertion is given through a prediction. The inquiry into the meaning and truth of reality assertions is therefore synonymous with that into the meaning and truth of probability assertions. The decisions as to the nature of physical propositions is thus shifted to the philosophy of probability theory, and it becomes easy to understand why this field has attracted such great interest in epistemological circles.

As to the question of the meaning of probability assertions, further study of the views expressed here shows that a probability assertion must be regarded as a prediction of a relative frequency. If the event under discussion is realised repeatedly, it will occur in a certain percentage of all cases. This percentage is called the probability of the event. This formulation is still quite imprecise; for a more precise formulation, the probability must be seen as a *relation* between two events O and P which we call *probability implication* and the degree p of which is designated as the probability of P with respect to O. I have introduced for this relation the symbolic form:

$$(O \underset{p}{\ni} P)$$

thereby opening up the possibility of a probability calculus that is linked to the logical calculus.[9] Suppose, for example, that *O* represents the throw of a die and *P* the showing of side 6. On the frequency interpretation, the quotient of the number of cases *P* and the number of cases *O* in a series of these events will tend toward a limit. This frequency interpretation has long been used by probability mathematicians. Its consistent extension in conjunction with the concept of a limit was first carried out by von Mises, although in a different form from that used here.

Attempts to interpret probability in a different way are doomed to failure, as they are unable to assign an objective meaning to the degree of probability *p*. This is particularly true of the so-called *Spielraumtheorie*, according to which the degree of probability always remains arbitrary.[10] The frequency interpretation is the only one on which it is possible to give a purely extensional interpretation of the degree of probability, i.e., by means of a report characterized solely by the occurrence or non-occurrence of *P*. In my view this is the decisive argument in favour of the frequency interpretation. But it must also be established that this interpretation is consistently used in physics.

The logical difficulties involved in the frequency interpretation are, to be sure, very great. We can never observe more than a finite number of cases, yet the stipulation regarding the limit is an assertion about an infinite series; thus it is never possible to confirm or refute a frequency prediction on empirical grounds. This prevents probability propositions, for we demand of strictly logical propositions that they can be shown to be true or false. These considerations have led various authors to deny that probability propositions are propositions at all. Schlick, for instance, speaks of a 'directive', viz., 'a directive for the construction of propositions'.[11] Insofar as this difference is merely terminological, this linguistic usage would be harmless. However, Schlick overlooks two vital circumstances pertinent to such 'directives' that render them closely parallel to strict propositions, so that including them under the general category of propositions appears justified. Because of these circumstances, I consistently use the term *probability propositions* [1930g, p. 169 ff].

The first point of similarity lies in the fact that we employ for probability propositions a scale corresponding to truth values. That is, we cannot require that probability propositions be true or false, but we can require that they be more or less probable.[12] In place, then, of the true–false alternatives of strict logic, a continuous scale of degrees of probability appears for probability propositions, and for this reason I have used the term *probability logic* for the theory of probability propositions. The existence of such a graduated scale is essential; without it, predictions would be of no use to us. It is this point that Schlick overlooks in his interpretation of predictions, which goes back to Wittgenstein. For Schlick and Wittgenstein, the prediction that the sun will rise tomorrow has exactly the same value as the prediction that a new comet will be observed tomorrow. According to Wittgenstein, we know exactly the same amount in the two cases – namely, nothing.[13] But our actual behavior belies this; whereas we respond to the occurrence of the former as to a highly probable prediction, we do not reckon with the second prediction, as it is highly improbable. Without this differentiation of degree, such 'directives' would be utterly without value. There would be, for instance, no justification for giving preference in the directives to just that case that is most probable; we might just as well issue a directive to always expect the least probable event. That we do not do this, that we would consider such a directive unreasonable, can be justified only if we assign to the prediction a degree of probability in the sense of an objective characteristic that exists just as much as the truth of a strict proposition. In the face of this fact, every attempt to deprive probability propositions of an objective meaning with respect to prediction is bound to fail.

Schlick wishes to escape the constraints imposed by this state of affairs by explaining that "directives [are not] true or false, but good or bad, useful or pointless" (Schlick, p. 155). But he fails to note that not even these *value* concepts can be applied to directives until we know whether the directives are probable or not in the theoretical sense. Why is it useful to consult a schedule before undertaking a journey? Because it is probable that the trains will depart at the times indicated in the schedule. Schlick cannot in any way show why it is *not* useful in planning trips to reckon consistently with a departure time two hours later than that indicated in the schedule; to be sure, it is very improbable that the trains will leave at the later time, but as this improbability of future events has, according to Schlick, just the same import as a high probability (viz., none), there is no reason why a basic principle that invariably reckons with the occurrence of what is most improbable should not be useful. Thus the introduction of value concepts cannot in any way ameliorate the problem of evaluating 'directives'. On the contrary, the value concepts presuppose that the 'directives' are theoretically confirmable, and it therefore seems more appropriate to leave aside entirely the question as to good or bad, which has no place in this theoretical problem.

Admittedly, we are not in a position to confirm our belief that the more probable event will occur on strictly logical grounds. This fact, familiar since the time of Hume, has been the ground upon which advocates of formal logic in particular have questioned the interpretation of probability propositions as justified predictions. However, I cannot believe that this conclusion can be seriously supported. If it is impossible to establish on strictly logical grounds that probability propositions are justified predictions, this just goes to show that strict logic is inadequate for the confirmation of reality assertions. We have no right to deny our assertions about reality all objective meaning as propositions about the future. Quite the contrary: our conviction as to the justifiability of probability predictions is such a fundamental fact that it is impossible to disabuse ourselves of it. Philosophy cannot consist in criticizing fundamental convictions on the basis of preconceived opinions as to the possibility of inferring certain propositions. Rather, we must simply accept such convictions, leaving to philosophy only the job of fitting them into a system. The justification of our belief in strict logic is, after all, in no better a position. Its justification consists in the assertion that "we cannot think illogically".[14]

The second point of similarity between probability propositions and strictly logical propositions is that the former are also confirmable, within their graduated scale, and confirmable as predictions, not just after the pertinent events have taken place. If, for instance, we consider the prediction that in the throwing of dice the relative frequency of side 6 will be 1/6, we will regard a given observed series of a finite number of elements as a probable confirmation or a probable refutation of this prediction according to whether the frequency 1/6 is or is not approximately achieved in the series. Thus we can speak of the *inductive confirmability* of probability propositions. We are not thereby introducing some arbitrary assumption; rather, we are simply formulating the approach taken by every one of us in science as well as in all our daily affairs. Thus confirmability presupposes the principle of induction, which is not, to be sure, logically demonstrable but from which we cannot be dissuaded. It can be shown, by the way, that the contradictory of the axiom of induction cannot be meaningfully conceived – that this axiom, that is, occupies a position within probability logic analogous to that of tautologies within pure logic.

The question might be raised as to the totality of presuppositions required by probability logic. In my afore-mentioned work on the axiomatization of the probability calculus, I demonstrate that these presuppositions may be reduced to a single presupposition, viz., the axiom of induction. No further assumption as to a particular form of regularity or irregularity in nature is necessary. All the other assertions used can be inferred from the axiom of induction by purely logical methods.

In these final remarks, we have been able to give only a brief summary of the state of probability logic. I see in this logic the only possibility of justifying from the epistemological standpoint the collective body of physical propositions as they are actually intended and actually applied. The conceptual constructs employed will undoubtedly seem implausible and uncertain to some of those encountering them for the first time, but a careful analysis of the problems will lead to the conclusion that these conceptual constructs are the only ones that are able to grant epistemological certainty either to the propositions of daily life or to the propositions of science. And this seems to me the only justification that can ever be given of a philosophical theory.

NOTES

[1] Schlick, M., *Naturwiss.* **19**, 145 (1931).

[2] LaPlace, P. S., *Essai philosophique sur les Probabilitès* (Paris, 1921), p. 3.

[3] By now it would be difficult to establish who first expressed this idea. Its appearance in Exner (*Vorlesungen über die physikalischen Grundlagen der Naturwissenschaften* [*Lectures on the Physical Foundations of the Sciences*], 1919, p. 701, which is cited by Schrödinger (*Naturwiss.* **17**, 10 (1929)), occurred at a time when the idea of this possibility had already been the common property of physicists for decades.

[4] Cournot, M. A., *Thèorie des chances et des probabilitès* (Paris, 1843), p. 104.

[5] von Mises, R., *Erkenntnis* **1**, 274 and 280 (1930), and Schlick, *op. cit.*, p. 151.

[6] Schlick, *op. cit.*

[7] von Mises, R., *Wahrscheinlichkeit, Statistik, und Wahrheit* (Springer, Vienna, 1928) pp. 61–78 [trans. as *Probability, Statistics and Truth*, 2nd ed., New York, 1957].

[8] Bergmann, H., *Der Kampf um das Kausalgesetz in der jüngsten Physik* (Vieweg, 1929) [Eng. trans. 'The Controversy Concerning the Law of Causality in Contemporary Physics', in *Boston Studies in the Philosophy of Science*, Vol. XIII, Reidel, Dordrecht, Holland and Boston, 1974, pp. 395–462].

[9] A more detailed presentation on this subject will be appearing shortly in the *Mathematische Zeitschrift* [1932f]. In this article I have carried out a logically rigorous construction of the probability calculus, in axiomatic form, whereby all the logical problems of this discipline are solved. I also demonstrate that the conception of probability implication as a two-member relation represents an abbreviation.

[10] Cf. Section IV above.

[11] Schlick, *op. cit.*, p. 151.

[12] This leads to an arrangement according to levels; we must distinguish probabilities of the first level, second level, and so forth.

[13] Thus Wittgenstein, in *Tractatus Logico-Philosophicus* (Kegan Paul, London, 1922), p. 180, 6.36311: "That the sun will rise tomorrow, is an hypothesis; and that means that we do not *know* whether it will rise."

[14] Wittgenstein, *op. cit.*, p. 128, 5.4731.

36. THE PHYSICAL CONCEPT OF TRUTH*

[1931g]

To make use of a philosophical concept in practical thinking is one thing, to render a pure and meaningful formulation of the concept quite another. Everyone faces the necessity of judging the truth of various assertions, and in daily life, as in science, methods have been developed in accordance with which we more or less consciously make such judgments. But the meaning of truth is not given along with the use of the concepts; it can only be grasped by philosophical reflection upon the sense and meaning of the processes of knowledge. To give you a glimpse into such epistemological research is my purpose today.

Let me make a preliminary remark, which I address particularly to the physicists. When I speak of epistemology, I have in mind something different from large circles of contemporary academic philosophers. That is, I do not understand by the term a discipline that produces knowledge from pure reason and then lays down prescriptions for scientists in the various specialties. Rather, I see epistemology as being inextricably bound up with the exact sciences. For the methods of epistemology can only consist in the analysis of scientific thinking; the sciences must be analysed with the help of logically axiomatic methods if questions regarding the meaning and presuppositions of knowledge are to be answered. The turning from an analysis of the knowing mind to an analysis of the crystallised product of knowledge is the characteristic feature of modern philosophy of science. This recent philosophical trend could only have arisen in a philosophical period of physics such as we are experiencing at the present time. Thus it is that the representatives of this movement who have invited you to this conference on epistemology come from the ranks of physics and mathematics. Yet despite this close connection, a differentiation of the spheres of operation has become unavoidable. The physicists are too heavily burdened with work on physical problems to undertake its epistemological analysis as well. Furthermore, an epistemological approach can prove a positive hindrance for the physicist in that it could, under certain circumstances, hold him back from concrete work – some of

* This paper was read at the Second Conference on the Epistemology of the Exact Sciences, Königsberg, 1930.

which, by the way, could not in its initial stages be justified from an epistemological standpoint. Admittedly, this is not always the case, for sometimes, and particularly at decisive points, the physicist needs epistemological insights in order to continue his progress in physics, a fact that has become familiar enough through the development of relativity theory and quantum theory. Mr. Heisenberg, to whom we are especially indebted for the epistemological investigation of quantum mechanics, will bear witness to this in his lecture, which is to follow mine, and it will turn out that the problem of characterising truth, which I wish to present to you today, is closely related to the research now being carried out in quantum mechanics.

There is one very simple and initially convincing characterisation of truth, according to which truth consists in agreement between the idea and the object. We find formulations along these lines as far back as Plato and as recently as Kant. Since then, epistemological critique has taken up this formulation: the names Helmholtz, Heinrich Hertz, and, more recently, Schlick, are particularly to be mentioned in this connection. That critique has made clear that the term *agreement* is meaningless in this characterisation of truth, as the contents of the idea on the one hand and the object on the other are incomparable. It is only in the mathematical sense that we can speak meaningfully of a representation, signifying thereby a univocal coordination between object and ideal models and asserting nothing as to a similarity between these two completely different elements. Hertz has given the best-known formulation of this idea.

> We create for ourselves mental apparitions or symbols of external objects, in such a way that the consequences of the images necessary in thought are always the images of the consequences of the depicted objects necessary in nature . . . . The images of which we speak are our ideas of the objects; they have an essential point of agreement with the objects, consisting in the fulfillment of the afore-mentioned requirement, but their purpose does not require that they agree with the objects in any further respects. Indeed, we do not know, and have no means of finding out, whether our ideas of objects agree with them in any points other than that one fundamental respect.[1]

Through the introduction of the mathematical concept of a model, the characterization of truth as agreement has been brought into a form which can be grasped by the logical tools of modern epistemological critique. After some investigation, however, it becomes clear that this characterization has severe drawbacks, which we will now compile.

Its first defect consists in joining the concept of truth to determinism. For univocal coordination between images and objects presupposes that there is, in fact, such a univocal agreement between images and objects. This stands out very clearly in the Hertzian formulation, which demands coordination not only between the presently observed images and the objects present but also between the logically necessary consequences of the images and the empirically necessary consequences of the objects. This amounts to nothing other than determinism. But since determinism is itself a theory requiring thorough-going philosophical scrutiny, it is very risky to make such a fundamental concept as that of truth dependent upon its validity.

Second, this characterization of truth is deficient in that it fails to offer any means whereby we can verify the truth of a statement before observation. If an astronomer announces that an eclipse of the sun will occur on a certain day, we wish to know *now* whether this proposition is true, yet this characterization permits us to make this decision only *after* the eclipse has been observed, for it is only then that we can establish whether the logically necessary consequences agree with the empirically necessary consequences. And this renders the characterization inadequate for all practical purposes, for the scientist consistently wishes to know whether or not his claims are true *before* they are confirmed by observation.

Third, given this characterization, universal propositions stand in a yet worse position than other propositions with respect to verifiability. For universal propositions make an assertion about all cases; but since it is never possible to observe more than a finite number of cases, it would never be possible to judge a universal proposition to be true. Thus universal scientific propositions would be completely unjustified.

To what do these objections appeal? It is easy to recognise that the difficulties occuring in them stem from the fact that the propositions of physics are predictions. It is only because these assertions involve the future that the three complications arise. Insofar as we are concerned solely with observation statements, the difficulties in question vanish. Physics, too, includes observation statements, for instance, in the experimental physicist's register. The truth of the statements is easily determined: they assert nothing more than the occurrence of perceptual experiences. However, science never rests content with observation statements, but always goes on to make predictive statements. How are we to verify these predictive statements? The experimental physicist's register may be accurate even though the prediction based upon it is false. The truth of predictive statements, then, is different in kind from the truth of observation statements, and herein lies the great epistemological problem of scientific truth. Up to now, philosophers have obscured the problem, transforming predictive statements into observation statements by imagining a later observer who makes his judgment only after the occurrence or non-occurrence of the perceptual experience. But this is a misconstruction of the problem, for it is a decision *before* confirmation that we require. We need to know how to judge predictive statements as predictions, and transforming predictive statements into observation statements pertaining to a later time does not supply an answer to our problem.

The situation becomes worse still, for closer study shows that, in fact, even the so-called observation statements in physics invariably contain predictions. Predictions are found throughout the reports in the physicist's register. For instance, when he speaks of an observed electric current of 2.4 amperes, he invariably also has in his mind the unspoken idea that the same value of 2.4 amperes would have been indicated had he used a hot-wire ammeter instead of a moving-coil ammeter. To be sure, this thought is not usually expressed, but it is nonetheless always tacitly presupposed. If the experimental physicist did not believe it, he would never be permitted to draw from his experiments those conclusions that he always does draw. To take another example, when the geologist reports that the northern range of the Alps is made of limestone, this 'report' always contains predictions as well, e.g., the prediction that, in boring through the layer of humus in an Alpine meadow, we would come upon limestone. In order to remove this element of prediction from scientific propositions, we would have to go a long way back, and all we would have left would be reports of observations in the epistemological sense, taking the form, "Green with white here and now" or "Tic-toc and light here and now." Even if it is true that all observation statements ultimately reduce to such epistemological reports, the scientist can never restrict himself to these pure reports. He invariably asserts more: indeed, he must, for he wants to include new predictions in his reports.

There is a radical solution to this situation, and some thinkers have actually tried to carry it out. According to this conception, the epistemological report constitutes the real contents of all scientific statements; everything else in them is added by the imagination, is an accompanying image. Thus there are no predictive statements in science. If we give certain statements the linguistic form of predictions, we are simply dressing observation statements in a curious disguise. On this view, the statement, "The sun will rise tomorrow," is tantamount to, "The sun rose yesterday, the day before yesterday, etc." (The meaning is actually supposed to embrace somewhat more, but it includes only assertions about the past, none about the future.) It is true that this conception leads to a rigorous solution to the problem of truth, but only, alas, at the cost of a gross distortion of physical thinking. The problem of predictive statements has indeed been eliminated, but so has the whole of physics. For physics will never content itself with merely stating past experiences. I can therefore see no solution to the problem in this direction. To deny the meaningfulness of a problem because no satisfactory solution has been found with the conceptual means available to date is not acceptable.

For this reason, I have always regarded the problem of truth as a problem of predictive statements. Certainly, the concept of truth for observation statements is also used in science, for it is ultimately grounded upon epistemological reports, but this concept of truth is not sufficient for science. The truth of predictions is the far more pressing issue.

How have past thinkers come to grips with the difficulties involved in the truth of predictive statements?

An auxiliary construction has been utilized, according to which the predictive proposition is true or false 'as such', although our knowledge of it is imperfect. We can, no doubt, approach this ideal truth, but we can never attain it. This is the usual twist given to the problem: the undecidability is transferred to the imperfection of man, while the ideal of strict truth is adhered to with respect to the propositions.

Widespread as this conception is, it is untenable, for in failing to offer any means by which the ideal truth may be approached, it reduces truth to a completely empty concept. If a proposition is supposed to be true in all cases, the fact that it holds true in one thousand cases does not bring it any closer to being true, for it is just as far from one thousand to infinity as from zero to infinity. As long as the concept of truth is not grasped from the standpoint of the concept of approximation, it will necessarily remain devoid of content. For we can never experience the ideal itself, but only a process of approximation; thus we can grasp the concept of truth only if the process of approximation has its own independent meaning, only if it defines the concept of truth. The ideal possesses only the significance of a limit, and just as the limit is not something that exists in itself, but simply takes on the meaning possessed by the process of approximation, the scientific concept of truth can only acquire meaning through the formulation of the method of approximation actually used in science.

This shows nothing other than the necessity of assigning to the concept of probability a position in epistemology superior to the concept of truth. For scientific approximations make use of the concept of probability. The physicist always designates his propositions as more or less probable, and additional experience renders a prediction more probable, but never true. Thus truth is only to be defined as a limiting case of probability. While, in the past, efforts have been made in probability theory to reduce the concept of probability to the concept of truth, the fruitless nature of these efforts has now become unmistakeable. Matters lie quite the other way around: a theory of truth can be given only through a theory of probability.

The reason this fact has been overlooked to date is that a great many physical propositions possess an extraordinarily high degree of probability. Their high probability has been identified with certainty, and many have lost sight of the fact that the equation of the two, quite justified in practice, must never be overlooked in epistemology. This oversight leads to the most wide-ranging effects. Thus philosophers have constructed a model of physical research with the intention of deriving from it an epistemological theory, but they have forgotten that this model is not identical to our actual knowledge and have read off from the model certain idealised features of the epistemological process that have no meaning with respect to the actual method of scientific learning. Past epistemological theories of truth have fallen victim to an hypostatization of the model, and they can only be rescued by a return to the original, actual process of knowledge.

We are faced here with a danger to which all scientific thinking is susceptible. Just as epistemologists have constructed for themselves a model of physical knowledge, physicists have constructed models the over-extension of which has led to erroneous ideas concerning fundamental features of natural phenomena. Newton's idea of a force of attraction between the planets was long regarded as the prototype of all propagation of effects, but we now know that, while Newton's mathematical formulas retain their validity as laws of approximation, the notion of an attractive force is only an accompanying image, having no scientific value. The formalised law attains scientific status only when it is detached from the model, and no matter how useful the model was as a subjective aid when the law was first proposed, continued adherence to it after it has rendered this service merely constitutes a danger, as it can lead to the drawing of erroneous conclusions. Similarly, Boltzmann derived his magnificent theorem from the idea that molecules collide like elastic balls; today we know that the general idea of a statistical solution to the second law of thermodynamics has been upheld, but not the mechanical model with which it was originally coupled. Present-day theories, too, depend along with their conceptual content upon apparent confirmation by means of a model. Thus Einstein's theory rests on the model according to which all the events in the universe are coincidences, the network of which constitutes what is genuinely objective, while the metrical dimensions of the network can be unrestrictedly expanded. This image fits magnificently for the universe as a whole, but the concepts of coincidence and space–time order have become highly problematical at the micro-level. No model possesses more than a limited validity, and there is nothing more dangerous than misjudging the boundaries, for the features of the model capture nature only with a certain degree of approximation, never in its totality. Thus we must be prepared today to forfeit even the last feature common to all previous physical models: the idea of determinism. It has become impossible to maintain even so much as the mere determination of one event by another, whether it be spatially located spheres or spatio-temporal coincidences. This determinacy, too, was merely an idealisation that was useful for many purposes; to postulate it as the ultimate epistemological truth would constitute a completely unjustified over-extension of the model.

The process usually evolves step by step. At first no one is aware of the idealization that has been brought about through the construction of the model. But after a while, curious difficulties begin to arise; the system displays contradictions and insoluble puzzles. In his despair, the physicist clutches at the apparatus of epistemological considerations, and it turns out that the physical difficulty arose from an epistemological bias. Einstein's famous criticism of the concept of simultaneity is one of the finest examples of such a development, and the current attempts in quantum mechanics to criticise the idea of causality have arisen from the pressures of a similar situation. In this latter case, to be sure, the epistemological critique preceeded the physical development, for philosophical analysis of the earlier physics had already established that the idea of causality was too narrow a concept, that a more general regular connection in nature, *probability relation*, needed to be introduced, of which quantum mechanics comprised only one particular, if highly interesting, form [1925d]. Through this anticipation of a later physical development, recent philosophy of science has already been able to demonstrate that it is on the right path in utilizing and expanding physical knowledge for purposes of philosophical discoveries.

But we must now indicate more precisely why the idea of strict causality constitutes an over-extension of the model. We can recognise that it does so when we consider that the laws of physics describe ideal connections that are never strictly realized in nature. We are invariably dealing with descriptions of nature in which we presuppose a certain ideal instance of the event and infer from it the applicability of other descriptions, but causality can only guarantee this inference for us if the ideal presuppositions underlying the conclusion are really fulfilled – and they never are. We must get quite straight the fact that the causal principle, in its strict form, would be a totally empty, because inapplicable, assertion if it could not be conceived of in some form other than that taken from the model. Within the model it is meaningful to infer from the presence of situation A the occurrence of situation B. In reality, however, the declaration of such an implication asserts nothing, really nothing at all, as we are not acquainted with so much as a single case in which situation A is strictly realised. It is of no help to amend the demand for causality by a continuity requirement to the effect that, at least, the approximate situation A determines the approximate situation B, for we do not even know if the situation A is so much as approximately realised. All we know is that situation A is realised *with a certain degree of probability*, and even if we determine situation A only within a certain margin of accuracy ϵ, we can still formulate the assertion in question only with the help of the concept of probability. It is highly probable that situation A lies within the margin of accuracy ϵ – that is all that we know. If we nonetheless infer with probability the occurrence of situation B, then our inference, of course, contains the concept of probability. We usually do not notice this because we generally choose the margin ϵ in such a manner that the attendant probability is very high; for all practical purposes, then, the probability assertions are indistinguishable from assertions possessing certainty. But if we therefore lay down the concept of causality in its usual rigorous form as the basis of physical knowledge, we become guilty of over-extending the model. Certainly, for many physical purposes we can forget the probabilistic character of natural laws and content ourselves with the model of rigorous laws of nature, but if, in epistemological investigations, we neglect the idealisation this constitutes, we thereby make a fatal error, leading to the total distortion of the theory of physical knowledge. Incidentally, even practical physics turns out, upon closer inspection, to contain the concept of probability; it has simply remained hidden within the narrow margin of error. Yet the very form of the Gaussian exponential curve, running, as it does, asymptotically in both directions, should show clearly enough that the falling of the result within the interval of precision can only be asserted with a high degree of probability, never with certainty.

The argument outlined could be objected to on the grounds that, while the concept of probability is always related to the occurrence of certain numerical descriptions such as are carried out in astronomy or geodesy, it nevertheless does not appear in the formulation of physical laws, as these contain no assumptions concerning the initial conditions of the phenomenon. Certainly, it is true that the assumption as to the course of a particular planetary path can be made only with probability; on the other hand, the objection would continue, we can assert with certainty that the planetary path will conform, for instance, to Poisson's differential equation, for this equation is true for every form of initial condition and at the same time accounts for the disturbing effects of the other planets. However, it is easy to show that this escape route fails, too. The physicist will never be able to claim seriously that his differential equations are strictly and precisely true, for the applicability of the differential equations is dependent upon the fulfillment of certain external circumstances that are never strictly present. We now know, for instance, from Einstein's theory of gravitation that even Poisson's differential equation is by no means applicable to every form of mass attraction, that it pertains, rather, to very special gravitational fields such as are never strictly realised. To be sure, we possess in the Einsteinian gravitation equation an instrument of far broader scope, applying to gravitational fields of a much more general nature. Nonetheless, no physicist would claim that this equation is the embodiment of an ultimate, unshakeable truth. We are warned off any such over-extension of the Einsteinian equation simply by the knowledge that the Einsteinian theory of gravitation is a phenomenological theory of matter the applicability of which can, for the time being, be guaranteed only for macroscopic matter, its validity within the sphere of quantum laws appearing highly questionable. We must therefore continue to maintain that even the functional form of the law invariably represents a schematization, just as a schematization is involved in the very choice of parameters and the determination of their quantity. Even the applicability of the laws of nature in the form of differential equations free of constants of integration can be asserted only with probability, for this applicability is connected with the fulfillment of certain conditions the presence of which can be asserted only with probability.

There is, therefore, no possible means of escaping the concept of probability in science. Every physical proposition, be it a proposition about a single particular physical system or the statement of the validity of a physical law in general, is a probability assertion, and thus only a theory of probability can supply an answer to the logical problem of the physical proposition. Some thinkers have objected, at this point, that the probability of physical propositions is of a different kind from the probability of statistical propositions, that a distinction is to be drawn between philosophical probability and mathematical probability, but here they have committed an error rife with consequences. I have presented the untenability of this standpoint in a series of earlier works [1916a, 1920c, 1920e, 1925d, 1929g, 1929k, 1929m, 1931i, 1932f, 1959a3], in which I also demonstrate in greater detail that the concept of probability in these two spheres is one and the same. For purposes of clarification, let me briefly indicate here that the appearance of a probability function in the theory of error as well as in the theory of games of chance reveals the identity of these erroneously separated concepts of probability. Indeed, it seems to me one of the most important facts in natural science that the very same concept of probability is found in the so-called statistical laws of physics as occurs in every other sort of physical proposition. Only the discovery of this fact enables us to frame a satisfactory theory of physical knowledge, for it reduces the problem of physical propositions to the problem of probability propositions. The question of the validity of a physical proposition is transferred to the philosophy of the concept of probability, and with the help of its methods, which have advanced greatly in the meantime, we can achieve the solution of the probability problem in physics.

The significance of this turn in the problem of physical truth will become clear if we now make use of the idea that has been developed to investigate the manner in which probability assertions are treated in physics. For during this process there will emerge the special conditions that led to adoption of the schema of strict causality and, thereby, to that of strict truth, upon the validity of which the justification of this older model is therefore completely dependent.

It turns out that the appearance of probability assertions is always an inconvenience for the physicist and that he has therefore thought up a method whereby he can avoid general probability assertions. This method consists in transforming, by means of certain devices, the probability of a given assertion into that of another, the degree of probability of which is close to 1; the resulting special probability assertion possesses a high degree of probability and can be treated as a proposition possessing certainty. This transformation can be carried out in two ways. The degree of probability can be brought close to 1 either by (1) precise analysis of the individual event, i.e., increasing the determining parameters and refining the functional form, which is the causal method, or by (2) making a transition to large numbers, which is the statistical method.

In these two methods lies the source of that dual character of physical laws which is formulated in the two types of laws, causal and statistical. Yet what has been falsely construed as two completely different forms of physical laws turns out upon closer inspection to be a duality of methods by means of which one probability proposition can be transformed into another one possessing a high degree of probability. The diversity of physical fields results in our choosing now one method, now the other. In the theory of gases, for instance, the second method is selected because the first cannot, in practice, be carried out, while the first is used in celestial mechanics, in which it can be carried out quite well and in which a transition to large numbers is frequently quite impossible. But we must bear clearly in mind that every physical proposition, no matter in what area, is always capable of being treated by either method and that the assertion of two fundamentally different types of regularity is untenable, as these two forms of regularity coincide in the single concept of a probability law.

Here the possibility arises of pointing the way along that path towards generalization upon which present-day quantum mechanics has embarked. For Heisenberg's indeterminacy principle asserts nothing less than that the first, or causal, method cannot be pursued without limit.[2] This is hardly surprising to the epistemologist, for there is no principle upon which the possibility of the first way could be guaranteed. A year before quantum mechanics came into existence, I had already indicated, on the basis of such considerations as have been outlined, the possibility that the future development of quantum theory could lead to the assertion of a limit in principle to the increase in probability in the analysis of individual events [1925d]. And indeed, it appears to be a quite natural development for physical knowledge as here conceived that the restricting of the idea of causality by quantum mechanics has now taken this direction.

The development described here has been criticized from some quarters on the grounds that the alleged limit to the idea of causality is only a limit to our knowledge of nature and does not infringe upon the ideal of a strictly causal event as such. Yet my treatment of causal propositions has surely shown adequately that this solution only seems to salvage causality. The idea of the strictly causal event is meaningful only insofar as it constitutes a proposition about a limit process; where this process is no longer possible, the assertion of strict causality has lost every meaning assignable to it. It is the confusion of the model with reality that is expressed in such conceptions; they overlook the fact that propositions about the model are meaningful with respect to our actual knowledge only insofar as they can be translated into propositions regarding the coordination of the model with reality. Let those who wish to rescue the idea of causality from quantum mechanics by means of so-called philosophical considerations take due note of this. Philosophers are fond of fleeing from precise investigation into the realm of vacuous propositions. Just as Carnap, in complete consistency with positivistic ideas, has recently taken up the battle against such philosophical shadows in other areas, so we must wage it in the sphere of causality. Here, in the sphere of predictive propositions, this means recognising that a meaning is to be attributed only to probability propositions, never to propositions of strict truth.

Some have chosen to regard this limitation of the causal method as a crisis of physical knowledge, but the consideratons presented here should be quite enough to show that the crisis exists only for those who lack an epistemological grasp of classical physics. What is actually taking place is a continuous extension of previous processes of knowledge in every direction. Physical regularity is not denied altogether: a transition is simply carried out to a regularity of a more general type. Anyone who has recognised this epistemological connection will regard the new development altogether as a therapeutic process in which the structure of knowledge is set forth far more clearly than in the earlier, more naive epistemological theory. For clarification, let me remind you of the theory of relativity, the physical and psychological consequences of which became historic once the theory became the common property of scientific research. Only those devoid of comprehension – so we may say today – could see in relativity theory a collapse of physics or speak of "the firm ground giving way under our feet" and "a rejection of objective knowledge". On the other hand, anyone who worked his way into the theory, thought through its basic concepts, and recognised the necessity of the questions it posed acknowledged that its effect was only to make physical knowledge clearer and more secure, for it offered clear-cut concepts and answers in places where the old theory overlooked the problems. The situation is just the same with quantum mechanics and the discovery of the probabilistic character of all knowledge. Only those who fail to think the epistemological problem through to the end and who confuse the constructed model with reality and therefore cling to the dogma that events exist 'in themselves' and unfold according to strict determinism can see in the probabilistic nature of quantum mechanics a renunciation of science. But anyone who has recognised that, while reality may be described by means of models, it cannot in other respects be thought of as having the character of the model, has comprehended that assertions about the real can only be made in the form of connections between the concepts describing it. And this knowledge leads to an insight into the nature of the problem of epistemology of a depth that the older physical theory cannot match.

For this knowledge means nothing less than a transition from the concept of truth used in the physics of models to the probability concept found in an epistemologically sophisticated physics, according to which assertions about the limit possess only the significance of convergence processes and can therefore be formulated only with the help of the concept of probability. There is no truth for physical assertions; probability is all that is attainable. If we nonetheless wish to use the concept of truth, it can play in physics only the role of the limiting case in which the probability is equal to 1. Truth, then, is a special case of the concept of probability. That is why the concept of truth can only be defined through the concept of probability, and not the reverse. More precise analysis of these questions leads from strict logic into a more general probability logic, a transition comparable to the relation of Euclidean geometry to Riemannian geometry. I presented my vision of this probability logic a year ago, at our first conference on epistemology, and may therefore refer to the report of that conference, which appeared in *Erkenntnis* [1930g]. It was my task today to show you the connection between these philosophical ideas and modern quantum mechanics. The intermingling of physical problems with epistemological problems is greater today than ever before – to the benefit, I believe I may say, of both.

NOTES

[1] Hertz, H., *Die Prinzipien der Mechanik* (Barth, Leipzig, 1894), pp. 1 and 2 [trans. by Jones and Walley as *Principles of Mechanics* (Macmillan, London, 1899; reprinted with an introduction by R.S. Cohen, Dover, N.Y., 1956].

[2] The indeterminacy principle has recently taken on a somewhat narrower form; it is always possible to set up experiments in such a way that a certain arbitrarily selected parameter of the event can also be predicted with unrestricted precision. But this invariably occurs at the cost of other parameters: the probability of the prediction cannot be increased towards 1 without limit for the combination of all parameters. As a result, the idea presented above concerning a limitation in principle of the prediction remains unaltered by this formulation of the indeterminacy principle.

37. HEINRICH SCHOLZ' *HISTORY OF LOGIC**

[1931d]

Advocates of a philosophy based upon the exact sciences have frequently been accused of not giving the history of philosophy its full due. This accusation demonstrates a total failure to comprehend the foundations of precise philosophical thinking. For those who uphold a scientific philosophy simply claim that its future development can be brought about, not through reflection on its historical antecedents, but solely through that immersion in the subject itself that is the mark of the exact thinker. That the history of philosophy can nonetheless also be a worthy object of scientific study has never been disputed. On the contrary, we would assert that it is only upon the basis of a familiarity with philosophical issues that it becomes possible to write a history of philosophy which is not merely a colourful series of philosophical fairly tales but is able to recognise a genuine thought whenever one appears amongst the variegated panorama of opinions. Only the author with a firm grasp of philosophical problems will be in the position to compose a real history of ideas. In Heinrich Scholz' brief *History of Logic*, we have proof that such a novel conception of the history of philosophy can be realised. For not only has Scholz shown impressive industry in assembling an extraordinary wealth of historical information, but he has gone beyond this basic requirement to become the first to scan the history of logic using the precise knowledge of modern mathematical logic as a searchlight. Through this method the occasional instances of genuine research and genuine logical knowledge stand out clearly amidst the musty odor of old books, and for the first time we begin to get some idea of what a history of philosophy will look like once we actually have a philosophy.

In an introductory section on the general outlines of logic, Scholz gives an overview of the various forms taken by the concept and contents of logic over the course of time. This is followed by two further sections about the classical form of formal logic and the modern form of formal logic in which two lines of development are distinguished, both extending all the way from Aristotle right up to the present day, nevertheless without occupying equivalent

* [The book reviewed is Heinrich Scholz' *Geschichte der Logik* (Junker & Dünnhaupt, Berlin, 1931), which has been translated by K. F. Leidecker as *A Concise History of Logic* (New York, 1961). – Ed.]

positions. For classical logic had long since reached its apex when Leibniz' inspired idea of a universal characteristic became the source from which mathematical logic was developed in the last century. The development of both classical and modern logic is presented by Scholz in a masterly fashion, and yet with such animation and human insight throughout that the perusal of this historical work is a pleasure even for the person not so concerned with the acquisition of historical knowledge. It is clear that Scholz' historical investigations have frequently led him to be deeply critical of academic philosophical logic, yet the earnest mood in which Scholz undertook these criticisms may be seen in the following passage from his book:

> For it is only what mathematical logic has to say about these matters that is of real consequence for formal logic. And let it be said here, once and for all, that the conclusions reached in this book have not been dictated by logical blindness, still less by logical arrogance. Instead, they find their source in that same sense of responsibility that forced Aristotle to judge as he did the actual logic of his great teacher Plato. It was not easy for him to undertake this critique; we beg the reader to believe that it is no easier for us.

It is also reassuring that this apologist for mathematical logic does not fall into blind adoration of its achievements.

> Nor are we asserting here that a command of mathematical logic as such is enough to make a thinking person into a philosopher. Again and again we come up against this deep-seated misunderstanding in arguing the cause of mathematical logic with non-logicians. We are, rather, simply asserting that a command of modern logic is at least desirable for any philosopher with an interest in scientific philosophical thinking and that adequate practice in the clarity and rigor of logical thought has even become a necessary requirement for such a philosopher. Necessary, without question, but not sufficient! Again, we do not mean to say that the new metaphysics may only be set forth in logical symbols. We are not about to be strangled with *this* noose, either. Rather, we demand of this metaphysics the strictness and exactitude of thought by which the logician is to be recognised even in his work *outside* logic. We demand no less than this, but also no more. That much that is still pompously proclaimed will disappear in the face of this rigor is a gain, not a loss. And no one today can overlook the quantity of new material that will come to light as a result.

To be sure, I should like to believe that this critique does not strike the advocate of symbolic logic in the way that might be supposed from Scholz' remarks. For logicians, too, recognise that along with the logical formula another sort of communication is required – otherwise they would stop writing books in which textual material appears alongside logical symbols. Thus the more important point in the quoted passage appears to me to be the indication that mathematical logic does not represent the ultimate level, even within the sphere of what can be logically formulated. For one vital conclusion of all reasonably rigorous work in the philosophy of physics seems to me to be that logic is not sufficient for the comprehension of physical thought in its totality, and I believe that probability logic, which Scholz mentions in a number of places in his book, unfortunately with some hesitancy, has already progressed a step beyond classical logic. Whether such logical systems of a more general nature are to be designated metaphysics appears to me to be a matter of taste, in which I personally would opt for the other side. But one point is certain: only he who possesses a precise mastery of mathematical logic and a comprehension of its accomplishments will ever succeed in progressing behond it – just as this history of logic could only have been written by a true modern logician.

38. AIMS AND METHODS OF MODERN PHILOSOPHY OF NATURE*

[1931j]

I

The term 'philosophy of nature' has recently acquired new significance. A new scientific philosophy, arising from the results of scientific research and dedicated to rigorous standards, is beginning to take shape. It has offered unexpected solutions to a number of old problems and has raised several novel questions whose existence had been overlooked at earlier times. The status of this new philosophy of nature is the topic of the following investigation.

What is new in this kind of philosophy is its method rather than its aim. Its aim is the solution of certain fundamental epistemological questions, some of which have played a role in the older philosophy, some of which, however, were only discovered in our times. Yet its method is fundamentally different from that of traditional philosophy. The new philosophy refuses to try to solve these epistemological questions by means of the abstract speculation, or pure contemplation, or analysis of reason resorted to by most preceding philosophers. The new philosophy is convinced that it can solve its problems only in close connection with scientific and mathematical research. It has elected as its method the analysis of actual scientific knowledge rather than the analysis of reason.

Any change as radical as that advocated in this programme will necessarity evoke the resistance of older philosophical schools. Those who see in philosophy a super-science that creates its own knowledge out of pure reason, intuition, or similar pretentious sources of truth, those who, owing to their faith in such superscience, look down upon the 'merely partial' results of the special sciences will voice objections which view the close association of philosophy and science as an impediment. Even though modern science has disproved many a venturesome result of traditional philosophy, this fact does not induce the super-scientist to revise his arguments, but only to become more intransigent towards the adoption of the method of analysing science. The defense of traditional ideas serves the purpose of deprecating the achievements of science. It has been contended that there exists a fundamental difference between the conception of knowledge in science and that in daily

*From *Modern Philosophy of Science: Selected Essays,* ed. and tr. by Maria Reichenbach, Routledge & Kegan Paul, London, 1959, pp. 79–108. Copyright © Maria Reichenbach 1959 except in U.S. Copyright © in U.S. by Maria Reichenbach 1959.

life, and that the proper starting point for epistemology is the common sense conception of knowledge. The procedure in science is said to be abstract, to be too far removed from genuine processes of understanding to enable the philosopher to construct a theory of knowledge by means of an analysis of the natural sciences. It is contended that only the methodology of the natural sciences, not the categories of nature, will be discovered in this manner. The product of such an approach will be 'philosophy of the natural sciences',[1] and not 'philosophy of nature'. True philosophy must be independent of science.

We mention this objection to the new philosophy of nature at the beginning of our presentation because it embodies a misconception of the scientific method characteristic of the trend in present day academic philosophy. The inability of traditional philosophers to comprehend the scientific method in its modern, highly complicated form has led them to distinguish the conception of knowledge in daily life from that in science. Of course, we can speak of the common sense method of solving problems, but this method, once stripped of its emotional components and analyzed solely with regard to its cognitive content, does not differ in principle from the method of natural science. The difference is one of degree, and the fundamental concepts that are used, such as the concept of existence or the concept of truth, are the same in both. This fact is not always clearly apparent because the epistemological structure of the complex theories of science is not as primitive as that of common sense knowledge.

Hence, we cannot accept the distinction between philosophy of the natural sciences and philosophy of nature. On the contrary, we believe that any philosophy of nature which refuses to integrate the concepts of science with the concepts of daily life deprives itself of the most reasonable way of gaining an insight into the character of the problem solving process. Precisely because the conception of knowledge in everyday life is relatively simple, it blurs certain essential relations which become obvious in the scientific process. The connections between percept and concept, observation and inference, become clearly discernible only in the scientifically formulated results; conversely, only the analysis of scientific method enables us to understand what kind of thought operations are employed in the solution of everyday problems. To retain the naive philosophy would be to blindly renounce the contributions to logical analysis which are implicit in the solutions of many significant mathematical and physical problems. It would mean renouncing something even more important; it would mean neglecting the results of a process of transformation which human thinking has experienced since the beginning of scientific research. In contrast to the conception of common sense knowledge, the conception of scientific knowledge has the advantage of having kept pace with the development which natural science has undergone from its inception two thousand years ago. It is not true that the general frame of theoretical reasoning is given once and for all, and that the progress of science consists merely in gradually filling this frame with more content. Reason is not a system of ready-made compartments in which the scientist deposits more and more subject-matter; this system of compartments is itself subject to historical change, and will constantly be transformed in such a way that it becomes progressively adjusted to the findings of experience. The change in such basic conceptions as those of space and time, causality, etc., the persistent emergence of epistemological questions in the midst of purely physical or biological research demonstrate the close connection between content and form in natural science. One would deny the philosophical maturation attained in centuries of scientific research if one were to divorce knowledge of nature from the theory of natural science, attaching it instead to more primitive levels of knowledge. If we look a little more closely, we shall notice that the conception of common sense knowledge goes through this same development, though at a somewhat slower rate. Today we count the view that the planets are revolving around the sun among the accepted facts of daily life although the scientists made this discovery only a few hundred years ago and immediate observation shows us the starry heavens as a vault dotted with little lights. The interpretation of these bright little spots as huge masses which have a three-dimensional order completely different from the two-dimensional arrangement in which they appear to the human eye must be regarded as a scientific abstraction. Yet we are so familiar with this interpretation that we have incorporated it unquestioningly into our common sense picture of the world. What the philosophers of the naive world view present to us as theory of knowledge is nothing but the analysis of those scientific findings which have already been assimilated by the educated person of our time. What such a common sense theory of knowledge supplies is, therefore, nothing but the scientific conception of a few hundred years ago. Thus Galileo and Copernicus are acknowledged, Einstein and Heisenberg are rejected. Under these conditions, we should prefer an epistemology which consciously takes its starting point from the natural science of the present and whose philosophic results compare qualitatively with the scientific results of our contemporaries.

The conviction that fundamental philosophic concepts are capable of development is the basis of the new philosophy of nature. Even though we agree with Kant and other philosophers that an analysis of the attainment of knowledge is essential, we differ in holding that only an analysis of scientific knowledge, not an analysis of pure reason, can furnish an answer adequate for our times. The introspection of reason is sterile because it only discovers earlier stages of the development of scientific thinking. Philosophy of nature, therefore, is not concerned with analyzing the potentialities of thought, but rather with analyzing the products of thought, its crystallizations embodied in elaborate scientific theories. Wherever thinking exists in this crystallized form, it exhibits a much greater differentiation, a much higher degree of adaptability, than a critique of pure reason has ever uncovered. The psychological explanation of this fact may still be open to discussion; the fact itself is indisputable. Man's reasoning capacity has attained much higher levels of theoretical achievement in close contact with the concrete objects of natural science than philosophers ever deemed possible on the basis of an analysis of pure reason. Therefore, there are no general presuppositions of knowledge, only presuppositions of particular hypotheses. The fact that there are presuppositions for *every* scientific system led philosophers to the fallacious conclusion that there are presuppositions for *all* of them. The distinction between every and all, which we have to make in this instance, is familiar to us from modern logic. Applied to our problem, it leads to the following formulation: For every scientific system there is a corresponding system of presuppositions, but, conversely, for every system of presuppositions, one can construct a system of possible experiences which would not be consistent with these presuppositions. In modern scientific philosophy, the above formulation replaces the old idea of a system of categories.

The attempt to preserve the concept of *a priori* knowledge in the face of such results, and to establish philosophical laws 'prior to all science' constitutes hopeless myopia. Academic philosophy reveals a complete misconception of the total situation by still clinging to the idea of the *autonomy of philosophy*. A philosophy which, independently of scientific research, engages in formulating laws for all fields of knowledge will eventually come into conflict with science. Today it is rather the *autonomy of problems* that must be recognized as basic in science and epistemology. It has always turned out that an unprejudiced analysis of problems, whether carried through from a philosophical, or mathematical, or physical point of view, furnishes the same results. The problems shatter the constraining frame of every system, and impose their own laws, independent of traditional views. In the wake of such an autonomy of problems, philosophy will inevitably be affected by the process of dissolution, a process which is characteristic of modern science and which has culminated in the extreme differentiation and specialization of the various branches of science. Some philosophers identify this process with a decline of philosophic knowledge; they demand a synthesis, the construction of a comprehensive system as the rationale of philosophy. To us, however, this tendency towards specialization seems to herald a process of recovery for philosophy. The process of differentiation constitutes a transition from the intuitive to the scientific method, from isolated speculation to scientific co-operation. Philosophy changes from being a sweeping world view to becoming a progressive science. These considerations do not exclude the possibility of a systematic synthesis of particular investigations; such an integration is only postponed. Once the analyses of specific problems have been carried through sufficiently, their integration into a system may become possible. But such a system will have quite a different range and depth from those that were constructed in too early an anticipation of systematic unity.

Because of its radical break with the philosophic method of the past, a method which, even in the last century, produced a number of systems, the philosophy of nature renounces any conscious connection with historical predecessors. Even the name 'philosophy of nature' is not meant to awaken historical reminiscences, neither of Schelling, nor of Haeckel, nor of the ancient Ionians. Previously, this name referred only to the subject-matter, not to the method of its treatment. The conception of the philosophy of nature held by men like Schelling or Hegel is diametrically opposed to the new philosophy of nature. It may be of interest to the historian to search for the expression of similar ideas in previous systems; for the new philosophy of nature this question is irrelevant. We do not intend, by this ahistorical procedure, to express contempt for the achievements of earlier thinkers. On the contrary, those who work productively in the field of philosophy have the greatest esteem for the philosophers of the past. Yet we do not seek our problems in the past, but in the living present. Compared to earlier times, the situation has changed so completely that the ancients can no longer be fruitful for us. Of course, we can read all kinds of ideas into the ancient philosophers; these men certainly knew more than their successors have credited them with knowing. But when we discover certain ideas in them today, we are able to do so only because we have rediscovered these insights in our own right. Let us mention an example. We know today that Leibniz had already formulated a causal theory of space and time. In our century, such a theory has resulted from the philosophical elaboration of the theory of relativity. Even though the relevant writings of Leibniz were known, their significance and theoretical content could be grasped only when the causal theory was independently developed on the basis of the theory of relativity [see 1924d]. Today we can look back and admire Leibniz' profound intuition; but even from this genius we can no longer learn anything, although, of all the older philosophers, he probably came closest to holding a modern conception of the philosophy of nature. Let those for whom, as Kant says, 'history of philosophy is philosophy itself' spend their time in studying the writings of the great philosophers of the past. We would rather emulate these great men in another respect; all of them were ahistorical thinkers, and did not care about the opinions of their predecessors. The fruits they have borne have had their effects, have become part of the tradition, and have certainly indirectly influenced the present in its way of thinking. But our modern task can only be performed without consideration of tradition, in close contact with the problems that scientific discoveries pose for the philosopher. It will be the prerogative of later times to decide whether, in spite of this attitude, there is an unconscious historical continuity; this question will probably receive an affirmative answer, yet the focusing upon such continuity is not the task of those who must first create the new philosophy of nature. For the creators, logical connections are important; historical connections will be established as a matter of course.

These general introductory remarks have been designed to characterize roughly the cultural position of the modern philosophy of nature. In the following, we shall indicate the problems with which the modern philosophy of nature is concerned. We shall proceed by treating the problems in relation to the special sciences from which they arose.

II

We shall begin with biological and psychological problems. The age-old philosophical problem of life is still the center of interest. This problem found its most decisive elucidation in Darwin's theory of evolution. Although this theory has become the accepted common property of biologists and is more or less taken for granted, the philosopher should remind himself of the far-reaching effect which Darwin's discovery had on the philosophic problem of life; the theory of evolution has reduced the problem of life to a single problem. All life is uniform because it has evolved in a continuous process from lower forms. There is no independent problem of life for plant, or animal, or man; there is only one problem, the problem of life as such. This simple consequence of Darwin's theory of evolution can at the same time be regarded as an example of how scientific results can influence the formulation of philosophical problems. Let us remember that medieval philosophy assigned to man a unique position among all living things, and that Descartes still made the distinction between man who has a soul and the animal which is a machine. In spite of the fact that Darwin's doctrine has encountered some justifiable criticism among biologists and that some of its shortcomings can no longer be denied, we must not forget what a tremendous discovery is expressed solely in the indubitable fact of descent and how this influenced the philosophical problem of life.

The shortcomings of Darwin's theory are connected with the part that is intended to explain the 'how' of evolution, his theory of survival. Survival through the struggle for existence, although always operative as a regulative factor, is not sufficient to explain the complicated and often useless forms of organisms. Darwin's famous examples, such as mimicry and others, had to be abandoned. Yet it seems to us that this criticism, however important from a biological standpoint, does not affect the fundamental philosophic idea expressed in Darwin's principle of selection. Philosophically speaking, Darwin's doctrine of the struggle for existence constitutes the attempt to explain the development of organisms causally rather than teleologically. The fitness of later forms is no longer regarded as the goal of the preceding development, but rather as the result of purely causal factors in combination with a statistical selection. The specific form which Darwin gave to this philosophic principle, stating it exclusively in terms of the struggle for existence, may be deemed inadequate today, but nevertheless it seems possible to retain the fundamental philosophic idea while correcting its biological content. A great number of the accepted biological theories can be made compatible with this point of view. It is the guiding principle of scientific biology to explain the regularities of organic evolution by causal factors and to reject the view that these regularities can be explained in terms of an anticipated future. In this context, the discovery that the concept of fitness had been defined too narrowly to be adequate for biology is an important one. Not only does organic evolution produce, in addition to useful forms, a number of highly useless ones, but for every evolutionary level we have to speak of a relatively greatest usefulness which is reached in its fittest representatives, and compared to which the so-called higher forms appear as a regression. Fitness becomes a relative concept and can no longer determine the direction of the evolution. It seems rather that a tendency towards differentiation, towards the production of more and more complicated forms, is inherently operative in the organic realm; the structure of these forms must be analysed before organic evolution can be understood.

In recent decades, the idea of vitalism has been greatly emphasized, and Driesch believed that he could demonstrate the impossibility of a causal explanation of life by means of biological experiments. These experiments were of great significance for biological research. When Driesch showed that half an egg of the sea-urchin will produce a whole sea-urchin, he refuted the narrow mechanistic doctrine of the preformation of all future organs at localized parts of the egg. We can still understand the surprise and admiration that was evoked by this epoch making discovery thirty or forty years ago. For the philosophic interpretation of the problem of life, however, these investigations did not have the importance that their author assigned to them. If Driesch believed that he had discovered the specific nature of organic life, he overestimated the developmental structure which he had found. Above all, we must reject as untenable the metaphysical interpretation which Driesch gave to his experiments by introducing the concepts of entelechy. 'Entelechy' means nothing but an imaginary entity that is supposed to account for the purposive tendency of organic evolution. Instead of postulating a determination of events by the future, Driesch postulates their determination by a metaphysical entity, an entity whose entire function exhausts itself in anticipating the future and guiding the organism towards its future aim. But this is only a teleological explanation in disguise. Subsequent biological research has shown that the inference from Driesch's experiments is unjustified. It has turned out that the development of the sea-urchin from half an embryo is in no way determined by any future aim; it is determined by causal factors alone, though this causal determination is more complicated than the mechanists had assumed. We have learned that the future of a single cell within the egg is not determined by this cell alone, but also by the biological environment of this cell and its location in the egg as a whole. Therefore, the same cell of the egg of a sea-urchin will produce something different depending on whether this cell is embedded in a whole egg or in the isolated half of an egg. Today, the biologist can predict exactly under what conditions half an egg will produce half an animal, and under what conditions it will produce a whole animal. We merely face a more complicated form of causal relations; an event in one part of an area is not only a function of the part but of the structure of the total area – this is the biological result that remains of the vitalistic theories. These relationships are not new to the physicist; he has known for a long time that determinations in terms of a totality can occur in stationary or quasi-stationary processes. Incidentally, this fact demonstrates that Driesch's formulations did not establish a difference between biological and physical events.

If we were asked whether by our rejection of vitalism we intend to assert that organic behaviour must be understood on the basis of physical laws, we would reply that the question is too vague to the answered precisely. What does it mean to say that living processes are determined by physical laws? The physicist would be the last to maintain that the known physical laws suffice for an understanding of living processes, since the discovery of a new scientific field has always required the formulation of new physical laws – witness radioactivity. Up to the present time, the phenomena of life have not been studied in physics, and it would be quite remarkable were the known physical laws to contain everything that is necessary for an understanding of life. We might reformulate the problem by saying that the laws governing organic behaviour should at least be of the same type as the laws of physics. But of what type are the laws of physics? As a matter of fact, and we shall return to this topic later, physical laws themselves have changed in type; witness the transition from causal laws to probability laws. There is no reason why the analysis of organic behaviour might not lead to a similar transformation in type. Perhaps we can regard the idea of a determination of all events by the respective present (or immediate past, i.e. by the time derivations of the parameters) as the general criterion of physical laws. Then the present situation in biology as we described it would favour the view that biological laws are also of this type. Of course, one cannot make any predictions in this respect; we certainly do not want to suggest that we wish to exclude *a priori* a determination of events by the future. We merely wish to say that the determination of events by laws of the causal type seems to us one of the properties of nature which has been highly confirmed. We should abandon the causal theory only on the basis of very compelling reasons, for to do so would entail giving up some of the most basic parts of our conception of the physical world, notably those involving the concept of time (see the following section).

Closely related to the biological problems are the philosophical problems of psychology, i.e., the questions concerning the nature of consciousness, a phenomenon encountered rather late in the evolutionary process. A recent development subsumes psychological events under the concept of Gestalt. 'Gestalt' refers to an organized configuration of experiences as represented, for instance, by a melody, which is actually given through a sequence of tones, but is not identical with this sequence (as can be seen from the fact that it is possible to transpose it into another key). In this connection, the principle of superposition is rejected by the psychologists. If one knows the laws governing the elements of a psychological process, one does not necessarily know the total process. The total process is not merely determined by the superposition of the elementary processes (as for instance a complicated sound wave is determined, according to Fourier, by the superposition of its harmonics); special effects occur which are determined by the configuration of the parts. The elaboration of such ideas in psychology is of great interest to the philosopher of nature. He recognizes the similarity of these ideas to concepts that have been used for quite some time in physics and mathematics where they have been analysed precisely. Thus the physicist knows that deviations from the principle of superposition occur in connection with the superposition of interdependent processes. The mathematician employs Gestalt concepts, such as the concept of topological shape, which are defined as invariants with respect to transformations. The concept of structure, used in modern logic, also belongs in this category. Developments in psychology have led to a generalization of the concept of law, a trend similar to that which has been visible for some time in mathematics and physics.[2]

This comparison raises some philosophic questions not yet answered by psychologists. The physicist knows that the occurrence of holistic properties is restricted to stationary processes, that the total structure of such processes can be explained in terms of action by contact, and that for such an explanation one need consider only the immediate neighbourhood which itself has no holistic properties. For instance, the Maxwell differential equations provide us with a description of the way in which a distant conductor causes the potential of a certain point in an electrostatic field to vary through a spreading of the field, and this description makes no reference to any holistic properties. It is interesting to speculate whether psychological Gestalt processes might not also be reducible to a theory of action by contact that would be independent of the concept of Gestalt. We must leave the solution of this problem to psychological research.

Let us briefly mention another problem, the problem of the individual, which has always played an important role in philosophy. Interest in this problem has been stimulated recently by various results of biological and psychological research. We may cite first the strange conclusions of psychiatric investigations, the distinction drawn between the conscious and the unconscious or the diagnosis of a split personality. Furthermore, the problem of the individual has been approached by new points of view biologically speaking. Two salamanders have been joined by an operation so that two individuals become one, having a common circulatory system and even a common nervous system.[3] Imagine what it would mean if such experiments were also successful with higher animals. Contrary to the philosophic pronouncement that the monads have no windows, there would be cases where two monads were completely united, possessing but one inner life, and there would also be all the intermediate stages in which there was a partial union, so that we would have to speak of a continuous range of individuality. It seems difficult to understand how one can still speak of the independence of philosophy from biology. Those who have had some access to biological research notice at every turn its close relationship to philosophic problems. A philosophy of nature that is not aware of this fact cannot do justice to the demands of the present.

III

We shall now turn to those problems in the philosophy of nature that have arisen from physics. We shall first consider the problems of space and time. Due to the theory of relativity and its philosophic elaboration, these problems have recently been subjected to an extensive analysis which has led to certain terminal results; in this area, therefore, we encounter for the first time a complex of problems whose analysis is relatively complete. The philosophic questions that could be raised have been answered. Of course, we say 'relatively complete' because we do not want to predict how later centuries will judge our results; but on the basis of our present knowledge, we can speak of a certain completeness.

We might summarize the results of the recent analysis of the problems of space and time as follows: Space and time are not forms of order which the knowing subject imposes upon reality as previous philosophers believed, but they are properties which indicate an objective order of the physical world and which are observed by the knowing subject in reality. This view constitutes a serious break with the philosophic conception which has been prevalent since Kant's time, but the developments in mathematics and physics have led so decisively to this view that we feel it is justified. We do not wish to disparage the historical achievement of Kant; on the contrary, Kant made an ingenious attempt to provide a philosophic justification for the theory of space of Newtonian physics. Kant was correct in considering the strange fact that geometry is, on the one hand, an absolutely certain science and that, on the other hand, this science presumably describes physical space to be a problem meriting philosophic analysis. The allegedly dual nature of geometry led Kant to his concept of the *synthetic a priori*, and he thought his view that space is a form of intuition arising from the subjective structure of reason explained this duality. According to Kant, space is a form of intuition which is *a priori* because it stems from reason, but which is also a necessary presupposition of all synthetic statements, and therefore descriptive of the physical world. We have to admit that this construction is an ingenious attempt to solve the problem as he saw it, but scientific conceptions have changed in the meantime, and Kant's proposed solution is no longer acceptable to us. The belief that Euclidean geometry is at once necessary and applicable to reality was due to the fact that only one geometry was known. During the last century, a plurality of geometries have been constructed, and we can no longer decide *a priori* which geometry holds for physical space. The various geometries express mere mathematical possibilities, but only experience can determine which of these possibilities is realized in the physical world. The problem of which geometrical laws hold in reality has found an empirical solution; in this respect, geometrical laws do not differ from other physical laws. We can always consider an empirical decision as a selection made from among several consistent formal systems in such a way that the choice itself is purely empirical. Mathematics tells us, for instance, how a mass point would move under the influence of a force of attraction which decreases in inverse proportion to the first, or the second, or the third power of the distance from its centre; experience decides that only the second of these possibilities holds for the motion of the planets. In a like manner, we have to regard the decision as to which geometry holds in the physical world as an act of selection based on experience.[4] Through this selection, i.e., by co-ordinating a certain conceptual structure to reality, physics describes a property of nature.

What sort of physical properties are formulated in geometry? Helmholtz gave us the answer: Geometry formulates nothing but the spatial relations between rigid bodies. We find these rigid bodies in nature, and we can make measurements with their help. We might, for instance, construct a large circle on level ground by fastening one end of a string to a stake and moving the other end like a compass. Then, if we were to take a meterstick and measure the circumference and the diameter of the circle, we would, in fact, obtain as the ratio of the resulting numbers, the number 3.14 .... Such a result is an empirical one; it is not logically necessary, but expresses a property of rigid bodies which has been formalized in the system of Euclidean geometry. If the ratio were different, we would not have had to wait for Bolyai and Lobatschefskij to discover non-Euclidean geometry; the ancient Egyptians would have found a more complicated relation between the circumference and the diameter of a circle, and Euclid would have axiomatized a more complicated geometry. Not only would the mathematicians have adjusted to that more complicated geometry, but so would everybody else; mankind would have long ago developed the ability to visualize non-Euclidean geometry. Our ability to visualize Euclidean geometry is a result of conditioning, not only from an ontogenetic, but also from a philogenetic standpoint, a conditioning which has its origin in the fact that our world of rigid bodies possesses a structure which approximates the Euclidean one to a high degree. If one understands the connection between the ability to visualize and the application of a geometry to reality, it is not difficult to visualize non-Euclidean geometries. Today we may consider it a philosophical result that non-Euclidean geometries can be visualized in the same way that Euclidean geometry is visualized (see [1928h], §9–§13). Incidentally, the description of physical space in terms of Euclidean geometry is only approximately correct; in astronomic dimensions, Riemannian geometry holds.

The properties of time are analogous to those of space. Time is not a form of intuition either, but a property of nature which we represent in a conceptual system. Causal chains are the physical phenomena which correspond (with regard to time) to the rigid bodies of space. We use the term 'causal chain' to mean any form of causal propagation, such as the transmission of a telegram by means of an electromagnetic signal, or even the transport of a material thing. All physical events form such causal chains and therefore exhibit a structure; this structure is a net whose lines we interpret as temporal order. It is possible to formulate the laws of this lineal structure precisely. It is one of the most important properties of these world lines, i.e., the causal chains, that they never return to their starting point. By 'return' we mean not only to the same place, but to the same point event, i.e., to the same place at the same time. It is not easy to imagine what it would be like if the causal chains were to return to their starting point because we are so accustomed to a different structure of reality. Closed causal chains would lead to very strange consequences; in such a world we might some day meet our former selves and enter into an embarrassing argument as to whether we are identical with each other or not. Fortunately, our world is so constructed that this event will never happen, but we cannot say that this impossibility is an *a priori* necessity; we are dealing with an observationally confirmable property of reality. Since there are no closed causal chains, the order of time is unique. Therefore, we can say that temporal order is determined by the causal structure of the world. If it is at all meaningful to ask what the nature of time is, the answer is given by this discovery.

We shall now turn to a group of physical problems, which, in contrast to the problems of space and time, are at present very much in the foreground of discussion; since their investigation is still underway, no final results can be reported even though some fundamental principles have been established.

These problems concern the nature of physical laws. In order to understand these problems, we must trace their treatment in the history of science.

The idea that there is a strict causal determination of all physical events has been regarded as the foundation of modern science, and this feature certainly constitutes a decisive advance beyond the science of ancient times. The ancients did not have a scientific conception of natural law; for them, laws of nature were mere rules of thumb, and therefore they were not astonished when once in a while an exception to a rule occurred. They would simply think that the law did not hold in that particular case. Modern science, on the other hand, has asserted that there are no exceptions to the laws of nature. What appears to us as an exception is rather the effect of the operation of more complicated factors, and we would be able to recognize the strict causal character of even the exceptional case immediately had we a comprehensive knowledge of all the parameters so that we could make our calculations accordingly. The principle of the strict causal determination of all events has celebrated great triumphs in science and has been the reason for the admiration and the often almost religious veneration which natural science has enjoyed among many of its adherents. Indeed, the idea of a strict causal determination of all natural events has a touch of superhuman sublimity about it which emphasizes, by contrast, the imperfection of human knowledge and action all too clearly. One must remember these emotional associations in order to understand why some eminent modern scientists of the older generation are negatively disposed towards the more recent developments in physics which have led to the abandonment of strict causality. They view this abandonment as the renunciation of a fundamental principle of physical research, and they would like to urge, almost as an article of faith, that belief in strict causality should be preserved even though, at present, the situation in natural science does not justify their faith. The younger generation does not agree with this point of view. In order to present the current conception of physical law, we must first enter into a logical analysis of that concept which is at the core of the problem of causality, the concept of probability.

The concept of probability was first formulated in a mathematical theory concerning games of chance. Soon, however, it outgrew this scientifically rather trivial subject and invaded physics where it initially appeared in the theory of errors. Even here it was regarded as a rather unimportant by-product since it apparently had significance only for determining the limits of exactness. The turning point came when the concept of probability was utilized in the kinetic theory of gases because then it acquired fundamental importance. Boltzmann's employment of the concept of probability in his analysis of the second law of thermodynamics is of major significance.

This law, which formulates the directedness of all thermal processes, had originally been of the strict causal type just like the first law, the law of the conservation of energy. In order to formulate the second law, the concept of entropy was introduced by analogy with the concept of energy, and the law was interpreted as expressing the fact that all thermic changes are accompanied by an increase in the entropy of the total system. Boltzmann showed that this law must be understood as a statistical law, as a law pertaining to the mixture of large numbers of particles. We will give an example to illustrate this point.

If we take a container divided into two compartments and fill one compartment with oxygen, and the other with nitrogen, and then remove the dividing wall, the molecules of the two gases will mix. Why does this happen? We cannot find a causal law to explain this process; rather, we explain it by a statistical law of whose correctness, however, we are firmly convinced. Of course, we cannot exclude the possibility that the opposite process might sometimes occur so that, for instance, the oxygen molecules of the air might accumulate on one side of the room and the nitrogen molecules on the other side. Although the consequences would be quite unpleasant for the people in the room, we do not know a law of nature that would exclude such an event. Its occurrence is extremely improbable; that is the only consolation which we can offer with respect to this possibility. This consolation is, however, sufficient for practical purposes. Anyway, our demands for certainty are not excessive in everyday life. We know, for instance, that every year there are a certain number of railway accidents; but we do not worry even though every time we go by train there exists a certain probability that an accident will occur. We identify low probability with impossibility, and high probability with certainty, and we forget completely that we deal, nevertheless, with probabilities. As far as train trips go, we never quite forget that we are dealing with probabilities as shown by the fact that many people deem it necessary to take out travel insurance. But the result of this is merely that the relatively high probability of a favourable event is replaced by a still higher probability. Of course, it is not impossible that the insurance company might go into bankruptcy because all of its clients suffered a train accident on the same day. When we consider statements of everyday life, we find that all of them neglect the low probability of a negative instance. Indeed, there is not one empirical statement which is certain – all of them are probability statements of more or less obvious character.

In Boltzmann's conception of the second law of thermodynamics, not only the mixture of molecules, but also the mixture of velocities is assumed. These two assumptions are fundamentally the same. Entropy is interpreted as the probability of a thermodynamic state; we can regard it as the 'degree of compensation'. At the time, Boltzmann's idea was considered a major physical discovery, and rightly so because it offers an explanation for the law of increasing entropy. What previously had to be accepted as a fact of nature, suddenly became comprehensible, being explained by a reduction to the principle of probability. But this explanation in terms of Boltzmann's principle turned out to be a dubious gift, for a law that had previously been regarded as strict was transformed into a law holding only approximately. Thus, uncertainty penetrated the most exact of all natural sciences. If physics acknowledged such probability laws, it could no longer claim the strictness for its laws which had so far constituted their greatest asset.

To avoid this consequence, a distinction was drawn between a strictly regulated physical nature, on the one hand, and the imperfect human knowledge of this regularity, on the other. The concept of probability was said to have its source in the imperfection of human knowledge, and to be a makeshift device of human ignorance which could be eliminated in an ideal physics. If we were in the position to ascertain precisely the initial positions and initial velocities of all gas molecules at a certain moment in time, and if we could calculate mathematically their mutual relations, we would be able to predict the behaviour of the gas with the same precision as the astronomer predicts an eclipse of the sun. Some time earlier Laplace had created the classical symbol for this view when he postulated a superhuman intelligence which could predict the future with certainty, and which would not have to resort to probability. This conception is called the subjective theory of probability because according to it, the concept of probability originates in human imperfection and has no objective significance.

But the subjective theory cannot do justice to the concept of probability. This interpretation neglects the fact that probability laws, in the form of frequency statements, stand the test of reality very well; their application to reality cannot be justified if probability laws are regarded as resulting from the ignorance of the knowing subject. If we formulate the laws of the theory of gases statistically merely because we do not know anything definite about the motions of the molecules, it is difficult to understand why these laws nevertheless describe the behaviour of gases. It is hard to imagine that nature takes human ignorance into such favourable consideration. This objection against the subjective theory of probability is decisive; therefore, an objective theory of probability has been constructed which attempts to show that probability laws hold objectively for natural events in the same sense that causal laws do. According to this interpretation, the regularity of such statistical processes as molecular behaviour constitutes an independent property of physical events and it is the task of science to formulate statistical laws just as it has formulated causal laws. It also seems inappropriate to regard the employment of statistical laws as provisional. Not even Laplace's superman would renounce the use of such laws – as the French mathematician Cournot remarked in the forties of the last century – since his exact calculations would lead to the same results for the computation of averages as do the statistical laws. His superhuman qualities would be apparent only in his ability to manipulate the mathematical apparatus of probability much better than we can.

With the help of such epistemological considerations, the statistical principle was established as a form of regularity on an equal footing with the principle of causality. Even at the time of Boltzmann, scientists debated whether the statistical principle might not some day supersede the principle of causality, and whether the fate which had transformed the second law of thermodynamics from a strict law into a statistical law might not affect all laws of physics in the same way. (It is hardly possible today to determine who expressed this idea for the first time.) It was possible to give a rigorous formulation to such ideas, however, only after the concept of causality had been subjected to critical analysis; we shall return to this development shortly.

The position of the subjective theory of probability was ultimately undermined by the investigations concerning the relation of causality to probability. The concept of probability turns out to be indispensable even for the so-called strict laws of physics as soon as one attempts to apply these laws to a particular case. Scientific knowledge consists in the ability to predict a given event. We know very well that the conditions with respect to which we calculate the course of an event are never completely realized in the physical world. For instance, we can never take all the causal factors into account when we calculate the motion of a material body such as a projectile. If in spite of this we are often able to make correct predictions, we owe our success to the fact that the concept of probability applies to the physical world. Probability takes care of those factors which were neglected in our calculations. In the theory of observational errors, the significance of the concept of probability is still obvious. It is, however, a mistake to believe that the probability in this case is restricted to a narrow interval of exactness; the difference is only that within this interval, the probability is noticeably smaller than 1, whereas outside the interval, it lies very close to 1. The asymptotic property of Gauss' exponential curve provides a precise analogue for this idea. We may only assert with high probability that the value of the predicted magnitude lies within the given interval, but we can never make this assertion with certainty. Every application of causal laws to physical things therefore depends upon the concept of probability. Even the statement that causal laws describe relations between real things rather than imagined things is a probability statement.

For the time being, physics has retained the idea that the probability of prediction will increase as the number of causal factors included in the mathematical description of the initial conditions increases. Yet this statement also rests upon the concept of probability. It is a convergence statement added to the statement about the probability relation between cause and effect. Therefore, strict causal statements must be replaced by the following two statements:

I. If one event is described by means of a certain number of parameters, the later event, also characterized by a certain number of parameters, can be predicted with a probability p.

II. The probability p converges towards 1 as more and more parameters are taken into account.

These two statements replace the assertion of strict causality; they incorporate the concept of probability. If one were to replace them by the statement 'physical events are strictly determined', one would be replacing a convergence statement by a limit statement. This procedure is harmless provided one understands that the limit statement cannot express anything other than what is formulated explicitly in the convergence statement. Usually, however, the definition of a limit statement in terms of a convergence statement is forgotten, and misunderstandings about the concept of causality arise; in particular, there arises the mistaken belief that the concept of probability can be eliminated. These mistakes resemble those made with respect to the concept of differential quotient defined as the ratio of two infinitesimal magnitudes. Misinterpretations of the concept of time which is closely connected with the concept of causality have a similar origin. But we cannot enter into a discussion of this matter at this point.

The formulation of the concept of causality given in statements I and II immediately suggests the possibility of generalizing the concept. We can envisage situations requiring the replacement of statement II by a more general statement asserting that the probability of prediction cannot be brought as close to 1 as we wish. This generalized version of statement II is the strict formulation of our previous conjecture that causal laws may have to be reduced to statistical laws. Of course, this reduction would alter the character of our scientific knowledge only if the reduction makes it impossible to describe a single event causally, i.e., if the probability of predicting the event cannot be brought as close to 1 as we wish by taking more parameters into account.

These considerations show that a rigorous epistemological analysis can eliminate certain preconceived notions stemming from a belief in some sort of *a priori* necessity. It is an empirical question whether statement II is true or false. We regard it as an auspicious sign of the agreement between modern philosophy of science and physics that the possibility of generalizing statement II (first given in [1925d], p. 138) was initially suggested by philosophic considerations concerning the principle of causality, and was subsequently taken up by modern quantum mechanics which has substantiated the applicability of the generalization. The factual assertion that there are limits to the predictability of an event could, of course, only be made in physics; the task of philosophy was exhausted by pointing out this possibility.

Heisenberg's Principle of Indeterminacy asserts that there are limits of predictability in quantum physics. We shall not discuss Heisenberg's justification in terms of the disturbance of the object by the means of observation since this is sufficiently familiar. Incidentally, the principle has a somewhat more restricted form than the denial of statement II. This restrictedness is very plausible to philosophers; physicists never make absolutely general assertions but formulate their generalizations as specifically as possible. According to Heisenberg, the probability of predicting the value of a single parameter (or even of several parameters of the same type) can be brought as close to 1 as we wish, but a simultaneous prediction of the value of a parameter of a different sort (or of several such parameters) has a correspondingly decreasing probability. The probability of predicting the values of all relevant parameters, computed as the product of the elementary probabilities, is restricted to a limit below 1; yet it is only the combination of all relevant parameters which characterizes an event exhaustively.

Physics has thus decided in favour of a more general type of law than that based upon the traditional concept of causality. Some physicists, among them a number of those outstanding scientists who are responsible for the major achievements of contemporary physics, tend to see this decision as a failure of physical science, as a renunciation of knowledge; and they have expressed the hope that the acceptance of probability laws will only be a passing phase in the development of physics. We think that their conception of physical laws is erroneous; presumably, this conception arose from a mistaken epistemological approach to the problem of causality originating in a false idealization of the possibilities of knowledge. It is not the task of epistemology to prescribe the structure of nature. If it turns out that strict causality in the form of statement II does not hold, then this result is an empirical assertion which must be accepted without regard to our emotional reactions. This result does not constitute a failure of physics, but represents an objective property of nature. Moreover, this discovery is of major importance and is as much entitled to our admiration as any other discovery of physics.

IV

I shall next consider those problems in the philosophy of science which concern the foundations of mathematics. These problems pertain primarily to the development of logic; epistemological considerations will be of secondary importance. Kant maintained that logic had not undergone any significant development since the time of Aristotle. We can agree with Kant's assertion if we take it as an eighteenth-century opinion, but since his time, logic has changed so radically under the influence of developments in mathematics that Aristotelian logic appears primitive by comparison. We shall now report on this new logic and its problems.

The new logic is characterized externally by the fact that it uses a symbolic notation similar to that used in mathematical formulas. This feature was at first considered inessential and pedantic. Some philosophers have said that symbolic logic could never accomplish anything more than the traditional logic formulated in ordinary language since everything that can be expressed in symbols can be expressed in words. This remark is in principle correct, but from a human and psychological point of view extremely naive, and it is hard to understand how it can still be seriously defended today. The employment of a symbolic notation is important for psychological reasons; symbols are indispensable for the clarification of our ideas. What would have become of mathematics without a calculus? Yet a mathematical calculus is also dispensable because in principle every mathematical formula can be translated into words. The introduction of a calculus in logic led at once to a considerable number of results which were unknown to traditional logic.

Let us mention a few examples. Traditional logic considered all statements to be of the subject-predicate form *S is P*. We know today that this is but one form of statement, the form of those statements containing one-place predicates, and that statements containing many-place predicates cannot be cast in this form. Predicates with more than one variable occur constantly, not only in science, but also in conversational language. Furthermore, traditional logic did not distinguish the meaning of 'is' in the sentence 'Socrates is a man' from its meaning in the sentence 'A negro is a man', a confusion which led to false interpretations of the concept of class and which may have contributed to the mistaken notion that concepts exist as independent entities. Modern logic has also revealed the fundamental distinction between a proposition and a propositional function; a propositional function is more general than a proposition, and contains free variables. The concept of a propositional function clarifies mathematical reasoning as well as the reasoning of everyday life; the structure of a large class of propositions, the class of general propositions, becomes clear only when propositional functions are exhibited as their constituent parts. In short, the use of symbolic notation has suddenly furnished a number of logical discoveries which, for psychological reasons, would never have been made by a logic tied to natural language.

Symbolic logic became an indispensable instrument in the foundations of mathematics due to the discovery of contradictions in this most certain of all sciences. In the theory of classes, one could formulate concepts which were apparently perfectly correct, but which led, by means of logically valid inferences, to contradictions. The occurrence of paradoxes in the theory of classes was very disquieting to the mathematicians. At first they tried to eliminate these paradoxes by considering the relevant expressions as inadmissible; but soon they began to worry whether they were not excluding too much in this fashion. A number of valuable and fruitful mathematical methods of demonstration were endangered by such restrictions. The pursuit of these investigations split the mathematicians into various groups. Brouwer's intuitionism is the most radical with respect to restrictions; he insists on ruling out even the unrestricted use of the Law of the Excluded Middle, and thus attacks the foundations of Aristotelian logic. Russell attempted to overcome the difficulties by means of his theory of types whose most important principle is the distinction of meaningless linguistic expressions from true and false sentences. A certain synthesis of the views of Russell and Brouwer can be found in the writings of Wittgenstein, who holds that a theory of language stands at the beginning of logic. Hilbert tried to save classical mathematics by providing it with an axiomatic foundation whose consistency he audaciously attempted to prove.

Although Hilbert's proof has not been carried through completely – recently, doubts have arisen whether it is at all possible to do so – the formulation of the problem itself, and the methods employed in connection with it constitute an extraordinary advance in philosophical analysis.

The most important philosophical discovery to be mentioned in this context is the distinction between form and content. The content of a thought can be expressed symbolically by means of a sequence of signs. This sequence of signs can be conceived purely formally if one forgets about the content of the sentence and regards it merely as a relation between signs. We can perform operations with these signs that can be compared to the moves in a game of chess. There are prescriptions about the elimination of parentheses, about the substitution of new signs for the original ones, etc.; there exists a genuine algebra of logic in which formulas are manipulated without regard to their meanings. The problem of the consistency of a formal system is treated in such a way that the occurrence of a contradiction is associated with the occurrence of a certain sequence of signs, and the system is proved consistent by demonstrating that this sequence can never occur. For various elementary parts of logic there are consistency proofs. Formalization enables us to establish the validity of sentences which have far-reaching importance; furthermore, these sentences remain valid when a meaning is assigned to the signs.

The earlier belief that logical consistency is an *a priori* postulate resulted from the confusion of two completely different assertions, as we can see today. The law of contradiction, the statement that a sentence and its denial cannot both be true at the same time, is logically necessary. But the necessity of this law does not imply that logic will never lead to contradictions. What can be required on the basis of the law of contradiction is only that contradictions *should not* occur; whether they *do* occur is a very different question. The assertion that no contradictions occur, that logic is consistent, is by no means self-evident. Whether a system of logical axioms is consistent requires special analysis – the consistency of a system must be proved. This discovery of modern logic is due in large measure to Hilbert, and it is the procedures used in carrying through a proof of consistency which suggested the formalist treatment of logical statements.

We must regard the distinction between form and content as a philosophical result of the highest order. It permits insights into the nature of reasoning that could not be anticipated. However, not all statements concerning logical relationships lack content. The statements of metamathematics have content since they are assertions about the formulas of mathematics. The methods of modern symbolic logic are not so very far removed from our reasoning in everyday life. We employ similar techniques in performing elementary arithmetical operations, for instance, when we multiply many-digit numbers and shift the row of numbers to the left before we add them up. This procedure constitutes a multiplication by 10, but during the performance of the multiplication nobody is aware of this fact. This is fortunate because only through such a process of formalization do practical calculations acquire their guarantee of correctness. Instructions for using household chemicals or machines further illustrate the prevalence of formal techniques. Directions are given for the handling of various devices, for instance, for the turning of a certain dial, without any explanation of the actual function of the device.

Many mathematicians and logicians have become interested in the problems of mathematical logic, and the symbolic logic which has been developed as a result of their interest plays an important part in the new philosophy of science. To characterize these investigations as purely 'mathematical' or 'unphilosophical' is to misconstrue the nature of philosophy; these studies deal with problems which have always been classified as philosophical, but for which no satisfactory solutions have hitherto been found.

V

In the following section, we shall deal with some problems that cannot be classified as belonging to any of the special sciences. It is quite plausible that the philosophy of nature should be interested in these problems since there is a gradual transition from the problems of the special sciences to these more comprehensive problems. The analysis of the laws of nature, for instance, cannot be restricted to physics; it is of concern to all the sciences.

We should like to mention first the problem of our knowledge of the external world. This is an age-old problem; during the last century, it was construed in an interesting manner by the positivist school of Ernst Mach, Avenarius, Cornelius, and others. Positivism maintains that sense data are the basic constituents referred to in any meaningful statement about reality. Whatever so-called things of the external world we may speak about are but logical constructions out of sense data. When an astronomer predicts an eclipse of the sun, the following process takes place: he observes the stars, i.e., he has certain perceptual experiences of bright and dark spots; then he interprets these perceptual experiences in terms of a theory referring to giant celestial bodies such as the sun, the earth, and the moon which are revolving around each other. From the laws governing these bodies he infers that there will be further perceptual experiences of bright and dark spots, experiences of the sort which we actually have during an eclipse. According to the positivist view, that part of the process referring to the external world is merely a device for relating the two sets of perceptual experiences; therefore, it is not necessary to speak about external things, for when we have described the relationship between the perceptual experiences, we have said all that can be said.

While positivism in its original form was open to serious objections which were never satisfactorily answered by the older positivist, Carnap has reformulated the position in such a way that these objections no longer hold. It is characteristic of Carnap's procedure that he applies the methods of mathematical logic in his analysis of this epistemological problem, and thus arrives at a logical neopositivism that stands on much firmer ground. Carnap primarily uses contextual definitions for the construction of concepts. This method does not define isolated concepts as traditional logic defined them by stating the genus proximum and the differentia specifica (for example, the lion is a predatory animal with such and such properties). Instead, a whole sentence containing the new concept is defined by a sentence (or a group of sentences) of a completely different logical structure. In this way, what is defined is not a concept, but a sentence in which the given concept occurs. Thus Carnap defines the sentence 'Material objects exist' by the sentence 'There are certain regularities between perceptions'.

The usual objections against positivism which contend that things would not exist for the positivist when they are not perceived are eliminated through this procedure. We do not wish to say that we therefore consider Carnap's analysis of the problem of the external world adequate; on the contrary, it seems to us that some fundamental objections can be raised even against Carnap's formulation. However, a discussion of these objections would go beyond the scope of the present essay. Independently of such considerations, we want to emphasize that Carnap formulates the problem of the external world for the first time in a meaningful way; any systematic attempt at a solution of this age-old epistemological problem will have to start from Carnap's analysis.

Another problem which we would like to mention briefly is that concerning the freedom of the will. Philosophers have always connected this problem with strict causal determinism. All of them have been concerned with the relation between determinism and freedom, no matter whether they affirmed or denied the freedom of the will, or whether they considered it a pseudoproblem. Since the analysis of the foundations of science has led to the abandonment of the conception of strict causal determinism, we must reformulate the problem of the freedom of the will. Of course, it would be shortsighted to interpret the replacement of strict causal relations by probability relations as an immediate justification of the freedom of the will. Freedom of the will cannot be equated with a mere denial of determinism. On the other hand, it seems just as shortsighted to deny that the problem of free will requires reformulation by arguing that psychological processes are molecular processes involving such a large number of particles that we can predict these processes with high probability and may therefore regard this high probability as certainty. Although there is practically no difference in everyday life between high probability and certainty, this difference must not be neglected for theoretical purposes. On the contrary, the solution of the problem of free will depends on this very distinction. Nor can we take it for granted that we are always dealing with high probabilities in regard to psychological events. We believe, therefore, that even this age-old philosophical problem will find a new solution on the basis of a modern analysis of the epistemological status of the laws of nature. Although we are not yet in a position to go into detail, we can say that the probability relations between events require us to view the difference between the past and the future in a new light; the analysis of this difference will be of great significance for the solution of the problem of free will.

VI

We have arrived at the end of our survey of the new philosophy of nature. It should be clear that this philosophy treats genuine philosophic problems by modern scientific methods. Let us summarize briefly the methodological principles involved.

In the first place, such a philosophy of nature constitutes a great triumph of rationalism. As far as we know today, there are no set limits to the capacity of human reason. But by reason (ratio) we do not mean the dogmatic reason of traditional rationalism, a system of categorical compartments into which all knowledge is forced; rather, we regard human reason as a flexible instrument capable of satisfying the relentless requirement of continuous adaptation to experience. This conception of reason seems to represent one of the most characteristic features of modern scientific philosophy.

However, modern philosophy of nature must also be regarded as a triumph of empiricism since experience is accepted as the final judge of all statements about the physical world. Experience is even employed in order to judge the ultimate categories which traditional philosophy had considered *a priori*. Modern empiricism does not contradict rationalism because knowledge is conceived as a system constructed by reason upon which experience exerts a regulative and selective influence. As explained above, the notion that there are necessary presuppositions of knowledge is given up; reason itself is subject to the regulative function of experience.

Finally, we should like to mention the technical aspect of the modern philosophy of nature which is one of its essential features and which has greatly contributed to its development. This philosophy has created such technical devices as symbolic logic, the axiomatic method, etc. It has developed into a marvellous instrument for systematic analysis by availing itself of the theoretical methods of the physical sciences. Some philosophers regret this development, and prefer the earlier intuitive way of philosophizing because they consider it more human. To us this attitude seems a kind of romanticism burdening the present with the emotional values of the past. It is unreasonable to judge the times in which we live by the value standards of former times. We cannot choose our fate; it is ours and we must accept it if we want to live. Those who have adapted themselves to the new philosophy of nature, who have felt its tremendous potential as a source of knowledge, have freed themselves from the nostalgia for the past, and have accepted the technical methods of analysis.

Behold the new philosophy of nature, still young, still in its beginning, and yet precise and progressive in its outlook. Still growing, it is already attacked. Some people see it as a continuation of the great epistemological tradition of the past; others consider it uninteresting, abstruse, and indifferent to those problems that appear essential to them, although they cannot contest its logical precision. How is it possible that there are such different opinions about the same thing?

We must realize that this controversy is not a matter of truth, but of personal attitudes and subjective evaluations. Two types of people oppose each other because of their difference in attitudes. The scientifically oriented person is one of them; he is concerned with knowledge, with the discovery of truth, and he regards a philosophy that is sustained by the same spirit which sustains scientific research as the only possible approach to philosophic analysis. The literary person is the other type; he is interested in human beings, in the understanding of emotional experiences. An objective analysis of logical and epistemological problems is alien to him. Those who attack the modern philosophy of nature and who would deprive it of any right to exist belong to the second type. I call this type of person the literary type because the writer is his prototype. But we can find him, less distinctly marked, among people of all professions, strangely enough sometimes even among physicists and mathematicians. At least some representatives of these disciplines take special pride in their literary approach to philosophical problems.

We do not intend to defend the view that an interest in understanding human emotions and experiences is either superfluous or without significance. What we are criticizing is the attempt of the literary philosopher to transgress the limits of his field, and to reject a scientifically oriented philosophy. I want to emphasize that this reproach is not directed against the literary person in general; fortunately, there are many far-seeing people who recognize both the legitimacy of the scientific approach and the limits of their own. Yet the majority of literary persons are hostile to the philosophy of nature. It is not merely a matter of criticism; rather they construct a philosophy of a very peculiar kind which they present to the world with the claim that it is the only true philosophy. One has to view this kind of philosophy as a rational superstructure based upon emotional experiences. These philosophers develop a theory of knowledge, a conception of being, sometimes even a kind of cosmogony all of which are evidently constructed without regard to systematic logical and epistemological analysis; their only purpose is to disguise emotional attitudes as a pseudo-scientific conceptual system. It may well be that this form of conceptual poetry satisfies its admirers emotionally in the same way that theological and mythological creations have satisfied believers for thousands of years. But it is the tragedy of all defenders of such modes of thinking that they do not recognize the limits of their activities, that they take the picture language in which they describe emotional experiences for scientific knowledge, and dare to oppose it to scientific philosophy. There are numerous philosophers of all ages who are guilty of this confusion of emotional, often religious, attitudes with theoretical knowledge. The most conspicuous example of this confusion has perhaps been the attempt to construct an ethics according to the geometric method. Incidentally, Spinoza's ethical principles are on a very high plane – only the geometric method is inadequate. In the camp of the literary philosopher, the geometric method is nowadays less fashionable; other methods are chosen instead. But the pictorial character of such a philosophy is obvious in spite of its scientific disguise. What we are offered today as a true metaphysics or ontology, be it even by writers with great human qualities, is nothing but such pseudo-science. What does it mean to say that being is founded on care, or that negation and the not are related to the phenomenon of anxiety? What is this but a kind of conceptual music reflecting emotional experiences? For the scientifically minded thinker, such assertions are devoid of meaning; yet today this sort of philosophy is widely acclaimed by those who hope in vain that it will give them an understanding of the nature of being and time.

Although modern philosophy of nature refuses to compete with literary philosophy, it does not want to characterize a concern with emotional or valuational problems as insignificant. What we repudiate is the possibility of treating such problems scientifically. If one looks a little closer, one will discover that modern social developments have demonstrated the futility of such pseudo-philosophical attempts. It has become ever more obvious that decisive new insights into the meaning of life, be it new visions of human society, or of the relationship between the sexes, or of the education of children and adolescents, or of the distribution of work and leisure in daily life, are found not by speculative philosophers but by people in practical life who discover new values in their activities and are able to make them acceptable to others through the impact of their personalities. Specifically, it is the academic philosopher's alienation from the revolutionary social processes of our time which explains why so much that is said in academic quarters about these matters seem strangely sterile and remote from life. Only from the soil of concrete activity can there spring new values of a sort never revealed by the pseudo-scientific analysis of emotional attitudes. It seems plausible to draw a parallel between this kind of philosophy and the development of epistemology as we presented it. We saw that new logical and epistemological insights are the products not of the analysis of pure reason but of scientific research. We may look at the development of the ethical values of modern man from a similar point of view; under certain circumstances, action is a better path to insight than contemplation, at least in those cases where contemplation becomes enmeshed in a pseudo-epistemological system. Anybody who intends to say something about the meaning of life should guard against conceptual analyses. He would do better to become a worker in a factory, or a teacher in a school, or a doctor in a hospital, or a participant in some social movement; then he will have something to say about the values of our times. But he should renounce rational constructions – he should leave those to theoretical research, to the value-free investigation of facts which alone constitutes scientific philosophy.

Only such a clean-cut delimitation of the scope of philosophy can decide the controversy about the nature of philosophy. Those who reproach the philosophy of nature with intentionally neglecting 'the ultimate questions' are demanding the impossible – that philosophy justify valuational attitudes by the construction of rational systems. Scientific philosophy rejects such an unreasonable task. If one therefore wants to deprive it of the name 'philosophy' because one is convinced that traditionally all philosophy has been explicitly concerned with justifying values, one misunderstands the philosophical attempts of the past. Such tendencies have certainly always existed in philosophy, but the term 'philosophy' designates such an inclusive complex of ideas that one cannot restrict it to just one goal. It seems that the pursuit of truth has been at least as powerful a trend in traditional philosophy, and perhaps a much more profound one, that traditional philosophy has been as much concerned with discovering theoretical solutions for certain fundamental problems of nature and life, be it by suggesting how questions may be formulated in a meaningful way, be it by answering them. Certainly, the great philosophers pursued this task, and to this pursuit they owe their historical influence. Many a traditional problem has turned up in our discussion of the modern philosophy of nature; let me mention only the problems concerning the nature of life, of space and time, of empirical laws, of our knowledge of the external world, and of the freedom of the will. Many technical problems which occupy modern scientific philosophy grew out of the traditional ones such as the problems connected with probability and symbolic logic, or those dealing with psychological phenomena. Only if one approaches philosophy as a theoretical endeavour which aims at the analysis of cognitive problems can one judge the philosophy of nature and its achievements.

Yet such theoretical philosophy need not justify its right to exist, even though it cannot offer a direct basis for practical ethical decisions. Theoretical knowledge has its own value; it satisfies a drive as vital as other drives. The exclusively literary person may perhaps never feel this way; yet he is like a blind man speaking about color when he calls epistemological pursuits and logical methods of analysis unimportant and remote from life. The life of the scientist is full of these matters, and the desire to work out precise methods is felt by all those who have at some time had their eyes opened to philosophy. Even two thousand years ago the philosophers knew this – if an ahistorical philosopher may be permitted a historical remark – and they tried to give answers in their way and with their logical tools. The answers that we must find today can only be discovered by the logical tools of our time, as they have mainly been elaborated in the mathematical and scientific methods. An approach geared to the present is the way, a reorientation of theoretical philosophy of nature.

NOTES

[1] See for instance Helmut Plessner, *Naturwissenschaften* 18, 1930, p. 869. Th. Haering (*Philosophie der Naturwissenschaft*, 1923) favours similar views.

[2] This idea has been emphasized by one of the leading representatives of the Gestalt theory, W. Köhler, in his book *Die physischen Gestalten*, Benary, 1920.

[3] See J. Schaxel, *Erkenntnis*, vol. I, no. 6, 1930.

[4] Actually, the problem is more complicated because the decision about geometry is uniquely determined only after certain definitions have been given. It can be shown, however, that geometry shares this characteristic with all other empirical statements. The lack of determination can be eliminated by supplying the requisite co-ordinative definitions. Conventionalism has led to misinterpretations and an unjustified defense of the geometric *a priori*. For a more detailed analysis [see 1928h, §4–§8].

39. KANT AND NATURAL SCIENCE

[1933b]

I. THE EFFECT OF KANTIAN PHILOSOPHY

The significance of Kant's philosophy reaches down into our own time. Not only have the theories of philosophers of a variety of schools been determined by Kant; scientists, too, have employed his philosophy in carrying out philosophical criticism of scientific theories or setting up systems of natural philosophy. No one who wishes to attain an understanding of present-day philosophical thinking can ignore his great influence. It is equally in evidence where scientists have attempted to perpetuate Kant's ideas by adapting them to the changes in science – as Helmholtz did with respect to geometry – and where they have totally rejected Kant's ideas precisely as a means of contrasting them with their own philosophical conceptions – as Mach did in his critique of the concept of the thing-in-itself. Thus Kantian philosophy does not only live today in the neo-Kantian system of thought, in the so-called Marburg school, which over the past fifty years has recreated a group of Kantians in the strict sense. Kant's ideas are to be found in other schools of philosophy as well, although some of these – the phenomenologists, for instance – place less emphasis on their connection with Kant. This sweeping influence does not apply only to German philosophy. Philosophers of other countries have also taken up positions for or against Kant in their basic writings, thus acknowledging his place as a cornerstone of philosophical thinking.

The beginning of this great influence lies far back historically. It started within Kant's own lifetime, for shortly after its appearance, Kant's principal work began to exercise a persistent effect, bringing fame to its author. We must bear in mind that the classical and romantic epochs of German literature were under his influence, that such men as Goethe and Schiller worked through his critiques of reason and referred to them in their writings, that in his day educated society discussed Kant with lively interest. Kant was not one of those men who is discovered and understood only long after his death. He himself experienced the success of his ideas, and even if interest in his theory has fluctuated since then, his philosophy has at the very least been at work in the background of all subsequent philosophical teachings. This is a fact that not even his opponents can deny.

What is the basis of this all-embracing influence? Why is it that a philosophical system has for one hundred and fifty years so largely determined the position of so many thinkers? It would be naive to suppose that it is because final and ultimate truths are to be found in this system. The relation between truth and success is not that simple. We know of cases in which the significance of very profound truths was not recognised until long after the death of their discoverer – and which even then have not attained the degree of recognition that other, more fortunate truths have received immediately after becoming known. And we know of others in which fame and recognition have been awarded to achievements of only moderate significance, without the public who discuss the great names of the day yet having realised how misplaced such fame is. Some of the great names in politics and war come to mind. Success, influence, is a sociological phenomenon, determined by two components, only one of which relates to the quality of the achievement; the other component is established by the sociological situation. It is determined by the extent to which the achievement in question answers to a sociological need and thereby encounters the conditions favorable for dissemination. There are situations in which the desire for a certain intellectual attitude lies dormant among the educated classes, without they themselves being conscious of it. The man who is able to express the spirit of his times – irrespective of whether he is aware of fulfilling this function – will ride high on a wave of approval and enthusiasm, experiencing the success that others have perhaps worked for in vain. To be sure, the achievement to which such a happy fate is allotted must display a certain minimum degree of quality, but it need not be superior to others of a similar cast. Above all, it need not represent accurate knowledge, the one truth among the many views contradicting the truth. The distinction of having succeeded is not itself a guarantee of accuracy. What the success justifies is the clarification of, the attitude toward, the system of ideas it presents. And that, too, is the justification for our attempt, in this era of a new scientific philosophy, to join the argument with Kant yet once again. In this discussion, we will touch not only upon Kant the individual but also upon Kant as the spirit of an epoch which upheld him because his theories became the expression of the prevalent intellectual orientation.

II. KANT'S SCIENTIFIC WORLD OF APPEARANCE

What was the historical situation out of which Kant's philosophy grew?

Among the sciences, mathematical natural sciences had undergone a tremendous upsurge. At the beginning of the century preceding Kant's, Galileo had, by means of his method of posing questions and the experimental method, given to mathematical physics that form which distinguishes it from ancient and mediaeval science and still constitutes its basis today. In a letter, Galileo once differentiated his method from the speculative Aristotelian philosophy of nature prevalent in the Middle Ages: he did not want, he said, to "penetrate into the true and immanent essence of natural substances by means of speculation," but wished instead to investigate "the empirical characteristics" of things, "their location, their movement, their shape and size," and other features which are likewise to be discovered by observation.

The method in question is the one he employed in establishing the laws of falling bodies and the one applied by his young contemporary Kepler in describing the regularity of the paths of the heavenly bodies. Later on Newton, who wrote just two generations before Kant, brought this same method its greatest success when he established the differential laws for the motions of heavenly bodies and demonstrated the fundamental similarity between the heavenly movements and the motions of freely falling bodies at the earth's surface. At the same time, the differential calculus, invented by Newton and Leibniz, transformed mathematics into a flexible instrument for use in the formulation of laws of nature. With these discoveries, mechanics became the model for all of the exact sciences, and it became a matter of course that all scientific knowledge had to conform to the ideal represented by mechanics. In contrast, the other sciences lagged way behind. Newton had laid down the first principles of optics; the dispute between the emission theory and the wave theory of light continued, but it was understood by the followers of both theories that in either case optics must be carried out in accordance with the model of mechanics, whether as the mechanics of light corpuscles or as a mechanical theory of ether. Nothing was known of electricity except for the phenomena of friction, to which the electrical theory of thunderstorms had been added since Franklin's time; the concept of the electrical current and everything attendant upon it did not yet exist. Heat was only pictured as a fine, penetrating substance. The principle of energy was still unknown. Lavoisier's theory of combustion coincides with only the final years of Kant's life. The construction of matter out of atoms was still construed according to the primitive notions developed by Descartes and his disciples from the atomic theory of ancient times. Chemistry still conformed completely to the conceptual world of alchemists, who had, to be sure, already assembled a fund of practical knowledge but were as yet unable to offer any clear conceptual explanation of chemical reactions. Biology was still more primitive.

The first version of the theory of evolution had not been introduced in Kant's day, and no one had yet attempted to establish biological laws.

Kant himself grew up with strong scientific interests. Among his writings we find a large number of purely scientific investigations, in which he reveals himself as an exacting expert in the handling of empirical facts, making great efforts to discover systematic connections underlying them. The very titles of his works betray the scientific orientation: 'Ideas Concerning the True Measure of the Vital Forces', 'The Question Whether the World is Growing Old, Viewed from a Physical Standpoint', and so on. The most important result of these works is the cosmological hypothesis, which ever since has been associated with the names of Kant and Laplace. Even if this hypothesis is no longer viable today, the way in which Kant establishes it shows a clear regard for the facts known at the time. For instance, he explains the great distance between Mars and Jupiter by the theory that the various bits of diffuse matter originally present within the space between their orbits were absorbed into Jupiter; this distance is therefore "the greatest between any of the planets" because the planet in question "has a greater mass than all others together." In connection with the ideas of the Englishman Thomas Wright of Durham, the notion that the Milky Way encircled our galaxy was developed; in this same connection, we find already expression of the idea that nebulae are independent galaxies.

Yet it is certainly striking that not a single work of mathematical content is to be found among his writings. He was, of course, familiar with the mathematical treatment of mechanics, and the discerning manner in which he explains, e.g., the elliptical motion of the planets – by demonstrating that the approach towards the sun on one side of the ellipse, after the fashion of a pendulum, must be followed by a moving away on the other side – shows his clear understanding of these relations. But he himself never undertook any strictly mathematical investigations, and he never uses formulas in his writings. This is all the more striking in that Kant himself coined the famous slogan that a science is only truly scientific to the extent that it contains mathematics. He apparently admired mathematics more than he mastered it. In this regard he differs from his predecessor Leibniz, who, like him, was a philosopher of the mathematical sciences, and perhaps this is the psychological root of the somewhat tense relations between Kant and Leibniz.

Despite this absence of mathematical works, it would not be possible to name many philosophers today with such a comprehensive knowledge of the science of their time or such an intense desire for participation in scientific thought. The opposition in which the academic philosophy of today stands to natural science would be least comprehensible to that philosopher who is its spiritual ancestor, for Immanuel Kant's philosophical work is deeply rooted in the scientific way of thinking and regards it as being, as a matter of course, the prototype of all knowledge.

III. EPISTEMOLOGY AS THE ANALYSIS OF REASON

And yet there is a profound difference between Kant's style of philosophical thinking and the working methods of natural science. This manifests itself in the fact that Kant invariably abandons the inductive method, which proceeds from the facts to their generalization, when he makes the transition from science to philosophy of science. Even in his day, Kant had the possibility of constructing philosophical analysis upon the basis of natural science in accordance with the available factual material, similarly to the way in which science itself rests in turn upon the facts of perception. But he fails to see this inductive route; he never calls his philosophy a philosophy of science. He wants more than that: he constructs a philosophical system that is grounded in pure reason and does not include any experience, thereby turning the inductive relation between natural science and philosophy into a deductive relation. His route does not proceed inductively from the epistemological concepts of the science at hand to philosophy, but the other way around instead – from the philosophical system he deduces the epistemological concept of science.

The features of the philosophic system derived in this manner are well known. It centers around the concept of the synthetic *a priori* judgment, i.e., a judgment that is certain and yet is not empty – not analytic or, as we prefer to say today, not tautological. Kant considers such judgments to be involved in the characteristics of space and time and also in certain universal scientific propositions, including, among others, the axioms of mechanics, the principle of causality, and the principle of the conservation of matter. Anyone well versed in mathematical physics in Kant's time had of necessity to come up against these principles, for they are central to scientific modes of thought. Accomplishments such as the reduction of mechanics to a few fundamental principles from which the whole of mechanics can be developed in the form of a spatio-temporal description possessing strict necessity, the universal implementation of the principle of uniform connection, and so on, inevitably raised the question of the source of the claim to certainty associated with these propositions. Kant develops the answer that these propositions have their origin, not in the world of experience, but in human reason. He believes that this solution is in no way incompatible with the idea that natural science is completely empirical in every other respect.

That all our cognitions start *with* experience does not mean that they arise *from* experience. For it could well be that our cognitions of experience are themselves composed of what we receive through impressions and what our own capacity for knowledge (which is simply occasioned by sensual impressions) offers of itself, we ourselves not distinguishing this addition from the basic material until long practise makes us aware of it and skillful at separating it out.[1]

With these words Kant develops his scientific program, for the abstraction of the rational element in cognition is the theme of his philosophical system.

But it is at this point that Kant does his characteristic about-face. He attempts to discover the rational element in knowledge, not by means of an analysis of natural science, but through a 'critique of pure reason'. In accordance with this approach, the work bearing this famous title includes no scientific observations and scarcely any scientific examples. Instead, it discusses in a very general way the most general properties of things as demanded by reason and even attempts to deduce by logic the categories, which is to say the most general forms of the scientific process. Thus the concept of knowledge originated by Kant does not, in the end, arise from comparative observations of the natural sciences but from an intuition of pure reason. The results of this analysis of reason have been presented by philosophers ever since as the logical prototype of the concept of knowledge.

The fact that the concept of knowledge so derived agrees so beautifully with the scientific concept of knowledge in Kant's era might seem suspicious to us, yet – and this is what is so surprising – this very fact was regarded, both by Kant himself and by many generations of his followers, as the pinnacle of the success of his philosophy. Kant believed himself to have shown that the concept of knowledge in mathematical natural sciences is grounded in the nature of reason; he fails to see either that he analysed only that aspect of reason that has been developed in mathematical physics or that this stage of knowledge need not be the final one. And just because he does not see this, because he constructs a deductive system that gives a justification of the contemporary scientific level as *a priori* necessary, he attained his historic success. For each generation of men seeks a philosophical system with a consuming desire; mankind has a profound need to conceive of what he experiences as being unconditionally necessary, and he who is able to show that the world as it is known could not possibly be otherwise is bound to be celebrated by many as the herald of the most supreme wisdom. The more self-sufficient his system, the more unified his epistemological construction, the more he will be esteemed. The architectonic of a philosophical system contains strong powers of suggestion that blind the faithful to the arbitrary elements or empirical conditions that form its basis. And the need for uniformity is connected to the aesthetic need; the belief that truth has to be constructed in the forms of simple regularity seems to be one of the deeply cherished primitive desires of mankind, and in many people the wish for an harmonious system of knowledge is stronger than the ability to test the facts without prejudice. We are dealing here with forces of sociological proportions; here are to be found the factors that determine the public success of a system of thought, and here, assuredly, are to be found the springs from which have flowed Kant's wide-spread success among the educated classes.

IV. KANT'S SCIENTIFIC PRINCIPLES REGARDED IN THE LIGHT OF THE SCIENCE OF TODAY

And today? How does Kant's philosophy look when viewed from the standpoint of the natural science of our own time?

One thing is clear: natural science has become something quite different since Kant's time. The mechanics of Galileo and Newton have been displaced by Einstein's theory of gravitation, in which space, time, and gravitation are combined in a uniform manner and for which the Newtonian law of attraction presents only an approximation. Geometry, which for Kant was the safest example of a synthetic *a priori* science, has dissolved into a multiplicity of geometries, and which of these geometries fits reality has become an empirical question. Electrical phenomena have appeared alongside mechanical ones, not only revealing a new sphere of physical regularities but also forcing mechanics from its privileged position. It turns out that the smallest particles of matter are electrical in nature and that mechanics presents only a relatively crude description of matter at the macro-level. The theory of light and radiation has led to the discovery of the quantum of energy, and the relation between the problem of quanta and the problem of the elementary particles of matter has led to those remarkable conceptions of the wave character of matter which we view today as an abandonment of the principle of strict causality and a transition to statistical regularity. It goes without saying that such a profound change must at the same time influence the results of any philosophical thinking that originates in natural science, and we must track down more precisely which positions in the Kantian system are affected by these changes.

The problem of space stands in the foreground of Kant's interest, for he believes it to be a sphere in which the appearance of synthetic *a priori* judgments is demonstrably inevitable. The certainty with which we pronounce the axioms of Euclidean geometry seems to him to prove that non-tautological propositions may also assume the character of *a priori* insights. And he see too, in our relation to geometry a special justification of this conception. For he believes he can demonstrate that the *a priori* character of geometry rests upon its specifically visualizable character; *pure intuition* seems to him a source of special insights that understanding alone could never gain. This Kantian idea of the intuitive necessity of Euclidean geometry has been picked up by other philosophical schools and subsequently used to defend the Kantian position against the discovery of non-Euclidean geometries, which took place in the post-Kantian era. In accordance with Kantian thinking these geometries have been viewed as being internally consistent and yet describing constructions that cannot be intuitively visualized; this, it is believed, offers a continued justification of the special position of Euclidean geometry.

And yet the development of mathematical physics does not concur with this interpretation. Helmholtz, in his day, recognised that the human capacity for intuitive visualization is not a fixed property, but is subject to adaptation and evolution, and that we have to view the intuitive preference awarded to Euclidean geometry as a product of ancient habit which finds its foundation in the Euclidean character of our physical surroundings. The physical entities by means of which we realize geometrical relations, i.e., rigid bodies and light rays, conform in their structure to the laws of Euclidean geometry, and this empirical fact is to be regarded as the source of such habituation. It is quite essential here to bear in mind that it is really a question of a fact, i.e., an empirically confirmable property of nature that is in no way to be conceived of as necessary. It might happen that our measuring instruments would realise quite different geometrical systems while being transported. The idea of conventionalism, according to which Euclidean geometry can always be maintained provided a suitable definition of congruence is selected, does not constitute an objection, for Euclidean geometry applies to our world through employment of a quite natural definition of congruence, namely, the definition by means of the rigid body. This is the fact of nature that led historically to the development of Euclidean geometry. In a more complex world, in which Euclidean geometry could only in a more complicated way be maintained by means of definition, it would surely not have come to have a favoured position; rather, mankind would quite early have constructed different geometries. And Helmholtz demonstrates further that on this basis it is possible to visualise intuitively non-Euclidean geometries.

> By the much-abused expression 'to represent' or 'to be able to think how something happens', I understand – and I do not see how anything else can be understood by it without loss of meaning – the power of imagining the whole series of sensible impressions that would be had in such a case.[2] *

As this idea was developed, not only was it demonstrated that non-Euclidean geometries can be visualised, but the reason Euclidean geometry appears necessary was uncovered. Actually, our Euclidean conceptions are not based on conclusions stemming from intuitive visualization but on purely logical inferences that derive their compelling nature from our having already unconsciously defined into our visualization Euclidean metrical relations, from which we then deduce the appropriate propositions [1928h]. Recently, attempts have even been started to demonstrate by cinematographical means that non-Euclidean geometries are equal from the standpoint of visualization.[3]

The Kantian conception that our intuition of space rests upon a special human *a priori* capacity thus collapses completely.[4] Contemporary natural science knows no intuitive necessity. This need not be interpreted as meaning – as, unfortunately, so many physicists think it does – that intuitive visualisation must be banned from physics altogether. It is just the other way around. All physical theories must always be intuitively imaginable, yet that does not mean that physics is subject to a certain compulsion exercised by intuition, but quite the reverse: that our intuitive capacity has the fundamental ability to make every theory that is not self-contradictory intuitively imaginable. In some cases the task involved may be difficult, but it must always be basically capable of fulfillment.

Kant's theory of time runs a parallel course to his theory of space. He develops it far less completely – in fact, it seems a mere repetition of his thoughts on the theory of space – but recent physics has developed, in the Einsteinian theory of time, a system the intuitive imaginability of which can be asserted with as much certainty as its incompatibility with the Kantian theory.

Kant's theory of substance met with a similar fate. Kant considers substance, too, to be an *a priori* concept, the nature of which is displayed above all in the law of the conservation of matter, and Kant's disciples wished to see a new vindication of his ideas in the later principle of the conservation of energy. Yet further developments, particularly in Einstein's physics, proved them wrong. We are not referring to the fact that the laws of the conservation of mass and of energy were, taken singly, proven false, only becoming viable again when they were joined together as a tensorial law in the theory of relativity. We are referring instead to the insight, which today is beyond

*But Kant bases his conclusions on the (theoretical construction) pure intuition just to avoid such accidents of empirical intuition.

doubt, that all such laws of conservation are of an empirical nature and that, in principle, we must be prepared to deal with the possibility that the principle of conservation is of a merely statistical character.

With these thoughts we arrive at the actual problem of the concept of physical knowledge today, at the problem of regularity, the development of which has also, in a peculiar sort of way, been in contrast to Kant's theory. Kant regarded the principle of causality as one of the most important tenets of his concept of knowledge. He sees this principle as expressing a synthetic *a priori* judgment, just as does the judgment of the intuition of space, but, according to his conception, of a more universal *a priori* form, for causality does not involve intuitive necessity. Here, he believes, we have an *a priori* that rests on pure concepts and yet is not, as he rightly recognizes, of a logical nature. That the principle of causality is not an analytic judgment, a tautology, had already been noted by Hume, and Kant, who, as he put it, was "awakened from his dogmatic slumber" by Hume's idea, constructed his theory of the synthetic *a priori* judgment with the object of solving this problem. He held that we are absolutely convinced of the necessity of the principle of causality, despite its non-tautological character, and he believed he could demonstrate that we will always adhere to the principle of causality, even if we are unable to establish the particular causal chain in an individual case. This version of his idea was for a long time accepted by natural science. The causal principle was regarded as a postulate that we invariably attempt to implement and which, even when faced with the most difficult situation, we will regard as at the most temporarily incapable of implementation, never as invalid, for (it was alleged) it is deeply rooted in the structure of human reason.

These ideas were overcome even before the advent of modern quantum theory. By that time, Kant's concept of causality had already been criticised on purely philosophical grounds, in connection with philosophical criticism of the concept of probability. Attractive as it appears at first glance, the Kantian theory reveals under closer scrutiny some basic faults of which it cannot rid itself. Two points in particular are to be mentioned.

If we examine the procedure by which we establish a special law of nature, we note that we are never dealing with certainty, but always only with probability. This stems from the fact that in every such case we are dependent upon the inductive inference, that is, upon the assumption that a previously observed regularity will recur in just the same way. The puzzling nature of this inference, which left even Hume in despair, is in no way dispelled by the postulating of a universal regularity in nature. For the assumption of such regularity, even if it conforms to the truth, says nothing as to the individual case; it cannot in any way justify our belief that the causal chain observed on one occasion reveals the permanent law. Thus Kant's *a priori* theory of causality does nothing to solve the problem of induction.

On the other hand, once we admit to the special position of inductive inference and its justification, the question of the principle of causality assumes a new aspect. For then we can present for discussion the question whether the principle of causality is valid at all, that is, whether the postulate of causality can be implemented in every instance. This question can in principle be decided with the instruments of inductive inference. That the decision itself has only the form of a probability statement makes no difference; all such cases essentially involve propositions asserting no more than probabilities [1932d].

The question of which form the assertion of causality is to take can now also be made more precise. Since all assertions about nature are merely probable, the assertion of causality is to be conceived of as an assertion about convergence; it signifies that, as progressive account is taken of more and more parameters, the probability of the prediction can approach certainty to any desired degree of proximity. And it is in just this form that the empirical verifiability of the assertion becomes clear, for it may well be that the laws of nature set limits to such precision that stop short of certainty. This is the logical form into which the causal concept of quantum mechanics must be fit; Heisenberg's uncertainty relation asserts that for the probability of predictions the limit lies this side of certainty.

With this approach to the problem of causality, the last and most important tenet of Kant's epistemological critique collapses. Now we must investigate the effects entailed by this shock to the Kantian epistemological system.

V. LOGICAL CRITIQUE OF THE IDEA OF THE *A PRIORI*

Not only did Kant assert the existence of synthetic *a priori* judgments to be a fact; he also tried to provide evidence for this fact. We must now investigate this proof, which culminates in the renowned idea of the necessary preconditions of all cognition.

Let us consider once again the passage quoted from Kant's principal work. In accordance with the program set forth, Kant investigates the means whereby knowledge is acquired, and he discovers that two different components are at work. Sense impressions supply the matter of cognition, the sensory content, and the understanding processes this matter, converting mere *sensation*

* but if infinitely many parameters are permitted, the principle becomes tautologous.

into *experience*, as Kant puts it. This is the process that he investigates in detail. He finds that the contents of sensation are ordered according to certain definite principles; these principles, however, are not supplied through sensation itself, but are, rather, additions originating in the human understanding. He classifies these *a priori* principles as *forms of intuition* or as *categories*, depending upon whether or not they are intuitively evident. These principles – so continues the Kantian train of thought – are on the one hand necessary, binding human reason, which cannot do otherwise than to think within the framework they provide; yet, on the other hand, they are not empty, not tautological. For that reason they are synthetic and *a priori* principles, and Kant believes himself to have adequately clarified the peculiar and initially seemingly incomprehensible existence of such judgments by means of the theory that they are necessary preconditions of cognition.

This argument is the focal point of Kant's entire philosophy and has been described by his disciples as irrefutable. But we have already seen that the very principles proposed by Kant – above all Euclidean geometry, absolute time, and the principle of causality – have been rendered invalid in modern physics. This leaves us with the question whether the physics of today is compatible with Kant's argument or whether we must reject it.

It can, in fact, now be shown that Kant's argument is untenable. It rests upon an erroneous overestimation of the participation of reason in cognition.

The following considerations make this clear. It is doubtless true that alongside the material supplied by sensation there appears in the act of cognition an act of transformation, that is, a cognitive process, in which certain principles are employed. But these principles have certain limitations. Since all cognition aims at inferring future data upon the basis of observed data and all cognitive transformation is directed at this one goal, this transformation is subject to a regulation: only those principles are applicable that lead to correct predictions concerning the future. If the understanding operates in accordance with a system of principles with which it is endowed by nature, then it cannot be established ahead of time whether conformity to this system of principles will lead to correct predictions. To be sure, the principles named by Kant are relatively broad, and it is therefore possible to some extent to structure the content of individual experiential propositions in such a way that the principles are preserved; for instance, it is possible to maintain the principle of causality even when it is not possible to indicate the causal relations for each individual case. But there are limits; under certain circumstances, the content of experience may assume a character that makes it impossible to maintain the principles. Such a collision could only be impossible if the principles involved were empty, i.e., analytic, in Kant's terminology. But precisely because they are synthetic, as Kant correctly realised, the possibility must be faced that Kant's system of epistemological principles may come into contradiction with our ever-increasing experience.

This is just what has happened in the course of the development of mathematical physics. The experiential contents, particularly as they are presented in relativity theory and quantum theory, cannot be reconciled with the principles put forth by Kant. This could be formulated more precisely by saying that, to be sure, only some individual principles must be rejected, while others can still be maintained, and that, furthermore, we have within certain limits a choice of which principles to retain and which to reject; but in any case it is true that the totality of all the principles can no longer be maintained. For these reasons, the pre-conditions of cognition named by Kant are not the preconditions of our knowledge of nature *today*.

What conclusions are we to draw from this fact with respect to the assessment of Kant's ideas? One approach would be to content ourselves with a generalization of the Kantian principles; even if the particular principles named by Kant are not the ultimate ones, the possibility remains that fundamental preconditions nonetheless do exist and must be gradually figured out. This idea has been developed primarily by the neo-Kantians, such as Cassirer, to whom we owe a debt of gratitude for their revision of the original rigid Kantian system. Yet this conception, too, leads to difficulties. For if we can never decide whether a system of principles is the ultimate system, the ideas of such an ultimate system become empty postulates that can be of no use in science. Indeed, this postulate cannot in any way be justified. For here, too, the argument outlined earlier, to the effect that it can never be said of any system of synthetic principles that it will be consistently compatible with all future experience, comes into play; it follows from this argument that no such ultimate system can exist. The relation between the totality of experience and a system of preconditions must, rather, be expressed in the following more complex form: for every totality of experience there exists a system of pertinent preconditions, but, conversely, for every given system of preconditions a totality of experience can be constructed that contradicts this system of preconditions. Therefore, there is no such thing as the most general system of synthetic preconditions.

In view of this situation, the only viable approach is a different one, rejecting completely the Kantian conception of a system of ultimate preconditions of knowledge. According to this alternative conception, the only question concerns the preconditions of the knowledge of a particular period. These

* (such systems of synthetic principles are epistemologically on a par with all inductively established empirical laws)
** this should be translated into model theory.

preconditions, which at the present time include, for instance, Riemannian space, the chain-like structure of causality, the energy principle, and the principle of quanta, represent empirical propositions of a high degree of generality. They present synthetic assertions about the world and are therefore *a posteriori*, precisely because our current experience is of such a nature that these principles are compatible with it. Thus the Kantian conception of necessary preconditions of knowledge is defeated; the preconditions of knowledge are not necessary, but are gained empirically and are permanently subject to the regulation of experience.

VI. THE HISTORICAL ROLE OF KANTIAN PHILOSOPHY

Kant himself did not regard the possibility of the development just described as being so remote as one might at first have supposed. From time to time he remarks that the applicability of the *a priori* principles he sets forth "require a certain organisation of nature," that is, that the principles cannot be applied to all possible experiential material without exception. "This harmony of nature with our cognitive faculty is presupposed *a priori* by the judgment . . . , whilst the understanding at the same time cognises it objectively as contigent . . . ." But Kant believes that, where this agreement does not occur, mankind must abandon its attempt at a scientific mastery of nature. "It might easily be thought that . . . it would be impossible for our understanding to detect in nature a comprehensible order".[5]

We cannot but admire the foresight contained in such words, yet we must note at the same time that it is the limitations he himself set up by means of his own method that compel him to pronounce such a judgment upon his own system. We have said that Kant's peculiar achievement was his clothing of the scientific cognitive concepts of his time in the form of an analysis of human reason, which is immutably given and the structure of which can be disclosed only in the form of a philosophical system. The practical application, the actual results of his philosophical methods are in marked conflict with this dogmatic rationalism. What Kant wanted was an analysis of reason; what he achieved turned out to be an analysis of the natural science of his time. It is in this unwanted contradiction that Kant's historical fate has its roots. His rationalistically systematic form gained for him the acknowledgment of the philosophical minds of his own day and the succeeding century; but it is to the scientific orientation of the contents of his work that he is indebted for his actual influence, the response from the ranks of the scientists, the penetration of his ideas into the thought processes of scientific experts right up to the present time. As so often in the history of sociological phases, the influence of a system of thought rests upon factors other than those the originator presented as its justification, with the result that the influence of the system takes on quite a different form from that corresponding directly with its content. Kantianism today has become essentially quite different from what Kant would have expected. His critique of reason has become an analysis of natural science through our consciously carrying out that of which he was not yet conscious. Those who subject natural science to philosophical analysis today are certainly not obligated to continue depending upon some form or other of Kant's conceptual construction, and the real error of the Kantians seems to me to be their desire to hang onto Kant's concepts and to salvage a few fragments of the system – a process that appears to lead inevitably to one collision after another with the developments of natural science. It would be entirely too presumptuous to continue an epistemological system into a new era in which scientific knowledge has undergone the most fundamental alterations. Historical development takes oddly circuitous paths to reach its goal, and we cannot naively demand that a line of development be continuous in the sense that its contents remain unchanged. A peculiar combination and transformation of the materials and forces contributed takes place in the melting pot of historical events; the results do not have the appearance that would have been expected by those who contributed the ingredients. But should it turn out that one of the ideas contributed has become really powerful, this can be accounted an historical success of the first order. The greatness of an historical achievement does not consist in its having *predicted* future developments but in having *produced* them. And this is the judgment of Kant that we must hand down today: his system has no validity for us, his theory belongs to the past along with the scientific vision of the world in the eighteenth century – yet Kant is certainly one of the few thinkers whose philosophical work forged the path along which philosophy of science progresses today.

NOTES

[1] Kant, I., *Critique of Pure Reason*, Introduction, I, A 1–2.

[2] Helmholtz, H., 'Über den Ursprung und die Bedeutung der geometrischen Axiome', *Schriften zur Erkenntnistheorie* (Springer, Berlin, 1921), p. 5 [Eng. trans. by M. Lowe as 'On the Origin and Significance of the Axioms of Geometry' in Hermann von Helmholtz, *Epistemological Writings*, R. S. Cohen and Y. Elkana (eds.) (Reidel, Dordrecht, Holland and Boston, 1977)].

[3] Gerhards, K., *Naturwissenschaften* **20**, 925 (1932).

[4] I cannot see that even the recent defense of Kant's theory of space offered by H. Driesch (*Philosophische Gegenwartsfragen*, Leipzig, 1933, p. 51) refutes these points, which I have presented more completely elsewhere. Quite the contrary: Driesch's attempt to separate 'pure' intuition from 'empirical' intuition seems to me a mere repetition of Kant's ideas, against which all the arguments I have presented are just as telling as they are with respect to Kant's.

[5] *Critique of Judgment*, Introduction, Section V, XXXVI (Akademie-Ausgabe) [trans. by J. H. Bernard, MacMillan, London, 1892].

40. CARNAP'S *LOGICAL STRUCTURE OF THE WORLD**

[1933c]

This extensive work by Rudolf Carnap represents a massive compilation of his ideas on logic and epistemology. Yet it is not Carnap's ideas alone but also those set forth by the Vienna Circle as a scientific conception of the world which have been fully presented for the first time in Carnap's major work. "The basic orientation and the line of thought of this book are not property and achievement of the author alone but belong to a certain scientific atmosphere which is neither created nor maintained by any single individual. The thoughts which I have written down here are supported by a group of active or receptive collaborators. This group has in common especially a certain basic scientific orientation. That they have turned away from traditional philosophy is only a negative characteristic. The positive features are more important; it is not easy to describe them, but I shall try to give a loose characterization. The new type of philosophy has arisen in close contact with the work of the special sciences, especially mathematics and physics. Consequently they have taken the strict and responsible orientation of the scientific investigator as their guideline for philosophical work, while the attitude of the traditional philosopher is more like that of a poet. This new attitude not only changes the style of thinking but also the type of problem that is posed. The individual no longer undertakes to erect in one bold stroke an entire system of philosophy. Rather, each works at his special place within the one unified science. For the physicist and the historian this orientation is commonplace, but in philosophy we witness the spectacle (which must be depressing to a person of scientific orientation) that one after another and side by side a multiplicity of incompatible philosophical systems is erected. If we allot to the individual in philosophical work as in the special sciences only a partial task, then we can look with more confidence into the future: in slow careful construction insight after insight will be won. Each collaborator contributes only what he can endorse and justify before the whole body of his co-workers. Thus stone will be carefully added to stone and a safe building will be erected at which each following generation can continue to work."

* [The book reviewed is Rudolf Carnap's *Der Logische Aufbau der Welt* (Berlin, 1928), which was translated under the above title by Rolf A. George (Routledge and Kegan Paul, London, and University of California Press, Berkeley and Los Angeles, 1967). – Ed.]

This ambitious program is carried out by Carnap in the present volume with the help of a precise logic which has developed during the last decades under the name of mathematical logic, a logic which has taken over from mathematics the language of formulas and, along with it, the rigorous construction of concepts and the profusion of results. To be sure, Carnap's book itself makes use of the logical symbolism in only a few parenthetical comments, no doubt with an eye to the wider circle of readers to whom the book is addressed and who at present would still be repelled by an excess of mathematical logic. Nonetheless, the textual material has, through the introduction of certain logical symbols, been conceived so clearly as to result in a lucid exposition of the foundations of modern logic. Indeed, we possess today scarcely any presentation of modern logic that can compare with this profound analysis of thought.

Yet more than logic is involved in this work; what has been constructed here is an epistemological theory which is to be described as the logical extension of the older version of positivism – a synthesis, as it were, of Mach and Russell – and which has already been accurately termed 'logical neo-positivism'. The basic idea is that all knowledge of reality stems from perceptual experiences, which underlie all science as a sort of register. They were called 'sensations' by Mach and subsequently 'data' by Russell and 'basic elements' or 'elementary experiences' by Carnap. As these elementary experiences are all we know of the world, there can be no other starting point for scientific knowledge. Scientific knowledge (and likewise the knowledge of everyday life) can only consist in logical transformations of elementary propositions referring back solely to the register. The real aim of Carnap's book is to describe the process of logical transformation. He terms this process 'constitution' and demonstrates how, in a succession of different levels, the concepts are constructed out of elementary experiences by means of constitution. The top level is made up of personal psychological objects, the second of physical objects, the third of psychological objects of others and mental objects; the levels proceed through this series in such a manner that propositions about the psychological objects of others and about mental objects will be constructed via physical objects. Carnap's process of constitution, which is an extension of the process introduced by Russell under the name 'definition in use' is to be conceived in such a way that intricate complexes of elementary concepts are replaced by abbreviations, or symbols, so that the assertions of both science and the workaday world are to be thought of as consisting altogether of such abbreviated propositions. If, conversely, we wish to comprehend the meaning of these abbreviated propositions, we must go back to the beginning and transform them back into elementary propositions. The meaning of every single proposition about real things consists, then, ultimately in a report about elementary experiences.

This is the powerful basic idea of positivism, presented in the logically rigorous form given it by Carnap. The usual objections to positivism – for instance, that objects must, after all, exist even when we are not perceiving them – thereby become pointless, for Carnap shows that a definition in use can be given for a proposition of this nature as well; it, too, will be nothing other than a report of elementary experiences. Here for the first time, then, there is developed an internally consistent system of positivism, the logical consistency of which must be admired by all who immerse themselves in its logical construction.

To be sure, I do not wish to assert that all possible epistemological criticisms of this system are thereby exhausted. For the epistemological question is whether such an interpretation of propositions about reality includes everything that we actually mean in asserting them. If Carnap's system were an exhaustive answer to this epistemological question, it would certainly have established a foundation for knowledge of the greatest security, for nothing is presupposed but the reports of immediate elementary experiences and the principles of pure logic. But it seems to me to be at least doubtful whether this reduction to perceptual reports and pure logic exhausts everything we mean to include in our assertions about reality. These doubts are principally aroused when we consider the use of the concept of probability in the natural sciences, for if we accept Carnap's reduction of scientific assertions, we forfeit the indisputable basic principle that such assertions are not merely reports of past perceptual experiences, but are also invariably predictions of future perceptual experiences. It is a puzzle to me just how logical neo-positivism proposes to include assertions of probability in its system, and I am under the impression that this is not possible without an essential violation of its basic principles.

But these are questions for further scientific investigation, and I believe that they, like a number of other questions, will be resolved within the confines of the scientific atmosphere of which Carnap's book is a part. Fortunately, this scientific atmosphere exists in various centers, of which the Vienna Circle is only one, and the discussions among these in some ways very different centers have already advanced to a high level. At any rate, the idea that philosophical assertions, like scientific assertions, are subject to the ultimate demand that they must assert something conceivable is a unifying principle upon the basis of which a further development of Carnap's ideas can be carried out and assent to which automatically excludes a large proportion of the so-called philosophical questions that still play a major role in other philosophical settings. Scientific philosophical knowledge has undeniably taken a great step forward with the appearance of Carnap's book, regardless of whether it constitutes an ultimate answer. For the path to further development can only proceed by means of the consistent formulation of conceptions, and the magnificently conceived form of positivism created and set forth in this work will remain in any case one of the most important statements of scientific philosophy.

41. THEORY OF SERIES AND GÖDEL'S THEOREMS

Sections 17–22

[1948e]

[These are the last sections of an unpublished manuscript which was circulated by graduate students and colleagues in the philosophy department at UCLA. We are publishing only that part which contains original work by Reichenbach. The manuscript was provided by Wesley Salmon. – Ed.]

17. THE PROBLEM OF AN INFINITE REGRESS IN THE ROOF OF CONSISTENCY

Since the time of their first publication, Gödel's results have been considered as a serious blow against Hilbert's program of proving the consistency of mathematics. If the consistency of a system cannot be proved within the system itself but only in the metalanguage, a proof of consistency seems valueless because we do not know whether the language within which the proof is given is consistent. It was Hilbert's program to secure the reliability of ordinary mathematical proofs by giving a proof of the consistency of the mathematical language. If it turns out with Gödel's second theorem that this proof can only be given in another language whose consistency is not proved, this seems to show that Hilbert's program cannot be carried through because it leads into an infinite regress.

We do not think that this criticism of Hilbert's program holds. It seems to us that the ideas just explained are based on a misinterpretation of the problem of consistency, and that a more profound analysis shows Hilbert's program to be unshaken, and independent of Gödel's results.

To prove the latter statement first, let us inquire into the significance of the theorem stating that the proof of consistency is only to be given in the metalanguage. What then would happen if the metalanguage should turn out to be inconsistent? This would lead to the consequence, not that our deduction of the statement of consistency is incorrect, but that the contrary of this statement could also be deduced; and this indeed would make our statement valueless. Now let us assume for a moment that Gödel's second theorem did not hold, or in other words, that Gödel had proved the contrary of his theorem.

This would mean that the proof of consistency of the language *L* could be given within L. A simple analysis shows that this would not improve the situation, since in this case our proof of consistency of L would be of value only if we were sure that L is consistent. In case L were not consistent, we could also deduce the statement of consistency of L, with the qualification that then the negation of this statement were deducible too. Thus if the consistency of L were deduced within L, this fact would not prove the consistency of L. This shows that the difficulties of the proof of consistency are not due to the fact that the statement of consistency is demonstrable only in the metalanguage; the infinite regress which we encounter in this case corresponds to the circularity resulting if the proof of consistency could be given in the language itself.

On the assumption that we can find an interpretation of the problem of consistency which avoids the infinite regress, or circularity, we shall put aside this problem for the present, and turn to the analysis of the consistency statement. We begin by repeating some results already explained in *Elements of Symbolic Logic* [1947c], p. 180. We explained there the distinction between the statement of consistency and the principle of contradiction. The axioms of logic contain only the principle of contradiction, which states that a formula and its negation are not both true. In reference to a logical system this principle turns into the *postulate* that a formula and its negation should not both be deducible. This however is a *demand*, not a statement; if we want to know whether the demand is satisfied we have to prove the *statement* that a formula and its negation are not both deducible. Until such a proof is given there is a blank left open in the system of logic. The discovery of this blank, of an unanswered but reasonable question, is the first great merit of Hilbert's analysis of the program of consistency. The second is the creation of methods to find an answer to this question. If the answer is given for any particular language L – and we know it has been given in the affirmative for certain elementary systems – this answer enriches our mathematical knowledge in the same sense as does the deduction of any mathematical formula. The theorem of Pythagoras, for instance, was assumed to be true, at least for a number of special cases, long before Pythagoras found its proof; but only after Pythagoras had deduced it from elementary axioms was the mathematical question of this theorem answered. In the same sense the question of consistency will be an open problem for the mathematician until it is answered by a demonstration of its truth, or its falsehood.

To come to a better understanding of the problem of consistency we shall therefore turn for the present to an analysis of the value of mathematical deduction, raising the question of value in general, and not restricted to the theorem of consistency.

18. AN EXAMPLE SHOWING THE VALUE OF MATHEMATICAL DEDUCTION

Let us imagine a man who does not know much mathematics but who likes to make experiments with numbers. He writes the third powers of consecutive whole numbers on a sheet of paper, takes the differences of these cubes, then the differences of the differences, etc. He thus comes to the following list of numbers:

third powers	1st difference	2nd difference	3rd difference
1			
	7		
8		12	
	19		6
27		18	
	37		6
64		24	
	61		6
125		30	
	91		6
216		36	
	127		
343			

Astonished that the third differences are always equal to 6, he continues his experiments and finds the same results for subsequent numbers. He will finally make the assumption that the same result will always hold, i.e., that the third differences of the cubes of consecutive whole numbers are always equal to 6.

This would mean proving a mathematical theorem by physical induction. Now physical induction is an inference which every mathematician abhors. Our man, though not a mathematician, does not feel satisfied with his proof either, and tries to find a better proof. He finally discovers the following reasoning:

$$\begin{aligned}(x-1)^3 &= x^3 - 3x^2 + 3x - 1\\ x^3 &= x^3\\ (x+1)^3 &= x^3 + 3x^2 + 3x + 1.\end{aligned}$$

By subtraction he finds

$$\begin{aligned}y_x &= x^3 - (x-1)^3 = 3x^2 - 3x + 1\\ y_{x+1} &= (x+1)^3 - x^3 = 3x^2 + 3x + 1.\end{aligned}$$

and then constructs

$$z_{x+1} = y_{x+1} - y_x = 6x.$$

Substituting here $x + 1$ for x, he gets

$$z_{x+2} = y_{x+2} - y_{x+1} = 6(x+1) = 6x + 6$$

and hence

$$z_{x+2} - z_{x+1} = 6.$$

Since this result holds independently of the special value of x, it holds for all whole numbers. Thus the theorem about the third differences is now proved, not by physical induction, but by mathematical deduction.

Our man will certainly feel much better with this result. We are ready to agree with him – but for what reason? Why is the man's second proof better than his first?

The first answer we should give to this question is, I think, as follows. We do not accept physical induction as a means of answering mathematical questions because it does not give us what we ask. What we want to know is not whether a man who operates with the cubes of whole numbers will always find the third differences equal to 6; rather we are concerned with the question of whether it is a *deducible theorem* that the 3rd differences are equal to 6. This question is by no means answered in the man's first proof; physical induction applied to the cubes cannot answer this question. I think this is the very reason that physical induction has no place in mathematics; it does not answer the questions which mathematicians ask.

Sometimes the rejection of physical induction is explained by the remark that a mathematician wants absolutely reliable results, and that therefore physical induction leading only to probability cannot satisfy him. I do not think that this is a good answer. Reliability is a matter of degrees, and in a case in which we have no absolutely reliable results it might be advisable to use results which are at least reliable to some extent. The question of reliability is a practical one, and I think that in all practical questions a mathematician will always be ready to apply physical induction. Imagine a mathematician telling another one: "I have here on this paper the solution of the *Fermat* theorem, but I do not tell you whether the result is affirmative or negative; if however you guess the right answer I shall give you a thousand dollars." I think every mathematician would then answer "It is negative," considering the fact that so far no triplet of whole numbers satisfying Fermat's equation has been found, and applying physical induction. As soon as a mathematical question is turned into a practical one, physical induction is applicable. If we do not consider physical induction a permissible method in mathematics, it is because we take it for granted that the aim of mathematics is to be stated, not as finding reliable results, but as finding deducible formulae.

If we do not want to decide one-sidedly for this conception of the purpose of mathematical proofs, we shall have to distinguish between two conceptions of the aim of mathematics. The first is a *formal conception* in which the aim is considered as being the deduction of formulae; the second is a *practical conception* in which the development of reliable statements is conceived as the aim of mathematical methods. We do not say that only one of these interpretations is the correct one; this would be a meaningless statement since we are concerned here with volitional decisions. What we want to say is that a criticism of mathematical methods will lead to different results according as we decide for the first or the second conception.

Turning back to our comparison of the two proofs concerning the series of cubes we shall now reconsider our results from the viewpoint of the second conception of mathematics, that is with respect to reliability. Is the second proof more reliable than the first?

It is obvious that we shall answer this question in the affirmative, thus constructing a second reason for the preference of the second proof over the first. However, the substantiation of this idea is not so easily given. To find it we shall have to leave the field of pure mathematics and turn to considerations which refer to the theory of physical induction.

19. RELIABILITY OF DEDUCED THEOREMS

Speaking of reliability of a language is meaningful only if we abandon the formal consideration of the language, and turn to its interpretation. Only an interpreted system can be reliable because reliability is a question connected with *truth*, and truth is a property only of interpreted statements. In what sense, then, are we to interpret mathematics if we speak of its reliability?

There are two different kinds of interpretation which we have to distinguish here. The first is an interpretation of the object language; the second is an interpretation of the semantical language. We shall begin with the consideration of the first, and postpone the discussion of the second to Section 21.

We know that the interpreted object language of arithmetic consists of statements about physical things with which counting operations are performed. The reliability of a deduced theorem then depends on the question of the validity of the axioms for the interpretation used; if the axioms are true for the interpretation, the theorem is also true. In other words, the question of the reliability of a theorem is reduced by the deductive proof to the question of the reliability of the axioms.

For a better understanding of this situation, let us turn from arithmetic to geometry. We know that here also the distinction between a formal system and an interpreted system must be made; we distinguish formal geometry from interpreted (or applied) geometry, the latter being a physical discipline. For the purpose of interpretation we coordinate to the undefined fundamentals of formal geometry, such as points, straight lines, congruence, etc., certain physical things, or relations of things; thus we coordinate to the term 'points' small particles, to the term 'straight lines' light rays, to the term 'congruence' the transport of solid bodies, etc. [see 1928h, Section 14] However, any such interpretation – and there are always a great many of them – is subject to the question of whether it is a *permissible* interpretation. To show this we have to prove that the physical things of the interpretation satisfy the axioms of the formal system. We have to show for instance that two small particles determine precisely one light ray as their connection, etc. Any such proof can be given only through empirical methods, and is therefore never absolutely reliable but restricted to empirical reliability. The reason for this restriction is that the proof can only be given by methods including physical induction. It follows that a deduced geometrical theorem, for instance, the theorem of Pythagoras, carries with itself the uncertainty originating from the question of the validity of the axioms, and is therefore not absolutely reliable.

The same holds for interpreted arithmetic. We here coordinate to the fundamentals of the axioms, such as the term 'equinumerous', physical operations; for instance we spread on the table the physical things belonging to a certain class, put close to each of them a thing belonging to another class, and interpreting *one–one-correspondence* by the physical relation *close to* we determine whether the two classes are equinumerous. Since we are concerned here with physical operations, the assumption that these operations correspond to the axioms of Peano is subject to empirical proof, and cannot be tested without physical inductions. This proof is given by very elementary methods, and may seem therefore practically certain; but we must never forget that we are here in the empirical field, and that, therefore, the validity of Peano's axioms for the physical operations of counting can be maintained only in the sense of an empirical statement of great reliability. It would not help us to say that if a deviation from the axioms should arise the mistake would be on the side of the interpretation, and that we would have to correct the interpretation. The statement of reliability is meant with respect to a given interpretation; we want to know whether this interpreted system is reliable, and not whether there is perhaps another one which is better. The implication "if the interpretation corresponds to the axioms it is a permissible interpretation" is without practical value since it does not tell us whether the interpretation actually used is permissible.

As in the case of geometry the rest of uncertainty, however small, is transferred from the interpreted axioms to the theorems, and thus we cannot say with certainty that in counting physical things we shall always find the theorems confirmed. Our theorem concerning the third powers of whole numbers (Section 18) may be interpreted as a statement about counting unit cubes in cubical volumes whose sides consist of whole numbers of units in length; since subtracting means dropping a sub-class of cubes which is in one–one correspondence to another class of cubes, our theorem translates into the statement that after performing certain operations with cubes composed of unit cubes we shall always have 6 unit cubes in our hands. Even if we assume that we make no mistakes in the arithmetical operations, this theorem is not absolutely certain because of the uncertainty originating from the physical interpretation of the axioms.

On the other hand we now see the reason why a deduced theorem is more reliable than a theorem obtained by direct physical induction such as the one used in our first proof of the theorem of the cubes (Section 18). It is also true that the reliability of the deduced theorem is ultimately based on physical induction, the latter being applied in the validation of the axioms; but this induction is more dependable than the direct induction applied in the independent establishment of the theorem. This is the reason that the second proof concerning the cubes is superior to the first from the viewpoint of reliability. A dubitable induction is replaced by a much better one – this is the progress attained in the deduction of the rule of the cubes from the axioms of arithmetic.

Strictly speaking, the situation is a little more complicated. The direct physical induction used for the support of the theorem is not entirely abandoned, but is *added* to the physical induction used in the establishment of the interpreted axioms; thus the evidence for the theorem is even a little stronger than the evidence originating from the isolated empirical testing of the axioms. Moreover, the evidence for the validity of the axioms is increased by the empirical confirmation of the deduced theorem as well as by the empirical confirmation of all other deduced theorems. This is an inference of concatenated inductions such as used in the verification of scientific theories (see [1938c], Section 41). With the deductive proof a great number of different inductions are united into an inductive network whose reliability is greater than that of each of the individual inductions of which it is composed.

It is important to realize that, as soon as we ask the question of reliability, the superiority of a deductive proof over an inductive one can be explained only within the framework of the theory of physical induction. The problem of reliability of statements constitutes a question which cannot be answered in deductive logic, but only within inductive logic.

20. CONSISTENCY OF EMPIRICAL LANGUAGES

We are now sufficiently prepared to raise, with respect to the reliability of deduced theorems, the question of consistency. If the consistency of arithmetic is not proved, we cannot regard it impossible that someone might deduce the contrary of a deduced theorem, for instance, of our theorem concerning the third differences of the cubes. Can we still maintain, in face of this possibility, the idea of the superiority of the deductive proof over the inductive one?

Before we answer this question let us state more precisely which kind of language we are here concerned with. A language consists of a set of axioms and a set of rules, the latter including truth rules, formation rules, and transformation rules [1947c], p. 16). If we say that a language is consistent, we mean that using the given linguistic rules it is impossible to derive both a formula and its negation; thus the statement of consistency refers, not to the axioms alone, but to the axioms in combination with the given rules. These rules, however, cannot be the source of contradictions if they are 'truth preserving' rules, i.e., rules of a kind such that every formula derivable from the true formula, is also a true formula. Considering now the set of axioms we may divide it into a subset of tautologies and a subset of synthetic statements. The tautologies cannot be the source of contradictions either, because every formula derived from a tautology by means of truth preserving rules, is a true formula. A language all of whose axioms are tautologies and whose rules are truth preserving must therefore be consistent; a contradiction cannot be derived because it would be a false formula, and false formulae cannot obtain through truth preserving rules applied to true formulae. Thus the source of contradictions can only be in the synthetic axioms. In our example of the cubes these are those axioms which we assumed as verified in the physical interpretation (cf. Section 19). The structure of the language to be considered may therefore be characterized as follows: the language will consist of tautologies and interpreted synthetic axioms, and possess truth preserving rules. Calling such a language an 'empirical language', let us ask the above question in the following way: is a theorem reliable if it is deduced in an empirical language whose consistency has not been proved?

This point may seem to be a serious objection; however the question can easily be answered if we realize that not only the synthetic axioms, but also the deduced theorem are interpreted, and that therefore the question of reliability concerns an empirical statement, which in any case cannot be maintained with absolute certainly. Only if the theorem about the cubes, in its meaning about physical things, were to be considered as absolutely certain, would the conception be seriously shaken by the possibility of contradictions in the mathematical language. However, if we admit a range of uncertainty in the validity of the deduced theorem, the possibility of contradictions can at most enlarge this range to a certain amount. Considered within the frame of the theory of induction, the problem of consistency will therefore be solvable in terms of degrees of probabilities, without leading to a break-down of logic.

Entering upon this analysis we immediately discover that the situation is even more favorable than we supposed, because the possibility of contradictions does not lower the chances of the deduced theorem being true. We see this as follows. If a system of axioms is inconsistent, it is impossible to find a physical interpretation for it; contradiction is always confined to language, and is never found in nature. This principle is nothing but an expression of the statement that a formula and its negation cannot both be true, i.e., of the principle of contradiction. Now the statement of the superiority of the deductive proof is based on the assumption that the interpretation underlying the theorems satisfies the axioms. We are not quite sure whether this assumption is true, and this is the reason that the truth of the deduced theorem cannot be maintained as certain. But from what we just said about contradictory systems, we infer that *if* the assumption is true, the system is consistent; if it were not, this system could have no interpretation, and the assumption therefore would be false. We see: with the same right with which we assume the validity of the axioms for our interpretation, we may also assume the consistency of the axioms.

It follows that the consideration of the problem of consistency does not change our previous results concerning deductive proofs; the possibility of a contradictory character of the axiomatic system does not diminish the reliability of a deductive proof. This result is due to the fact that reliability concerns only interpreted mathematical statements, and is therefore to be discussed within the theory of empirical knowledge. Since the latter theory can be given only within the frame of the theory of probability, let us apply the symbolism of probability theory for a clarification of the problem under consideration.

Let "*b*" be the assumption "the axioms are true for the interpretation in question", and let "*c*" be the statement of consistency of the object language. We then know that if our further postulates of empirical languages are fulfilled, we have

$$b \supset c. \tag{1}$$

For probabilities we use the symbol

$$P(x, y) \tag{2}$$

which means: probability of "*y*" with respect to "*x*", or more precisely: probability of the statement "y if x" [we write quotation marks here because we consider probability here as a predicate of sentences, like truth. – H.R.]. Probabilities are always relative in that they correspond to relations between two statements; we have therefore in (2) two terms separated by a comma. [the quotation marks are omitted; this may be considered as a convention that the parentheses of the *P*-symbol include among their functions that of quotation marks. cf. [1937b], pp. 321–322; and [1939b], pp. 56–57.] If we speak of the probability of "*b*", we have to add, therefore, a first term relative to which this probability is meant. We shall consider the totality of our knowledge as this point of reference of the probabilities under discussion; since both "*b*" and "*c*" belong to the semantic language, we shall understand by "*a*" the sentence "the totality of our knowledge is true," thus having all sentences of the same level. We are concerned here with probabilities of the form

$$P(a, b) \qquad P(a, c) \qquad P(a, b.c) \tag{3}$$

where the latter probability is the probability of the logical product "*b.c*" relative to "*a*". Now because of (1) we have the relation. [As to this relation see [1935h], p. 116, (4). As to the possibility of a frequency interpretation of probabilities like (3) cf. [1935h] and [1938c] Chapter V.]

$$P(a, b) = P(a, b.c). \tag{4}$$

This means: the probability that the axioms are true and the language is consistent, is the same as the probability that the axioms are true. We see from this that the assumption of consistency does not introduce a new source of uncertainty into the problem.

Now let us consider a theorem deduced in the object language; denoting by "d" the sentence "the theorem is true" we have, because of the elimination rule (see [1935h], Section 19 or [1937b], p. 282)

$$P(a, d) = P(a, c) \cdot P(a.c, d) + P(a, \bar{c}) \cdot P(a, \bar{c}, d). \tag{5}$$

The term $P(a.\bar{c}, d)$ means the probability that the theorem is true if the system is inconsistent (and the axioms, therefore, not all jointly true); this case is of course possible, although we may consider this chance as small. Independently of any assumption about the degree of this probability we derive from (5) the inequality

$$P(a, d) \geqq P(a, c) \cdot P(a.c, d) \tag{6}$$

which follows because $P(a, \bar{c})$ and $P(a.\bar{c}, d)$ cannot be negative. The equality sign holds in case $P(a.c, d) = 0$.

We know further that if the object language is consistent, the probability of "d" is greater, or at least equal to the probability of "b", the uncertainty of "b" being the only source of uncertainty of the deduced theorem; we therefore have

$$P(a.c, d) \geqq P(a.c, b). \tag{7}$$

With (6) and (7) we derive

$$P(a, d) \geqq P(a, c) \cdot P(a.c, b) = P(a, b.c) \tag{8}$$

where we have used the rule of multiplication. With (4) we get

$$P(a, d) \geqq P(a, b). \tag{9}$$

We see: if we consider a deduced theorem as not absolutely reliable its probability is greater, or at least equal to the probability of the assumption "b" that the axioms are true in the interpretation used; moreover, this value already takes into account the uncertainty due to the assumption "c" of consistency, since the probability of both assumptions "b" and "c" together is equal to the probability of the assumption "b".

This result may also be stated as follows. If we deduce from a system of axioms an empirical statement concerning future observations we are not quite sure whether the expected observations will occur. Now it is irrelevant whether we attribute the remaining uncertainty to a possible failure of an inductive inference used in the verification of the axioms, or to a possible contradiction arising from the deductive operations used in the derivation of the statement. The chances of the latter possibility are included in those of the first.

We see that in an interpreted language we have good inductive evidence for the consistency of the language. This evidence is determined by the equation (following the rule of elimination)

$$P(a, c) = P(a, b) \cdot P(a.b, c) + P(a, \bar{b}) \cdot P(a.\bar{b}, c). \tag{10}$$

Because of (1) we have $P(a.b, c) = 1$; as the terms $P(a, \bar{b})$ and $P(a.\bar{b}, c)$ cannot be negative we have

$$P(a, c) \geqq P(a, b). \tag{11}$$

This means: the inductive evidence for the consistency of a language is greater than, or at least equal to, the inductive evidence for the truth of the axioms of a language. We see that in an empirical language deductive proofs need not be accompanied by a proof of consistency of the language.

21. LOGICAL EVIDENCE

After these results we turn once more to the objection of the infinite regress. We reached our conclusion by deductive operations in a language L whose consistency is not proved; is it then permissible to consider our result as reliable?

It is reliable if L is an empirical language; but to prove this we would have to make statements in the metalanguage L'. We would have to state in L' that L consists of empirically verified axioms and of tautologies, and that the deductive operations in L are truth preserving. All these are empirical statements in L'. However, they are not obtained by immediate empirical evidence, but by deductive operations in L'. Thus proving that a formula is a tautology presupposes the assumption that the first two columns of the truth tables exhaust all possible combinations of truth values (which means assuming the principle of *tertium not datur* in L'). It follows that in order to prove L to be an empirical language we first have to prove that L' is an empirical language; only after such a proof would the statement "L is an empirical language" be reliable. Repeating this consideration for L' we see that we are led here into an infinite regress.

The infinite regress to which we come here is somewhat different from the regress pointed out in Section 17. We do not come to the result that the consistency of the metalanguage should be proved before we use this language; instead we demand only that some statements about the nature of this language should be proved before we apply the language. These statements are of a simpler kind than the statement of consistency, and thus far our result means an improvement of the situation. However, the infinite regress is not yet eliminated.

We do not think we can avoid this regress as long as we keep to the principle that knowledge about a language has to be formulated in another language before it can be used as the basis of deductive operations. This principle will always lead to an infinite regress for the following reason. If we formulate conditions under which a language is considered as *acceptable*, i.e., can or shall be used for deductive operations, the principle under consideration will require that the fulfillment of these conditions is to be formulated in statements belonging to another language. Before we can determine whether these conditions are fulfilled we shall therefore have to ask the question whether this second language fulfills the conditions, etc. Although we could formulate such a *rule of acceptance* for languages we could not apply it, as its application would be possible only after we had already run through an infinite number of cases of applications of the rule.

It seems to us that the only way out of this difficulty is to admit the existence of an *immediate knowledge* about the language in which we are talking, a knowledge which, although it can be formulated, need not be formulated in order to be applicable. *Logical evidence*, i.e., the logical constraint which we feel in using tautologies and applying truth preserving rules, is the psychological criterion which tells us whether the language in which we are talking satisfies the operational conditions of an acceptable language. If, furthermore, *perceptive evidence*, i.e., evidence through sense observation, shows us that the synthetic axioms of the language are true, we know that the language we are talking is an empirical language. Both logical and perceptive evidence constitute the guides which we rely upon when we accept a deduced statement of an empirical language as reliable. In relying upon this immediate knowledge we avoid the infinite regress into which we would be led if we were to make the acceptance of a statement dependable on the acceptance of another statement.

We do not want to say that every tautology is logically evident; most of the tautologies, such as the higher theorems of mathematics, are of course too complicated to be immediately seen as tautological. Mathematical proof therefore consists in the decomposition of complex inferences into small steps which, taken individually, are logically evident. We may furthermore refer to the fact that the tautologies applied as evident can all be reduced to interpreted tautologies, that is to formulae which are special cases of general tautologies, and therefore contain no propositional variables. Thus the use of logical evidence may be confined to very simple deductive steps. It is, however, impossible to entirely eliminate logical evidence from deductive operations since without the use of logical evidence as criterion of validity we would be led into an infinite regress.

If we admit with this that logical systems cannot be constructed without reliance upon logical evidence, we do not want to say that there are *absolute truths* in which we have to believe. Although an interpreted tautology appears logically evident when we think it, we do not know whether it will again appear logically evident to us when we think it tomorrow. We even know of cases in which a formula previously evident seemed to us not evident at a later time. The history of mathematics has shown that demonstrations accepted at one time as valid have been abandoned later, that accepted theorems have been later restricted in the domain of their validity. We shall take account of this situation by saying that, although we shall consider tautologies as certain, the statement that they are certain is only empirical. Thus the uncertainty of empirical knowledge is transferred to logic, with the only difference that uncertainty comes in here on a higher linguistic level.

With this shifting of the empirical character of logic to the higher linguistic level we repeat considerations explained in *Elements of Symbolic Logic* [1947c] p. 186 in which we pointed out the empirical character of a sentence stating that a given formula is tautological. We gave as a reason for this that we are not quite sure whether the empirically given formula actually has the structure which would make it tautological. In the preceding consideration the same idea is expressed by saying that we do not know whether the formula will appear evident at a later time.

Since the uncertainty of the higher level statement makes the certainty of the logical formula practically valueless, we see that the ultimate reliance on logic is to be analyzed in the theory of empirical knowledge. We cannot develop this theory here (see [1938c], Ch. V). We may, however, add the remark that we need not speak here of a belief in empirical statements; we can interpret them as *posits*, or *wagers*, i.e., statements dealt with as true statements as long as they have not turned out to be false. If the latter should happen, they are replaced by new posits. Empirical knowledge follows a method of posits and corrections, of trial, error, and new trial; its justification derives from the fact that following this procedure is the best we can do.

Following the guide of logical evidence is in the same sense a procedure of trial and error, with the difference however that the posits we make are not in the language itself, but in the metalanguage. (The tautological statements of the language L may also be considered as posits, but they are *appraised* posits whereas the corresponding statements of the metalanguage are *blind* posits (cf. for these terms [1938c] pp. 352–353). Logical evidence may be conceived as an index of the highest possible *initial weight* (*ibid.* p. 277).) Thus we posit that an interpreted tautology is certain. We see that the difference between analytic and synthetic statements is not dropped in this empirical interpretation of logic; there are analytic statements, but they are not 'absolute' since their reliability is restricted by the uncertainty of the metalanguage sentence stating that they are analytic. It follows that, although mathematics is not denied the analytic character of its propositions, ultimately even mathematical procedure is justifiable only as a method of trial and error.

22. THE VALUE OF A PROOF OF CONSISTENCY OF MATHEMATICS

Although interpreted mathematics is an empirical language, the results of Section 20 cannot be immediately applied to the problem of consistency of mathematics. The reason is that the relation between the formal system of mathematics and its interpretation is of a rather complicated kind.

The axioms of mathematics constitute a system which is not fully covered by the interpretation; in particular, the concept of the infinite does not actually enter into the interpretation, since all physical observations concern only a finite number of objects. We refer here to the distinction between *precision mathematics* and *approximation mathematics* introduced by F. Klein.[1] To explain this distinction let us use the example of irrational numbers. For all physical purposes the system of rational numbers is sufficient, because physical measurement confirms mathematical statements only within certain limits of approximation. It would therefore be meaningless to ask, for instance, whether the specific weight of copper is a rational or an irrational number. The system of mathematics splits therefore into the two parts M_p and M_l, where M_p is the subsystem used in physical interpretations, and M_l an additional subsystem of a purely logical kind which is not included in the physical interpretation. This uninterpreted subsystem M_l is not without value; it is a fact that mathematical deductions within the comprehensive system M are technically much simpler than deductions restricted to a physically meaningful subsystem M_p. Thus we determine a velocity as a differential quotient instead of using a quotient of small differences, which would strictly correspond to what we actually measure; this *idealization* simplifies the use of mathematics considerably. However, as soon as we turn to the question of consistency the distinction explained leads to important consequences.

Since all theorems of applied mathematics are deducible from the subsystem M_p, it is only this subsystem which is verified by the interpretations. The inductive inference leading from the verification of the system to the truth and, therefore, consistency of the axioms concerns only the subsystem M_p, leaving open the question of the consistency of M_l, and therefore of the comprehensive system M. In spite of that, if we consider the whole system M as consistent, our inference is based only on the fact that so far no contradictions have been discovered. This means that we infer the consistency of the system M only by inferences of the kind used in the first proof of the theorem of the cubes (Section 18); the statement of consistency of M is therefore obtained only through direct inductions of a very questionable type. This is the reason that mathematicians are anxious to find a deductive proof of consistency.

We may state the situation as follows. Since not all axioms of M are interpreted, we have $P(a, b) = 0$, if "b" is to include all axioms of M. The inequality ((10), Section 20) therefore does not furnish any utilizable lower bound for $P(a, c)$. Thus the latter probability is determined only by inductive generalization based on the fact that so far no contradictions have been observed. This shows that the inductive evidence for the consistency of M is different from the inductive evidence we have for the consistency of M_p. If we are dealing with an interpreted system, like M_p, the inductive inference concerning its consistency runs through the verification of the axioms; this is a satisfactory induction, because it is a case of concatenated inductions (cf. the end of Section 19). In an uninterpreted system, or a system only partially interpreted, like M, the inductive evidence for the statement "c" of consistency is unsatisfactory since this evidence consists in nothing but direct inductive generalization; such an inductive inference would be an *isolated* induction, and therefore not very dependable. Thus our criticism of the assumption of consistency of mathematics is based, not on a refusal to accept inductive evidence, but on the distinction between good and bad inductions.

Let us analyze the situation in order to find out what would be reached with a deductive proof of consistency of mathematics. We know that the discussion of consistency can be given only in the metalanguage. Speaking in the metalanguage we consider mathematics as a system of signs, disregarding any interpretation of these signs, and looking at nothing but the operations which we perform with the signs. The signs are physical things, but although the system of axioms is represented by a group of these physical things, namely the corresponding signs, these signs are not interpretations of the axioms. This is seen from the fact that a contradictory system of axioms can be written, whereas it cannot be interpreted.

While the object language is not interpreted, the metalanguage *is* interpreted. We have to know, for instance, what the term 'the same sign' means, i.e., we have to apply a relation of similarity to individual signs. In making substitutions and inferences we have to understand the meaning of the rules controlling these operations. We have to know the difference between a variable and a constant, in order to apply the rule that what is proved for a free variable holds for all constants.

We can, for instance, state what the theorem of the cubes (Section 18) means in the metalanguage. It means here that in performing certain operations with signs of whole numbers, namely the operations we perform when we 'figure out' third powers and their third differences, we shall finally arrive at the number-sign '6' whatever be the signs of whole numbers with which we start. We see that this *semantical interpretation* of the theorem is very different from the *object interpretation* of the theorem explained in Section 19, in which the theorem refers to operations with physical cubes, and not with number-signs.

In a similar way we can express the statement of consistency in the semantic interpretation. It then means that if a formula "a" is deducible by operations with the signs of the axioms, the formula "$\bar{a}$" is not deducible. Seen from the metalanguage this is a theorem of the same kind as the semantically interpreted theorem of the cubes; both theorems state that, in performing operations with signs, only combinations of a certain kind can obtain. Both theorems therefore need a deductive proof.

As for the theorem of the cubes in the semantical interpretation, this proof is immediately given with the deductive proof of the theorem in the object language. For the statement of consistency, the situation is different since we have no theorem corresponding to it in the object language. We have an indirect proof of a statement of consistency based on the empirical verification of the interpreted object language; but we found that this proof concerns only the consistency of the subsystem M_p. Thus we have no proof for the consistency of M. We see that the construction of this proof, which is the objective of the formalist school of mathematicians, would fill in a blank left open in the theory of mathematics.

This result shows that, from the viewpoint of the formal conception of the aim of mathematics (cf. Section 18), a proof of consistency is to be required, because it answers the question about the deduction of a theorem which otherwise is derivable only by means of direct physical induction. However, we see that, as in the problem of the cubes, the deduction is desirable also for the practical conception of the aim of mathematics, because it would improve the reliability of the interpreted comprehensive mathematical system M, for whose consistency we have no proof based on a satisfactory physical induction.

We come to the result that we need a proof of consistency for classical mathematics, because mathematics is not a fully interpreted language. Thus it is the special nature of classical mathematics which makes a proof of consistency necessary; it would be unjustified to demand that for every kind of mathematics a proof of consistency should be given. It is therefore possible that, although we demand a proof of consistency for classical mathematics, we need not demand such a proof for the metalanguage in which the proof of consistency is given. Applying the results of Section 20 we can indeed show that such a proof is unnecessary.

The language in which the proof of consistency is carried through is the interpreted semantical language, and therefore an empirical language. We found in Section 20 that the reliability of a deduced theorem in an empirical language is never of the type of absolute certainty, and that it does not depend, on the other hand, on a proof of consistency of this language; this consistency is assumed with the same right as the validity of the interpreted axioms. We now have simply to transfer those results to the metalanguage. The proof of consistency of the object language *M* is the deducible theorem "*c*" whose reliability is under consideration. The axioms "*g*" from which this theorem "*c*" is deduced, and which are formulated in the metalanguage, include syntactical descriptions of the axioms of the object language, and descriptions of the rules of deduction used in the object languages. The reliability of the deduced theorem "*c*" will depend on whether the axioms "*g*" are true, i.e., on whether the syntactical descriptions of axioms and rules of the object language correspond to the actual structure of the object language. With the same right with which we assume "*g*" to be true we may also assume "*g*" to be consistent. It is this consistency of "*g*" which we mean when we speak of the consistency of the metalanguage, and hence we come to the result: The inference leading to the consistency of the metalanguage is based on a satisfactory physical induction, since this induction is of the type running through the validity of interpreted axioms, and not of the type of a questionable direct generalization.

It is an essential premise of this conclusion that the statement of consistency is to be considered as an empirical statement. To meet eventual objections to this conception we shall add some remarks. We explained in [1947c], p. 186, that syntactical statements can be made in the two different forms (2) and (3). They may be distinguished as the *assertory* and the *implicational* form. The assertory form refers to a sentence which is written on paper, or otherwise given, and states some properties of it; of this form is the syntactical statement, "this sentence is deducible." The assertory statement is always empirical since it tells us some properties of an empirical object, in this case a sentence. Only in the implicational form can a syntactical statement be tautological. It then describes a sentence to which it refers without asserting that any actually given sentence satisfies this description, such as in "a sentence of this and this structure is deducible." For the statement of consistency we have the same distinction of two forms; but what is called the problem of consistency always concerns the assertory form of the statement, because we want to know whether a given language is consistent. It would not help us to know "if a language has this and this structure it is consistent," or "if the language *M* has this and this structure it is consistent;" what we want is a statement of the form "*M* is consistent." This is an assertory syntactical statement, and therefore empirical. It can be proved in the metalanguage only if the axioms "*g*" from which it is deduced are interpreted axioms, i,e., are asserted with respect to the given language *M*. Using for the mathematical metalanguage a terminology which has been developed for the mathematical object language, we may say: the statement of consistency belongs to applied mathematics, not to pure mathematics.

We summarize: for an empirical language a proof of consistency is not necessary, because the assumption of consistency is included in the assumption of the truth of the axioms. For mathematics we need a proof of consistency, because only a subsystem of mathematics is interpreted in the object language, and thus the consistency of the whole of mathematics cannot be inferred from empirical verifications of the axioms and theorems of mathematics. Since the statement of consistency concerns an empirically given language, its proof is to be given in a metalanguage which is interpreted and empirical, for which a proof of consistency therefore can be dispensed with.

We see that the problem of consistency does not lead to an infinite regress if it is conceived as a problem of inductive, not of deductive logic. Although the proof of consistency is to be given in a deductive way, the statement of consistency is an empirical statement whose truth depends on the truth of the axioms of the interpreted semantical language, and whose reliability, as well as the reliability of the language to which it refers, is a matter of degrees to be discussed within the frame of the theory of induction.

In mathematical expositions of the problem of consistency we sometimes read that a proof of consistency is of value only if the metalanguage is simpler than the object language, to the effect that the metalanguage is more likely to be consistent than the object language. The chances of reaching this aim seem imperiled by the Gödel theorem. With this way of stating the problem it is at least admitted that the problem of reliability of mathematics is not a problem of deductive but of inductive logic, since the term 'simpler' refers to degrees of reliability. It is, however, not very clear what is meant by the term 'simpler', or why a simpler language is to be more reliable. It seems to us that the correct formulation would rather be as follows:

The semantic language M' in which the proof of consistency of M is given should be a language which does not contain a subsystem M'_l which is not included in the interpretation of the semantic language. If it has one, the consistency of this uninterpreted subsystem M'_l would not be demonstrable through the inductive evidence for the validity of the interpreted axioms of M'; therefore, if the subsystem M'_l would be used in the deduction of the statement of consistency of M, the reliability of this statement would not be equal to that of the interpreted axioms of M' but might be smaller. The tendency to give the proof of consistency in a language simpler than the object language is justified if the word 'simpler' here is understood to mean a language which contains no uninterpreted parts, i.e., whose axioms are fully covered by the interpretation. It seems as though this aim should be reachable, because the statement of consistency is meant as an interpreted statement; however, we have no proof that this aim can be reached, and all we can say is that the Gödel theorems at least do not exclude its attainability. It seems to be the advantage of intuitionist mathematics that it comes nearer to this program than does general mathematics. This would justify the desire to give the proof of consistency within a language satisfying the postulates of intuitionism, and we may add here the remark made that Gödel himself has made it clear that his theorems do not contradict the possibility of an intuitionist proof of consistency of formal mathematics. This aim has already been reached for important parts of mathematics. Recently G. Gentzen has published a proof of consistency for the theory of whole numbers; a corresponding proof for analysis has so far not been achieved, but we are entitled to hope that this aim will one day also be reached.[2]

NOTES

1 F. Klein, *Elementarmathematik vom höheren Standpunkte aus* (Berlin, 1924), p. 39.

2 Cf. Hilbert-Bernays, *Grundlagen der Mathematik* (Berlin, 1939), Volume II, pp. VII, 360, 373.

PART IV

ETHICAL ANALYSIS

42. THE FREEDOM OF THE WILL*

[1959a–7]

The following essay is based on an unfinished manuscript left by my husband. The original manuscript consists of a treatise on the freedom of the will which contains several fragmentary inserts, and a separate fragment on the freedom of action. These papers were written in 1951 and antedate the manuscripts of the French lectures he gave at the Sorbonne in June 1952 under the title of 'Les fondements logiques de la mécanique des quanta' [1952d], and of his books *Nomological Statements and Admissible Operations* [1954e] and [1976a], and *The Direction of Time* [1956b].

I have tried to combine the various fragments into one continuous essay. As far as possible I have preserved the original manuscript, but I have made some stylistic changes, partly reorganized the order of presentation, provided a number of new illustrations, and elaborated on the text wherever I felt it was necessary. I was guided frequently by the later manuscripts mentioned above which clarify some of the issues touched upon in the paper on the freedom of the will. Although I have not changed any of the philosophical content of the paper, the responsibility for its final version is mine.

Anyone who wishes to consult the original manuscript may ask for a photostatic copy from the Department of Special Collections of the Library of the University of California at Los Angeles where the original manuscript has been deposited. – M.R.]

I

The problem of the freedom of the will has held the attention and taxed the ingenuity of philosophers of all ages. While in general everybody is convinced that his will is free, or at least behaves in such a fashion that this conviction can be inferred from his behavior, it has frequently been maintained that a strict causal determinism seems to exclude the possibility of free will. If in principle every event is predictable, then human behavior can be predicted like any other natural event, and it is difficult to understand in what sense it can be called free.

Philosophers have offered various solutions to this problem, with the result that free will has been defended as well as denied on the basis of a strict determinism. The defence usually pursues the line that freedom presupposes strict causality; otherwise, it is argued, we could not be the causes of our actions, and would be incapable of doing what we want. The denial of freedom on deterministic grounds has sometimes been tempered by a compromise, as in the case of Kant whose pragmatic considerations can only be called a *tour de force*. Although he is persuaded that an epistemological analysis

*From *Modern Philosophy of Science: Selected Essays*, ed. and tr. by Maria Reichenbach, Routledge & Kegan Paul, London, 1959, pp. 151–192. Copyright © Maria Reichenbach 1959 except in U.S. Copyright © in U.S. by Maria Reichenbach 1959.

shows that the will is not free, he believes that this conclusion contradicts our conception of a meaningful ethics. Kant therefore makes the freedom of the will a postulate of Practical Reason in his attempt to rescue morality. Spinoza admits a certain freedom by distinguishing between causes inside and outside our bodily system. If no outside event interferes, if the causal chain is confined within our organism, we feel that we have voluntarily initiated this chain, that is, we feel free and are free in the sense that we are not compelled by an external cause to act against our will. Actually, however, only God is free because he exists and acts solely by the necessity of his own nature, whereas all other things, including human beings, are dependent on external causes. We have the feeling of freedom because we are aware of our volitions but often ignorant of the causes that have determined them. Accordingly, the feeling of freedom is merely an epiphenomenon. By analogy, a stone which was moved by an externally caused impact and which continued to move after the impact had ceased would feel itself free if it had the capacity to think and if its motion coincided with its wishes to continue on its path. Spinoza's presentation of the problem of free will is much more consistent with the rest of his system than are Kant's wishful postulates with his epistemological investigations.

We propose to discuss the problem of free will in general rather than to take issue with the interpretations of individual philosophers. In order to clarify the scope of this essay, we should like to distinguish between the *practical* and the *theoretical* side of the problem of free will. It has often been argued that common sense knows when we are free and when we are not. I am free to lift my arm unless somebody has tied it behind my back. Usually, a person is free to smoke a cigarette or drink a glass of wine, but if he is the victim of an addiction, he is not free when he resorts to his favorite drug. A kleptomaniac is not free with respect to actions that normal people control by their will.

This informal presentation of familiar examples from everyday life seems to indicate an agreement that, on the whole, we can make the distinction between freedom and lack of freedom for practical purposes. It appears plausible that the practical problem has centred mainly around the concept of responsibility as faced in ordinary life and specifically in legal decisions. Age, illness, and various situations are sometimes regarded as extenuating circumstances which absolve a person from the responsibility for his actions. The legal profession seems to be increasingly aware of these considerations; this is borne out by the fact that, due to psychiatric influence, the notion of punishment is being replaced by the notion of rehabilitation or cure, and help or therapy is given with the aim of making the person morally free. The law formulates the general conditions for responsibility (freedom) and non-responsibility (lack of freedom), and empirical evidence determines which rules apply in a particular instance. The practical problem, therefore, is not in principle too difficult to solve.

Yet it is well known that a feeling of freedom cannot be regarded as a guarantee that we are free; an action due to hypnotic suggestion is not the result of a free will, although the person feels that he is acting in accordance with his own free deliberations. Freud has shown that many actions felt as free are actually determined by unconscious motives but that psycho-analysis may help to make us once more masters in our own house.

In this context, it is assumed that freedom can be determined by empirical evidence. Practically minded people, therefore, have often been inclined to dismiss the theoretical problem, which arises on the basis of logical and semantic considerations, as unimportant and irrelevant for social situations – as one of those philosophical nightmares created by an overdose of abstract reasoning. They would delegate the solution of the problem to sociologists, physicians, and psychologists rather than to philosophers who, in their eyes, are often too aloof from realistic and practical concerns.

Although we believe that the core of the problem of free will must be sought in the practical problem and is to be decided on empirical grounds, we would not be willing to disregard the theoretical problem. On the contrary, we think that the theoretical problem of free will can be solved along lines of reasoning that start with the practical problem. From the philosophical point of view, such a theoretical investigation deals with a clarification of the meaning of 'freedom', with an explication of this term on the basis of a precise epistemological, semantic, and logical analysis which makes the usage of the term compatible with conversational and scientific language. We have the impression that a solution of the theoretical problem was impossible as long as it was sought within the framework of a strict causal determinism. To anticipate briefly, we believe that the theoretical problem of free will can be solved if deterministic considerations are replaced by probability considerations. The present essay is intended to outline such a theoretical solution.

We know that this is a touchy subject. 'Quantum physics has abolished causality; the electron is free, because its path is unpredictable.' So some writers have argued; others have pointed out how silly a conception of free will would be if human behavior were considered to be of the same type as electron behavior. It is not this kind of argument that we wish to discuss. It is obvious that a substitution of probability relations for deterministic relations does not *eo ipso* solve the problem. Yet it would be a serious error to believe that such a replacement has no bearing upon the problem of free will. We wish to show that, in fact, it does lead to a solution if it is done in full view of the specific relationships that characterize human volition and that are expressed in the practical distinction of free and unfree action.

II

Since the terms 'determinism' and 'indeterminism' have historically been used in various senses, we should first like to give a more precise formulation of the difference between them – a formulation that will, at the same time, constitute a definition of these concepts as employed in this essay. [See [1925d], [1932d], p. 47, written before Heisenberg's discovery of quantum mechanical indeterminacy; and [1952d] and [1956b], Section 11].

The causal hypothesis can be formulated in two different ways: in a conditional form and in a categorical form. In the conditional form we say: If $D^{(1)}$ describes the total physical state of the universe at the time t_1, i.e., if $D^{(1)}$ describes the cross section of the universe at the time t_1, we can predict with a certain probability $p^{(1)}$ the total physical state E of the universe at a later time t_2. Of course, the conditional form does not guarantee the truth of the prediction. We therefore supplement the conditional form of causality by the assumption that it is, in principle, physically possible to know at the parameters of the initial situation and to know the ultimate physical laws. Thus we arrive at the categorical form which constitutes an empirically testable statement and reads: There exists a description $D^{(1)}$ of the total physical state of the universe at the time t_1 such that the total physical state E of the universe at a later time t_2 can be predicted with a certain probability $p^{(1)}$. The prediction is regarded as true if observation shows that it is satisfied within a certain exactness ϵ. (It would be demanding too much to require $\epsilon = 0$.) The hypothesis is based on physical, not on technical possibility. [1938c, Section 6].

The hypothesis of causality has to be conceived as a limit statement because of its reference to ultimate initial conditions and ultimate laws. It must therefore be translatable into convergence statements involving the concept of probability. Using the categorical form of causality, we assume that the description $D^{(1)}$ which predicts the future state E with a probability $p^{(1)}$ can be replaced by a more exact description $D^{(2)}$ which raises the probability of E to a higher value $p^{(2)}$. This procedure can be continued, and the descriptions $D^{(i)}$ will furnish higher and higher probability values $p^{(i)}$ for the prediction of E. The hypothesis of causality can thus be formulated as the assumption that the probability p of E can be brought as close to 1 as we want by means of more and more exact descriptions $D^{(i)}$. This assumption was made in classical physics.

The hypothesis of determinism goes beyond this assumption by asserting that there exists an ultimate precise description D towards which the approximate descriptions converge and which permits us to predict the future situation E with certainty. It should be noted that this assertion does not follow from the assumption made in classical physics, although it does not contradict it. It may be possible to increase the probability values for the prediction of E so that they come as close to 1 as we want, but there need not be an ultimate description which supplies certainty for the prediction. In other words, if the series of probability values for the prediction of E converges towards a limit, it is not logically necessary that there exist a limit in the corresponding series of descriptions. The definition of causality by convergence statements is therefore more general than the deterministic definition, and includes the latter as a special case. In order to avoid overlapping categories, we shall call the conception which assumes convergence of the probability values for the prediction of E, but denies the existence of a limiting description, the *probability interpretation of causality* or *indeterminism*.

Instead of speaking about a cross section of the total universe, one can restrict the statement to descriptions of parts of the universe enclosed within a certain spatial volume V. In this case, it is not sufficient to know the condition of V at the time t_1; in addition, the conditions of the boundary surface of V must be known for the time interval $t_1 - t_2$. This requirement is necessary because the rest of the universe has an influence upon events in V; it is assumed that this influence spreads through action by contact, and therefore that it has to pass through the surface of V. It can be shown that this device would help us to define determinism for an infinite as well as for a relativistic universe. However, such additional knowledge is not needed if the total cross section of the universe is assumed to be known; in that case, the influence of the environment on any volume V is computable from the initial cross section. For the sake of convenience, we shall use the formulation referring to the total cross section.

The deterministic conception can be illustrated by the following schema:

$$\begin{matrix} D^{(1)} & D^{(2)} & D^{(3)} & \dots & D \\ p^{(1)} & p^{(2)} & p^{(3)} & \dots & 1 \end{matrix} \tag{1}$$

The first row represents the descriptions of the initial situation at the time t_1

which achieve progressively greater exactness; the second row gives the corresponding probabilities of the situation E at a later time t_2. The probabilities converge towards 1. In addition, the deterministic interpretation asserts that the first row has a last term D to which the probability 1 corresponds. The existence of this ultimate description is assumed in the arguments of the determinist, although perhaps with the tacit qualification that human beings may never be able to reach this state of knowledge.

The probability interpretation of causality is similar to determinism with respect to the second row since it also assumes that the probabilities converge towards 1. It differs, however, from the deterministic interpretation because it denies the possibility of an ultimate description D, that is, it denies that the sequence of more and more precise descriptions converges towards a limit. The following schema, illustrating indeterminism, therefore omits the last term in the first row:

$$\begin{matrix} D^{(1)} & D^{(2)} & D^{(3)} & \dots & \\ p^{(1)} & p^{(2)} & p^{(3)} & \dots & 1 \end{matrix} \tag{2}$$

It may be objected that although it does not lie within human capacities to furnish an ultimate description of initial conditions, the universe certainly has a definite state at a momentary cross section; so instead of speaking of an 'ultimate description', we would do better to speak of an 'ultimate state' or of 'ultimate conditions'. But the indeterminist does not deny that the universe is, at any given moment, in a definite state. He does deny that the universe has, at any moment, a state from which any state of the universe at a later moment follows with physical necessity.

Schemas (1) and (2) represent the formulations of determinism and indeterminism presupposed in the subsequent analysis of the problem of freedom. Evidently, both determinism and indeterminism are hypotheses going beyond observables; but we shall postpone the discussion of the empirical evidence that can be adduced for each to the end of the essay.

III

We should like to divide our analysis into two stages: we shall first concern ourselves with the problem of free and unfree action, and secondly, with the problem of free and unfree will. For both problems, the analysis will be carried through first on the basis of a probabilistic, and then on the basis of a deterministic conception of causality.

It is important to keep in mind that, in the practical meaning of the terms, freedom is not opposed to predictability. An action may be predictable and yet free. A few simple examples will illustrate this point.

Joe Smith is an outstanding member of the Democratic Party, and therefore we can predict that he will vote for the Democratic candidate. But his vote is a free action, in any practical sense of the word 'free'. Moreover, whatever his party affiliations may be, we can predict that Joe Smith will have his breakfast tomorrow morning; yet nobody would doubt that he could just as well not have it. The latter is an example of so-called conventional behavior which is always both predictable and free.

How can we express this fact in logical terms? The answer is easy if we formulate our predictions by means of probability implications with values close to 1, that is, if we assume a probability interpretation of causality. As we shall see later, a deterministic interpretation which is expressed by formal, non-truth-functional implications indicating the relation between cause and effect leads to paradoxes. [We are using Russell's definition of 'formal implication'.] We shall first present the probability interpretation, and then turn to an analysis of the deterministic interpretation.

Some preliminary remarks concerning our terminology and the introduction of symbols will be in order. We shall assume that there is a familiar psychological experience called *volition*. Our behaviour is usually directed towards a goal; we intend to do a certain action, we want to perform it, and we know from immediate experience when this is the case. In other words, we shall regard 'volition' as a primitive term in the course of this analysis. We do not say that we are free when we experience volition; on the contrary, we should rather maintain that we can be mistaken in this belief. Instead, we wish to define the term 'free', and we shall do so by regarding the term 'volition' as having an ostensive meaning.

With respect to an action B, a person may have three different attitudes:

1. The person has the volition to do the action B. We shall denote this case by 'VB'.
2. The person has the volition not to do B. We shall denote this case by '$V\bar{B}$'.
3. The person is indifferent with respect to the action B. We shall denote this case by 'IB'.

Obviously, IB is the same as $I\bar{B}$. However, $\bar{V}\bar{B}$ is not the same as $V\bar{B}$; $\bar{V}\bar{B}$ means having no volition to do the action B, that is, it means either having the volition not to do the action B or being indifferent with respect to the action B. This case would be denoted by '$V\bar{B}$ or IB'.

Given that a person has a volition *VB*, whether he can perform the action *B* depends on further conditions. A person may want to buy a car, but may lack the money to do so. He may want to climb a mountain, but may not have the physical strength or the skill to do so. The description of these conditions, which belong partly to a person's environment and partly to his bodily or psychological state, is therefore a conjunction of statements. We will let the single letter '*A*' summarize this conjunction. '*A*' will thus denote the class of situations in which a volition occurs or an action is performed; however, for convenience we shall use the same letter to refer both to a single situation and to the class containing such situations.

Psychologists might argue that it is sometimes difficult to distinguish between a volition *VB* and a situation *A*. It might be possible to describe a particular instance of volition or parts of such an instance of volition (which, as we know, is a rather complex affair) in such a way that no volitional terms are used. For instance, we might refer to certain bodily reactions generally associated with an instance of volition, such as shining eyes, or a smile, or a tightly clamped mouth, and we might include the description of these reactions in the description of a situation *A*; then *A* is so defined that, physiologically speaking, it covers that part of the brain which we reserved for the term '*VB*'. Such a description of a situation *A* would practically exclude a volition $V\bar{B}$. We do not deny this ambiguity. But we shall assume that it is possible, in principle, to draw a satisfactory line of demarcation between the description of an instance of volition *VB* or $V\bar{B}$, and the description of a situation *A* which does not include parts of the volition or equivalents of it in nonvolitional terms. Practically speaking, the situations *A* in which we speak of freedom of will or lack of such freedom are of this kind; and they are the kinds of situations referred to in the following analysis. In the formulas to be developed, the term '*A*. *VB*' is understood to presuppose the clause 'If it is possible to distinguish descriptively between the phenomenon of a volition *VB* and the nonvolitional phenomena of a situation *A*'. The construction of the term '*A*. *VB*' is therefore governed by the rule 'select a description *D* of such a kind that *A* and *VB* are distinguishable'.

There are various ways to achieve this distinction and split off *VB* from the rest of the description. It might be done in terms of physics, or by means of introspection or depth analysis, or by giving an account of behavioristic symptoms. However, those dualistic descriptions which distinguish physical events from 'mental' events that do not have a physical correlate are excluded from our considerations.

Persons other than the one having the volition never directly observe the volition, but only see behavioristic symptoms of the kind mentioned above from which the existence of a volition can be inferred. For the observer of these symptoms, the existence of a volition in the other person would be an illatum. [1938c, pp. 212, 277] But the observation of behavioristic reactions represents only one method of inferring the existence of a volition in another person. A different way would be to observe the person in a situation which contains, in addition to the goal of the action B, a stimulus or goal C that conflicts with the original aim B and that may lead to the volition $V\bar{B}$. For example, a child may want to snatch a piece of sugar (action B); but if he suddenly sees his mother watching him, it may be advisable for him not to take the sugar in order to avoid punishment (goal C). If in the presence of goal C the behavior changes from B to $\bar{B}$, we may conclude that VB has changed to $V\bar{B}$, and thus infer that there exists a volition in the child. In this way, behavioral changes under the influence of an additional stimulus C permit us to assert the existence of a volition. It may be argued that even the person himself often infers the existence of his volition in this indirect way. The following analysis is independent of epistemological considerations concerning the various methods of confirming an instance of volition. In the subsequent formulas, the term 'VB' is to be interpreted as an abbreviation for that part of the description which can be regarded as referring to the volitional state of a person, and is thus distinguished from the rest of the description denoted by 'A'.

IV

With this introduction to our terminology and definitions of symbols, let us turn to the problem of free action. When we call an action free, we want to say that a volition has a causal influence on the action. This assertion means that in a situation A, a volition VB will lead to the action B, whereas a volition $V\bar{B}$ will lead to the action $\bar{B}$. In the case of indifference IB, we leave it open what action will result. Regarding the causal influence of volition upon action as a probability relation, we characterize freedom of action by the following three relations [Using the probability notation developed in [1949f] where the sign '$\sim$' means 'equals approximately'.]:

$$P(A.VB, B) \sim 1 \tag{3}$$
$$P(A.V\bar{B}, \bar{B}) \sim 1 \tag{4}$$
$$P(A.IB, B) = u. \tag{5}$$

In these probability expressions, 'A' does not stand for a single situation A, but for the class of such situations; '*VB*', '*B*', etc. must be interpreted in a similar fashion. The probability *u* in (5) may have any value between 0 and 1. For conventional actions like putting on the left shoe after having put on the right shoe, or answering the question 'How are you?' by the reply 'Just fine, how are you?', the degree *u* will be close to 1. In this case, *B* is predictable with respect to an indifferent attitude.

If we wish to regard relations (3)–(4) as a definition of freedom of action, we must be able to interpret them as conditionals contrary to fact. The use of conditionals contrary to fact is indeed the nucleus of the freedom problem. For example, I have decided to visit a friend and I do so; was I free to do the contrary? We would answer: yes, because if I had decided not to visit my friend, I would not have gone to his house. In this case, we are making a statement about what would have happened if the volitional decision had been in the opposite direction. The logical analysis of freedom, therefore, involves a study of conditionals contrary to fact.

In other publications, [1947c, Chapter VIII] and [1954e], a theory of conditionals contrary to fact has been developed within the framework of a general theory of nomological statements. This theory applies to the use of those conditionals contrary to fact which can be schematized by formal implications of an analytic or synthetic kind. For instance, the statements 'If today is Sunday then tomorrow is Monday' or 'If a person touches a high voltage wire he will be killed' can be used in the counterfactual sense whenever the antecedent is false. There is, however, a second kind of conditional contrary to fact which does not lend itself to a schematization by formal implications. These counterfactuals can be schematized as probability implications which cannot be transformed into formal implications because the probability of the conditional prediction is not high enough; yet even if this probability is rather high, certain other conditions for the introduction of formal implications may not be satisfied [For a formulation of these conditions see [1949f], pp. 435–6.]. In the present essay, we wish to show that if the probability implications (3)–(5) are schematized as formal implications, they cannot be interpreted as conditionals contrary to fact asserting freedom of human behavior in a meaningful way. The proof will be given in Section 5.

As mentioned above, certain probability implications like (3)–(4) can be used to assert conditionals contrary to fact even if they cannot be schematized as formal implications. For this purpose, a number of requirements must be introduced which, if satisfied, permit us to interpret relations (3)–(4) as conditionals contrary to fact. When these requirements are added, relations (3)–(4) can be used as a definition of freedom of action.

The situations, volitions, and actions referred to in relations (3)–(4) constitute events which have certain time positions and whose causal relationships we want to describe. Since we are not concerned here with a general theory of causality and time in which time order is defined in terms of causal order, we shall assume a world in which time position and time order are known for these events. We thus restrict ourselves to the simpler problem of defining specific causal relationships in such a world. Within this frame, that is, within the familiar macroscopic world of our environment, we wish to lay down requirements that allow us to construe relations (3)–(4) as formulating a causal influence of volition upon action.

We shall assume that events x, y, z, ... are observed as separate spatio-temporal units, and that these events can be classified into certain classes G, H ... The events belonging to a class may succeed each other in time. However, we impose the following requirement upon the definition of the classes:

Requirement α: The classes must be codefined. [1956b, p. 187]

A class G is codefined if it is logically possible to classify an event x as belonging to G coincidently with the occurrence of x. In other words, observing x we must be able to say whether x belongs to G, and it must not be necessary to know for the purpose of this classification whether certain other events y, z, ... happened earlier, or later, or simultaneously at distant places. For instance, the class term 'red' is codefined because looking at x we can say whether it is red. But the class term 'day following a rainy day' is not codefined; in order to know whether a day belongs in this class, we have to know whether it was preceded by a rainy day. Logically speaking, a codefined class term is a one-place predicate which is not contracted from a many-place predicate. A definition of the kind that permits the act of classification to coincide with the event that is to be classified will be called a *codefinition*. Many class terms are codefined, for instance, 'house', 'sweet', etc.

The following example will illustrate the need of this requirement for the freedom problem. If we define VB as the class of volitions to do B which always precede an occurrence of B, and $V\bar{B}$ as the class of volitions to do $\bar{B}$ which always precede an occurrence of $\bar{B}$, then (3)–(4) will be trivially satisfied, but we would not for this reason regard the action B as free. In this case neither VB or $V\bar{B}$ is codefined. For instance, given the above definitions of VB and $V\bar{B}$, relations (3)–(4) will hold both if B is smoking a cigarette and if $\bar{B}$ is throwing face 6 on a die, but we would normally call the former action free and the latter unfree. We shall see further reasons for requirement α in connection with requirement δ, stated below.

The next requirement concerns the probabilities of the codefined classes.

Requirement β: The probabilities must be genuine [1954e, pp. 127, 128]

This requirement is needed because a probability expression has a unique numerical value only if its reference class is not empty [1949f, p. 55]. But the reference classes used here may be empty; for instance, it may never happen that a certain member of the Democratic Party will want to vote for the candidate of the opposite party. In order to rule out unreasonable probability values for such empty classes, we must indicate the conditions which furnish genuine probabilities.

One condition is the openness of the reference class. A reference class G is *open* with respect to the attribute M if it is not empty and if those of its members for which we have confirmed the occurrence of the atrribute M form only a small fraction of the total class. A probability $P(G, M)$ is genuine if its reference class G is open with respect to the attribute M, or if it is inferred from a set s of true statements from which the nonopenness of G cannot be inferred.

Let us use an example. Assume that G is open with respect to M and that we have

$$P(G, M) = p. \tag{6}$$

The usual statistical classes are of this kind. Now assume that K is a class of classes such that for every class C_i which is a member of K we have

$$P(G.C_i, M) = P(G, M). \tag{7}$$

This relation can be established by having it confirmed for some members of K and then extended inductively to all members of K. Now it may be true that the class $G.C_1$ is empty although C_1 is in K; but this fact cannot be inferred from the above statements. Hence, $P(G.C_1, M)$ is a genuine probability.

For instance, let G be the class of human beings falling from a high tower, M the class of people who die, K the class of professions C_i. Let C_1 be the class of prime ministers, C_2 the class of generals etc. Then

$$P(G, M) = 1 \tag{8}$$
$$P(G.C_1, M) = 1 \tag{9}$$
$$P(G.C_2, M) = 1 \tag{10}$$

and so on. Thus we find by induction that for any class C_i which is a member of the class K of human professions

$$P(G.C_i, M) = 1. \tag{11}$$

But it may never happen that a prime minister falls from a high tower. $G.C_1$ will therefore be empty, but $P(G.C_1, M)$ is a genuine probability of the degree 1 because it is derivable from (11) and because we cannot also derive the emptiness of $G.C_1$ from (11). (Not all genuine probabilities express laws of nature. For instance, mortality statistics hold only for specific social conditions. They contain individual terms, and for this reason we do not regard them as laws of nature.)

We shall now relate this requirement to the problem of freedom of action and show why the probabilities in relations (3)–(4) have to be genuine. Consider the situation in which a normal child of English speaking parents is brought up by them in the United States. Let A be the class of such situations, and let B denote the class of actions of understanding English. Intuitively, we would not regard the child's understanding of English as a free action. Nevertheless, relations (3)–(4) hold, but (4) is not a genuine probability. The class $A.V\bar{B}$ is not open (we presume that it is empty) and we have no grounds for assigning a unique value to $P(A.V\bar{B}, \bar{B})$. Thus we see that requirement β excludes from the class of actions certain cases which we would intuitively not call free.

Since causal influence concerns only the future we need:

Requirement γ: The reference class must precede the attribute class

The elements of the classes used in probability expressions are given in a certain order; the events $x_1, x_2, \ldots$ progress in time and stand in a one-to-one correspondence to the events $y_1, y_2, \ldots$ of any other class co-ordinated to it [1949f, p. 46]. All events x_i have the property G, that is, they belong to the reference class; some of the events y_i have the property M, that is, they belong to the attribute class. If every member x_i of G precedes temporally every member y_i of the corresponding sequence, we say that the reference class precedes the attribute class. (However, x_{i+1} may succeed y_i.) Applied to relations (3)–(4), this condition means that the volitions VB precede the actions B. The requirement is necessary because otherwise volition could not be the cause of action. We should also realize that an analysis of the problem of freedom of action presupposes a theory of the direction of time.

Finally we need:

Requirement δ: The volition must be causally relevant to the action

This requirement is more precisely formulated as follows: Physical conditions must exclude a codefinition of a subclass A_1 of A that would make one of the following relations false:

$$P(A_1.VB, B) \sim 1 \quad (12\text{A})$$
$$P(A_1.V\bar{B}, \bar{B}) \sim 1. \quad (12\text{B})$$

The import of this requirement may be illustrated as follows. Suppose a person insists that day and night are controlled by his will. Whenever the sun is low on the horizon in the east, he has the volition that there be day, and whenever the sun is low on the horizon in the west, he has the volition that there be night. Relations (3)–(4) would then be found to hold. In order to show that there is no causal relationship between the person's volition and the occurrence of day and night, we would divide the reference class A of situations in which the sun is low on the horizon into two subclasses A_1 and A_2; A_1 would be the class of situations in which the sun is low on the horizon in the east, and A_2 the class of situations in which the sun is low on the horizon in the west. If A_1 as defined above is used in (12B), and we take VB as the volition that there be day, then (12B) would turn out to be false. With respect to the situation A_1, a volition towards the event $\bar{B}$ would not change the existing probability $P(A_1, \bar{B})$; that is, a volition $V\bar{B}$ is not causally relevant to the occurrence of $\bar{B}$ in the situation A_1. Let us even suppose that the class $A_1.V\bar{B}$ is empty, i.e., that the volition $V\bar{B}$ never occurs in the situation A_1. Nevertheless, we can find, by means of inductive inferences based on other observations concerning human beings and the occurrence of day and night, that the probability of the coming of day (B) following the sun's being low on the horizon in the east (A_1) remains the same, no matter what the person's volition in the situation A_1 may be.

That is,

$$P(A_1, B) = P(A_1.VB, B) = P(A_1.V\bar{B}, B) \sim 1 \quad \text{and}$$
$$P(A_1, \bar{B}) = P(A_1.V\bar{B}, \bar{B}) = P(A_1.VB, \bar{B}) \sim 0.$$

Still, $P(A_1.V\bar{B}, \bar{B})$ constitutes a genuine probability; therefore, we see that requirement δ is necessary.

It should be noted that requirement δ is tied to requirements α and β. Otherwise, it would always be possible to define a subclass A_1 of A which violates (12A) or (12B). If we defined A_1 as that subclass of A consisting of situations followed by the event B, then (12B) would be false. The reference class $A_1 . V\bar{B}$ of (12B) then need not even be empty since relations (3)–(4) are merely statistical; it may occasionally happen that a volition $V\bar{B}$ is followed by the event B. But this class A_1 is not codefined because the classification of a situation x_i of the class A into the class A_1 cannot be achieved at the time of x_i, but must be postponed until the event referred to in the definition of A_1 has been observed. (Instead of referring to the occurrence of the event B in the definition of A_1, one might refer to an event C which is causally between VB and B and which is of such a kind that it practically entails the event B. A_1 would then be defined without explicit reference to B, and yet (12B) would be violated. This definition again would not be a codefinition. In the example of day and night, C might be the disappearance of the stars.)

Let us remember that it was our aim to use relations (3)–(4) for a definition of freedom of action, that is, we wanted to be able to construe them as formulating a causal influence of volition upon action. This aim has been reached by the addition of requirements α–δ. Requirements α–δ enable us to interpret relations (3)–(4) as conditionals contrary to fact of a particular kind; due to these requirements, relations (3)–(4) express a causal conditional contrary to fact, that is, they tell us what would be the effect if the cause were different. They formulate a statement of causal relevance according to which volition is causally relevant to action. We therefore give the following definition:

Definition 1. An action B is *free* in a situation A if relations (3)–(4) hold and requirements α–δ are satisfied.

We should like to add some psychological remarks. The feeling we associate with a probability statement that satisfies requirements α–δ may be called *inductive confidence*. It is an intuitive feeling we have when accepting beliefs that are based on inductions. We do not maintain that inductive confidence can be used as a basis for logical arguments, nor do we think that the theory of probability and induction presupposes inductive confidence for any of its conclusions. But the psychological phenomenon of inductive confidence can be employed as a criterion when we are searching for an adequate explication of conditionals contrary to fact; if we succeed in defining in logical terms a class of statements which coincides with the class of statements associated with inductive confidence, we have reached correspondence with the usage of conversational language. The justification of this usage is derivable from the logical properties of those statements of this class which satisfy the following condition: if the antecedent of the statement were true, its conclusion would be warranted by an inductive inference. To this condition, requirement δ

adds the provision that the volition must not be redundant for the conclusion, that is, that its function as antecedent cannot be taken over by other factors existing simultaneously with it.

If the probability relations (3)–(4) (relation (5) is empty and can be omitted) satisfy requirements α–δ, they can be interpreted as asserting the following:

> *A volition would have influenced the action if this volition had occurred.*

This statement, which is a conditional contrary to fact, will be called the *freedom conditional*; if it is true, the volition is called *efficient*. In contradistinction, the definition laid down in relations (3)–(4) in combination with requirements α–δ will be called the *freedom condition*. According to this terminology, the freedom condictional can be asserted by means of the freedom condition which consists of a formal condition, formulated in relations (3)–(4), and metalinguistic conditions of interpretation, given by requirements α–δ. The complete freedom condition stated in definition 1 can therefore be stated as follows:

> *An action is free if it would have been influenced by a preceding volition.*

We can now ask whether we can make any prediction when nothing is known about the volitional attitude of a person. There are many instances in which we would predict the result B even in such a case. For example, we do not know whether Mr. X has any deliberate plans to answer Mr. Y's question "How are you today?", nevertheless, we would be willing to predict that he will give the conventional answer. In this case, the probability expression would read:

$$P(A, B) \sim 1. \tag{13}$$

This relation is compatible with relations (3)–(5) for any value of u, in particular, with a value u close to 1. It is merely required that the case $V\bar{B}$ does not happen too often; but the probability of $P(A, V\bar{B})$ need not be 0.[1]

Relations (3)–(5) and (13) express the fact that an action can be free and yet predictable, and thus they fully account for the practical use of the term 'free'.

We turn now to the definition of unfree action. This definition differs from that of free action with respect to requirement δ, whereas requirements α–γ remain in force. We define:

Definition 2. An action B is *unfree* in a situation A if the class of situations A can be subdivided into two exclusive and exhaustive subclasses A_1 and A_2 such that the following relations hold:

$$\begin{aligned} P(A_1.VB, B) &= p \\ P(A_1.V\bar{B}, B) &= p \\ P(A_1.IB, B) &= p \end{aligned} \tag{14A}$$

$$\begin{aligned} P(A_2.VB, B) &= q \\ P(A_2.V\bar{B}, B) &= q \\ P(A_2.IB, B) &= q \end{aligned} \tag{14B}$$

The following illustration may help. Let A be'birth of babies'; let A_1 be 'single birth'; let A_2 be 'multiple birth'; let B be 'boys'. It can easily be seen that the volition of the mother will have no causal influence upon the birth of a boy, that is, the probability in the three cases of (14A) and the three cases of (14B) will remain the same. (14A) and (14B) must however satisfy requirements α–γ. The class A_1 or the class A_2 may be identical with A; the first or the second set of relations (14) will then be omitted.

If Definition 2 is satisfied, volition is causally irrelevant to the action. It should be noted that causal irrelevance is not merely the negation of causal relevance; the negation would only require that one of the relations (12A)–(12B) be false, whereas causal irrelevance requires that the degree of probability p or q be the same in relations (14A) or (14B), respectively.

Between free and unfree actions there are actions which are free to some degree. In this case, relations (3)–(5) have to be replaced by the relations

$$\begin{aligned} P(A.VB, B) &= p_1 \\ P(A.V\bar{B}, B) &= p_2 \\ P(A.IB, B) &= p_3 . \end{aligned} \tag{15}$$

For example, if we take the theory of psychosomatic interdependence seriously, we would say that the patient's will to recover has some influence on his recovery, and that the degree of probability of his recovery is a certain extent a function of his attitude towards it.

Requirements α–γ must again be satisfied. Requirement δ is to be replaced by the requirement that no subclass A_1 of A can be codefined in such a way that one of the following relations is false:

$$\begin{aligned} P(A_1.VB, B) &= p_1 \\ P(A_1.V\bar{B}, B) &= p_2 \\ P(A_1.IB, B) &= p_3 . \end{aligned} \tag{16}$$

The values p_1, p_2, p_3 in (16) are meant to be the same as those in (15).

If $p_1 > 1 - p_2$, volition is positively relevant to B to a certain degree; if $p_1 < p_3$, volition has a negative influence on B. An example of the latter case is the familiar experience of the deterrent effect which a deliberate effort to remember a name that has momentarily been forgotten usually has upon its recollection. This essay, however, will not be concerned with situations in which volition has a partial influence.

Let us briefly summarize the results of the foregoing section in which we attempted to give an explication of free action within the framework of a probabilistic causality. We formulated the different attitudes a person might have towards an action, and we distinguished the volitional attitude of a person from the general situation in which he finds himself. We then proceeded to set up the probability relationships that describe the causal influence of volition upon action. We showed that these relations must be interpretable as conditionals contrary to fact if they are to be used as a definition of freedom of action. For this purpose, we added a number of requirements that have to be fulfilled by the above relations. On this basis, we formulated the freedom conditional which, if true, expresses the efficiency of volition.

We have also shown that actions may be predictable although nothing is known about the volitional attitude of a person. At the end of the section, we set up the probability relations that would hold in the case of unfree action, and of the partial influence of volition upon action.

V

We shall now turn to a discussion of the solution which a deterministic interpretation of causality offers for the problem of freedom of action. On this interpretation, statements expressing high degrees of probability are replaced by two-valued formal implications. Since in an instance of free action we want to assert that volition has a causal influence upon action, the implications which state the relations between volition and action must be interpretable as conditionals contrary to fact. They cannot, therefore, be conceived as adjunctive [1947c, pp. 28–32] or material implications because such implications are always true when the antecedent is false. Hence, we shall express the causal relationship by means of *nomological non-truth-functional* implications, and a few words about implications of this kind will be in order. (For a detailed presentation see [1954e] which is a revised version of the account in [1947c] Ch. VIII. The discussion of contrary implications is carried through only in the later publication, the results of which are presupposed for the present analysis.)

First we define a class of *original nomological statements* which contains laws of logic and laws of nature. From the original nomological statements, the class of *derivative nomological statements* or *nomological statements in the wider sense*, is constructed; it comprises all those statements that are deductively derivable from sets of statements of the first class. Only those synthetic statements that lend themselves to inductive confirmation satisfy the requirements by means of which these two classes are defined. It can then be shown that for the definition of 'reasonable implication', the class of original nomological statements is too narrow, whereas the class of derivative nomological statements is too broad. Within the class of derivative nomological statements, therefore, a subclass of *nomological statements in the narrower sense* or *admissible statements* is defined. For the purpose of explicating the notion of admissible implications, that is, implications that can be interpreted as predictions or as counterfactuals, the following terminology is introduced:

1. Merely factually true statements are *true of first order*.
2. Synthetic nomological statements are *true of second order*.
3. Analytic nomological statements are *true of third order*.

If S is true of order K, the negation of S is false of order K. A set of statements has the order of the lowest-order statement contained in it. A statement has the order of the highest-order set from which it is derivable. The major operation of a nomological statement is called a *connective operation in the wider sense*.

One of the requirements for admissible implications is that the antecedent must not be false of the same order as, or of a higher order than, the order of which the implication is true. Such implications are called *connective in the narrower sense*.

For connective implications in the wider sense we shall use a grave accent. The statement '$a \grave{\supset} b$' means the same as the metalinguistic statement ' "$a \supset b$" is nomological in the wider sense'. There are two forms of connective implications in the narrower sense. The first, called *admissible implication*, is written with an acute accent. The statement '$a \acute{\supset} b$' means the same as the metalinguistic statement ' "$a \supset b$" is an admissible statement'. The second form, called *proper implication*, is written with an arrow. The statement '$a \rightarrow b$' is true if '$a \supset b$' is admissible, and it is false if '$a \supset \bar{b}$' is admissible. In all other cases it is meaningless.

Both forms of connective implications in the narrower sense are adequate for the expression of conditionals contrary to fact. They permit the assertion of such a conditional only if the implication is true of an order higher than that of which its antecedent is false. If an implication is synthetic and true of second order, it can be interpreted as a conditional contrary to fact only if its antecedent is false of at most first order. Of this kind are the usual conditionals contrary to fact. For instance, the following is both an admissible and a proper implication: 'If this telephone wire were not supported, it would fall down.' This implication is true of second order because it is derivable from a law of nature, but its antecedent is merely factually false and thus false of first order. The implication is therefore admissible and can be maintained as true. The contrary implication 'If this telephone wire were not supported, it would not fall down' is false if it is regarded as an accent implication or an arrow implication, although it would be true if it were construed as a material implication, since the antecedent is false. This material implication, however, is true only of first order, whereas the original implication is true of second order. If two contrary implications are true of the same order, it can be shown that neither one is an admissible or proper implication. This result renders admissible and proper implications suitable for conditionals contrary to fact, and guarantees that a conditional contrary to fact is unambiguous, i.e., that it does not lead to contradictory assertions if the antecedent is true.

In contrast, let us consider the following implication: 'If masses attracted each other in inverse proportion to their distance, this telephone wire would fall down.' Considered as a material implication, it is true of second order because the antecedent is false of second order; since the order of which the implication is true is the same as that of which the antecedent is false, the implication is not admissible. Considered as a proper implication, however, the statement is meaningless because the contrary implication 'If masses attracted each other in inverse proportion to their distance, this telephone wire would not fall down' is not admissible either. Calling this implication meaningless seems to be closer to conversational language. We would hesitate to state what would happen in such a case because we cannot tell how the mechanics of solid bodies would change if Newton's law were false. We shall therefore make use of proper implications for the following discussion. The results are easily transcribed for admissible implications.

VI

We now return to the problem of free action. For a deterministic interpretation of causality, relations (3)–(5) are transcribed into formal non-truthfunctional implications representing nomological statements in the wider sense:

$$A.VB \overset{\cdot}{\supset} B \tag{17}$$
$$A.V\bar{B} \overset{\cdot}{\supset} B \tag{18}$$
$$A.IB \overset{\cdot}{\supset} ?. \tag{19}$$

These relations must be satisfied since determinists wish to state that a volition causes an action according to a law of nature. However, in order that these relations be interpretable as conditionals contrary to fact, it must be required that the corresponding proper implications hold, i.e., that these relations can be written in the form

$$A.VB \rightarrow B \tag{20}$$
$$A.V\bar{B} \rightarrow \bar{B} \tag{21}$$
$$A.IB \rightarrow ? \tag{22}$$

These relations assert: if there were a volition towards B or $\bar{B}$, then B or $\bar{B}$, respectively, would occur. This conditional obviously expresses the meaning of freedom of action. As before, a volition may be called *efficient* if it satisfies relations (20)–(22). We put a question mark in (22) because, as in (5), we cannot say what, in general, will happen in the case of indifference. In a particular instance, say in an instance of conventional action, it may very well be true that a 'B' can be put in place of the question mark.

This interpretation seems to be reasonable and is often referred to by adherents of a deterministic causality. In cases of free action, volition is a part of the cause, or, in other words, deliberate decision is one of the links of the causal chain. So far, therefore, the deterministic interpretation of freedom appears permissible. A serious difficulty arises, however, as soon as relation (13) is taken over into the causal schema.

This relation would have to be transcribed into the form

$$A \overset{\cdot}{\supset} B \tag{23}$$

or

$$A \overset{\cdot}{\supset} \bar{B}. \tag{24}$$

Relations (23)–(24) mean that B or $\bar{B}$, respectively, follows from the situation A alone. If one of these relations is true, we can derive by means of (17)–(19) that relations (20)–(22) no longer hold. In order to prove this conclusion, let us assume that (23) is true. Using the rule (which we shall call the *invariance principle* of implication) that an implication remains true upon further specification of the antecedent, we derive from (23)

$$A.V\bar{B} \mathrel{\dot{\supset}} B \tag{25}$$

which is the contrary implication of (18). Since (25) has been derived from (23), it is true of second order, and therefore true of the same order as (18) is. Yet we stated above that if two contrary implications are true of the same order, neither is admissible.

We can also give the proof directly as follows: Given (18)

$$A.V\bar{B} \mathrel{\dot{\supset}} \bar{B}$$

and (25)

$$A.V\bar{B} \mathrel{\dot{\supset}} B$$

we can derive

$$\overline{(A.V\bar{B})}$$

that is

$$A \mathrel{\dot{\supset}} \overline{V\bar{B}}. \tag{26}$$

This result means that the antecedent of (18) and (25) is false of second order, or, in other words, that in a situation A a volition towards $\bar{B}$ is physically impossible. Expressed differently, the order of which the implications are true is the same as that of which their antecedent is false, and they are therefore not admissible. It follows from the definition of proper implication that relations (20)–(22) are meaningless. If, in a deterministic world, the action B is predictable on the basis of the situation A alone, without reference to volition as stated in (23), then relations (17)–(19), although still true, cannot be transcribed into the form (20)–(22), and therefore cannot be interpreted as conditionals contrary to fact. In such a case, it is meaningless to say that a volition in the opposite direction would have caused the opposite action.

It may help to consider an analogous example from physics. Let A be: There is a refrigerator whose internal temperature is 4°C and in which a glass of water has been left for some time. [If one prefers a description of A without reference to a time interval, one would have to replace the phrase 'left for some time' by a description of the state of the molecules of water in the glass] Let C be: The water in this glass is at 4°C. Then we can predict E: The water in the glass is at its greatest density; that is

$$A.C \mathrel{\dot{\supset}} E \tag{27}$$

is a derivative nomological statement. Now we can formulate the counterfactual: If the water in this glass were not at 4°C, it would not be at its greatest density: that is

$$\bar{C} \dot{\supset} \bar{E} \tag{28}$$

is also a derivative nomological statement which is admissible because it is true of second order whereas its antecedent is false of first order.

But the following is also a derivative nomological statement: If there is a refrigerator whose internal temperature is 4°C and in which a glass of water has been left for some time, then the water in the glass will be at its greatest density; that is, we have

$$A \dot{\supset} E. \tag{29}$$

Using the invariance principle of implication, we can now derive two contrary implications. From (28) we can derive

$$A.\bar{C} \dot{\supset} \bar{E} \tag{30}$$

and from (29) we can derive

$$A.\bar{C} \dot{\supset} E. \tag{31}$$

The two contrary implications are true of the same order; therefore, neither is admissible. We can also say that the antecedent in (30) and (31) is false of second order, that is, that '$A.\bar{C}$' describes a physically impossible state of affairs and that therefore (30) and (31) are not admissible. It would be meaningless to say: If in the situation A the water in the glass were not at 4°C, then it would not be at its greatest density.

This example affords us the opportunity to make some remarks concerning modalities [1947c, Section 65]. By means of analytic nomological statements, we can define logical modalities; by means of synthetic nomological statements, we can define physical modalities. We shall restrict our discussion to physical modalities because of their importance for the freedom problem.

Modalities are used with reference to facts or situations. Ordinary language uses modal terms both in an absolute and in a relative sense. Correspondingly, we shall distinguish between absolute and relative modalities. We say, for instance, 'It is necessary that water at 4°C be at its greatest density', or 'It is possible to bring water to the boiling point'. Since no reference to other facts or special conditions is involved in these statements, the modalities are absolute. An analysis of the environmental conditions, however, may lead us to say 'In the situation A, that is, in a refrigerator at 4°C, it is impossible for water to boil.' Here we are employing a relative modality because we mean 'relative to the situation A, it is impossible for the water to boil'. Since we are making use of modalities in our explication of reasonable counterfactuals with reference to freedom of action and freedom of will, it is important to realize that both kinds of modality are used in conversational language as well as in science, and that it would be a mistake to believe that one of them can be discarded. It is not the task of the logician to legislate with respect to the usage of such concepts; his only concern is to clarify their meanings. A very important use of relative modalities occurs, for instance, in the court-room when the defence attempts to establish an alibi. The attorney does not claim that it was impossible for the defendant to commit a murder, but rather that it was impossible for thim to commit a murder in Los Angeles at the same moment that he was in Washington.

In the following definitions of absolute and relative modalities, 'nomological' means 'nomological in the wider sense'.

Definition of absolute physical modalities

p is physically necessary $=_{Df}$ the sentence describing p is a nomological statement.

p is physically impossible $=_{Df}$ the negation of the sentence describing p is a nomological statement.

p is physically possible $=_{Df}$ neither the sentence describing p nor the negation of the sentence describing p is a nomological statement.

Definition of relative physical modalities

q is physically necessary relative to $p =_{Df}$ there is a nomological statement n such that the sentence describing q is derivable from the conjunction of n and the sentence describing p.

q is physically impossible relative to $p =_{Df}$ there is a nomological statement n such that the negation of the sentence describing q is derivable from the conjunction of n and the sentence describing p.

q is physically possible relative to $p =_{Df}$ there is no nomological statement n such that the sentence describing q or the negation of the sentence describing q is derivable from the conjunction of n and the sentence describing p.

When we say '$V\bar{B}$ would have been possible', we are using an absolute modality. But the case in which the action B is predictable on the basis of the situation A alone requires relative modalities because we mean that $V\bar{B}$ is impossible relative to the situation A if determinism is assumed. In such a case, it is not meaningful to say 'Instead of VB the volition $V\bar{B}$ could have occurred.'

In order to deal with the difficulty discussed on pp. 451–3, and to ensure that relations (17)–(19) be interpretable as defining freedom of action and be translatable into relations (20)–(22), the following requirement must be added. For the deterministic interpretation of causality, this requirement assumes a function similar to that of requirement β for the probabilistic conception.

Requirement ϵ: The combinations $A.VB$ and $A.V\bar{B}$ must not be physically impossible

In other words, there must not exist nomological statements of the form

$$A \mathbin{\dot{\supset}} \overline{VB} \quad \text{or} \quad A \mathbin{\dot{\supset}} \overline{V\bar{B}}. \tag{32}$$

This requirement, introduced here in order to render relations (17)–(18) interpretable as conditionals contrary to fact, conforms to common sense; to say that in the situation A a volition would have influenced the action is reasonable only if a volition in either direction was possible in this situation.

It should be noted that a statement about absence of freedom also presupposes requirement ϵ. If the action B is *unfree* we have

$$A.VB \mathbin{\dot{\supset}} B \tag{33}$$
$$A.V\bar{B} \mathbin{\dot{\supset}} B \tag{34}$$
$$A.IB \mathbin{\dot{\supset}} B \tag{35}$$

From these statements, relation (23) is derivable; relation (23) is therefore compatible with absence of freedom. Nevertheless, if relation (33)–(35) are to be interpreted as lack of freedom of action, they must be interpreted as conditionals contrary to fact. We must be able to say 'If there were a volition VB in the situation A, then B would occur'; 'If there were a volition $V\bar{B}$ in the situation A, then B would occur'; 'If there were indifference towards B in the situation A, then B would occur'. This interpretation requires that the antecedents not be physically impossible. Otherwise, the contrary implications could likewise be set up as nomological statements.

The determinist must therefore distinguish between three kinds of actions; free, unfree, and neutral. *Neutral* actions are those for which one of the conditions (32) holds, that is, for which requirement ϵ is violated. For neutral actions, the freedom conditional (i.e., the statement 'If the opposite volition had occurred, it would have caused the opposite action', symbolized by the arrow implications (20)–(22)) is meaningless because for such actions no confirmation of the conditional is possible. We mentioned earlier (p. 449) that synthetic nomological statements must lend themselves to inductive confirmation. Although relation (18) may still be true, the violation of requirement ϵ prevents us from interpreting it as a conditional contrary to fact.

The *freedom conditional* (20)–(22) was distinguished from the *freedom condition* which, for determinism, is given by relations (17)–(18) in combination with requirement ϵ. Since ϵ is violated for neutral actions, the freedom condition is not satisfied for such actions. To say that neutral actions are not free is therefore meaningful and even true. But to say that volition is efficient or not efficient for neutral actions is meaningless because efficiency is expressed through the freedom conditional.

Neutral action must be carefully distinguished from indifferent action, symbolized by 'IB'. If relations (32) do not hold, that is, if requirement ϵ is satisfied, it is meaningful to say, with respect to indifferent action, that instead of indifference a volition VB or $V\bar{B}$ could have occurred and would have caused the corresponding action. This conditional contrary to fact is then expressed through relations (20)–(22).

We have shown that it is the strict causal law (23) or (24) which makes it meaningless to speak of volitional efficiency. If the situation A is of such a kind that either $A \dot{\supset} B$ or $A \dot{\supset} \bar{B}$ holds, then we cannot speak of volitional efficiency in that situation, and therefore there is no freedom of action in that situation. Conversely, if it is meaningful to say that a volition is efficient, that is, if the action is not neutral, the action cannot be determined by a strict causal law that does not refer to volition, such as relations (23)–(24). Freedom of action excludes a strict causal law statable without reference to volition.

This result has been derived on the assumption that efficiency of volition is to be expressed by a proper implication symbolized by the arrow. But our result is not restricted to this particular form of reasonable implication. If we use admissible implications in (20)–(22) instead of proper implications, that is, if we replace the symbol '$\rightarrow$' by the symbol '$\dot{\supset}$', we arrive at the conclusion that the statement about efficiency of volition is false instead of meaningless. This interpretation is not essentially different from the first one. However, there is no way of constructing an interpretation for which efficiency of volition can be asserted if relations (23)–(24) hold. Although relations (17)–(19)

remain true if (23) is true, the two contrary implications (21) and (25) are true of the same order. Whatever interpretation is used for a conditional contrary to fact, it will have to satisfy the requirement that such a conditional be unambiguous; the two contrary implications cannot both be admitted for this interpretation. If there is agreement with respect to this requirement, our result is independent of the special properties of proper implication assumed for the above discussion.

VII

In order to anticipate some objections to our explications, we should like to probe a little further into the possibility of representing the causal relationship between volition and action by means of formal implications in terms of the arrow connection.

We said that relations (17)–(18) must be interpretable as conditionals contrary to fact if they are to serve as a definition of freedom of action. They must, therefore, be translatable into the proper implications (20)–(21). Relations (20)–(21) constitute the freedom conditional in the deterministic interpretation of causality. Let us see under what conditions it can be maintained in this form.

There is a continuous transition from wishing, to wanting, to willing an action B. When we denote wishing, wanting, and willing by 'W', we have merely

$$P(A.WB, B) > P(A, B). \tag{36}$$

This probability need not be high. However, we may finally reach a stage where we decide on the action B, and the volitional decision VB is then followed by the action B.

This statement requires the qualification that the action B may be of such a kind that we cannot always perform it. We may decide to go to the theatre, but an accident intervenes; we may decide to run a race, but are completely exhausted after we have covered half the distance. In all such cases we have merely, as above,

$$P(A.VB, B) > P(A, B). \tag{37}$$

Under these circumstances, we are free to a certain degree to do the action B. Yet, in all such cases, we can at least take some step towards the action; we can leave our house in the direction of the theatre, or start running on the track. When we select for B a suitable initial link in the causal chain that is distinct from the volition VB, we can always achieve the action B. On these conditions, we can replace the high probability of (12A) and (12B) by an arrow implication, and assert the freedom conditional (20) and (21)

$$A.VB \rightarrow B$$
$$A.V\bar{B} \rightarrow \bar{B}.$$

If neither VB nor $V\bar{B}$ exists, we speak of indifference. Actually, the special symbol 'IB' for indifference is not necessary, because it would be equivalent to '$VB \vee V\bar{B}$'.

Since the deterministic version $(A \rightarrow B)$ of the case in which an action B is predictable on the basis of a situation A alone, without reference to volition, leads to contrary implications due to the invariance principle, and has to be given up if the freedom conditional is to be retained, we can replace it by (13)

$$P(A, B) \sim 1.$$

(13) holds, for instance, for conventional actions that can be predicted without any knowledge of the volitional state of a person. In addition to (13), and in contrast to $A \rightarrow B$, however, we can have

$$A.\ \overline{VB \vee V\bar{B}} \rightarrow B. \qquad (38)$$

Relation (38) says: Given a situation A and indifference towards B, i.e., neither a volition towards the action B nor a volition towards the action $\bar{B}$, the action B will follow. For example, under ordinary circumstances (A), if I neither want to breathe (VB) nor want not to breathe ($V\bar{B}$), I breathe.

From (38) and (21) we can derive respectively

$$A.\overline{VB}.\overline{V\bar{B}}.V\bar{B} \rightarrow B \qquad (39\text{A})$$
$$A.\overline{VB}.\overline{V\bar{B}}.V\bar{B} \rightarrow \bar{B}. \qquad (39\text{B})$$

But we find that the antecedent in both cases constitutes a contradiction; it is false of third order, and the implications are therefore not admissible. Yet this result does not affect the admissibility of (21) and (38), and relation (38), in contrast to (23), is compatible with (20) and (21), that is, with freedom of action. It would be no contradiction of (38) to say that if in the situation A I did not want to breathe for a certain interval of time, I would not breathe.

Now the determinist might say that freedom of action is possible in the situation A if relation (38) is admissible.

We must realize that we are using relative modality because we mean that the action B is physically possible relative to the situation A. If we define

$$A' =_{Df} A, \overline{VB \vee V\bar{B}} \qquad (40)$$

we have, if (38) holds,

$$A' \rightarrow B. \tag{41}$$

If we now replace A in (20)–(21) by A', the antecedents of these conditionals become contradictions, and (20)–(21) are no longer admissible. Relative to A', therefore, we cannot maintain freedom of action.

The case of indifference (38) must be carefully distinguished from the relation

$$A \rightarrow B \tag{42}$$

If freedom of action in a situation A is so defined that the addition of VB or $V\bar{B}$ to A produces B or $\bar{B}$, respectively, then A cannot strictly determine B, but must stand in a probability relation to it. Relation (42) would therefore have to be replaced by (13).

Let us make another attempt to save determinism. It may be objected that in all such cases we do not use relative modalities, but absolute ones. If $\bar{q}$ is necessary relative to p, then if p happens, q is not possible. But actually we say: instead of p, q could have happened. For instance, on a rainy day (p), the air is not dry (q). Yet we say: today the air could have been dry; it was not impossible. Obviously we are using an absolute modality in this case because we do not mean 'possible relative to the weather situation of today'. We do not want to maintain the possibility of $p.q$ as an *addition*, but the possibility of $\bar{p}.q$ as a *replacement*. In other words, we do not want to say that although p occurred, q could have occurred in addition to p. Rather, we assert that $\bar{p}.q$ could have occurred instead of p.

If we apply these considerations to the treatment of free action within a deterministic conception of causality, we might say that the possibility of $V\bar{B}$ does not mean the addition of $V\bar{B}$ to A' since '$A'.V\bar{B}$' is a contradiction, but the replacement of A' by $A.V\bar{B}$, because $A.V\bar{B}$ could have occurred instead of A'.

The determinist, however, will have difficulties in defending this position because he is not using absolute modalities in a legitimatic way in this instance. As we pointed out on p. 435, the hypothesis of a deterministic causality is based on the following argument:

There is a set L of fundamental laws; there is a complete description D_0 of the state of the universe at the time t_0 which assigns precise values to every parameter; there is a complete description D_1 of the state of the universe at a later time t_1. Part of the description D_0 refers to a situation A in which a volition towards an action B occurs, and this part is denoted by '$A_1.VB$'.

Then D_1 is physically necessary relative to any preceding state D_0 if the following relation

$$D_0 \rightarrow D_1 \tag{43}$$

can be asserted. If determinism holds, this relation is true and excludes replacement of D_1 or parts of D_1 by their negations. It would therefore be a mistake to use an absolute modality and say: instead of $A_1 . VB$ the volition $V\bar{B}$ could have happened.

If probabilistic causality holds, relation (43) is not true. In the probability interpretation, if D_0 is separated from D_1 by a sufficiently long time interval we have

$$P(D_0, D_1) < 1 \tag{44}$$

and $\bar{D}_1$ is therefore physically possible. Here we can use an absolute modality because the probability of (44) is so low that we cannot replace it by a nomological statement.

If we accept the probability interpretation, we can say, with respect to the problem of free action that even if the situation $A.\overline{V\bar{B}}$ is observed at the time t_1, $V\bar{B}$ could have occurred, that is, $V\bar{B}$ was physically possible as a replacement rather than as an addition. Then $V\bar{B}$ is not a possibility relative to the simultaneously existing situation A, but the volition $V\bar{B}$ is possible relative to earlier situations and might have occurred instead of A. Yet, obviously, this interpretation is compatible only with indeterminism. This result rules out a deterministic interpretation of relation (43) in terms of replacement or absolute modalities. We can maintain an interpretation of 'possibility' in terms of replacement or absolute modalities rather than in terms of addition or relative modalities only within the probability conception of causality. In the latter interpretation we can say: $V\bar{B}$ was possible at the time t_1 although $A.\overline{V\bar{B}}$ occurred at the time t_1. While still excluding (42), we can derive from (20) and (38)

$$A.\overline{V\bar{B}} \rightarrow B. \tag{45}$$

Relation (45) reads in words: If in the situation A a person does not decide against the action B, he will certainly perform it. And we may even confirm through observation that he does not decide against B, and then predict B with nomological certainty. But we could still say that the person was free at the time t to decide against B since the possibility of his deciding against B is not asserted relative to the situation A at the time t, but relative to earlier situations.

VIII

As a consequence of the foregoing analysis, we may either admit that, if relation (42) is true, relations (17)–(19) can no longer be interpreted as conditionals contrary to fact, and thus regard an assertion of free action as meaningless; or, in order to maintain that it is possible for a person with a volition towards $\bar{B}$ to be free to do what he wants, we may give up relation (42) and substitute relation (13) for it, in which case the volition $V\bar{B}$ may occur in rare instances. This dilemma points up a particular difficulty for the deterministic interpretation of causality; it explains why predictability has been viewed as excluding freedom of action. Indeed, if the action B is predictable without exception on the basis of a situation A alone, without reference to the volitional state of a person, then the action B can be called neither free nor unfree, since (23) is incompatible with the interpretation of (17)–(18) as freedom conditional. The invariance principle of implication confronts the determinist with the alternative of *predictability versus freedom*, or, more precisely, with the thesis that predictability without reference to volition makes a statement about freedom meaningless.

It is with respect to this problem that the probability interpretation of causality manifests its superiority. For probability implications, no invariance principle holds. These implications admit of exceptions; that is, from $P(A, B) \sim 1$ we cannot derive $P(A.V\bar{B}, B) \sim 1$. [This cannot be done even if $P(A, B) = 1$. In this case we have $P(A, V\bar{B}) = 0$; but the fact that the limit of the relative frequency of $V\bar{B}$ is 0 does not mean the impossibility of the occurrence of $V\bar{B}$ (see [1949f], pp. 54, 72, 118). The value of $P(A.V\bar{B}, B)$ is not determined by the values $P(A, B) = 1$ and $P(A, V\bar{B}) = 0$; it is therefore left open, and may even be = 0.] A high probability in (13) is, therefore, compatible with a high probability in (4) without the requirement that the reference class in (4) be empty. It follows that even if (13) is true, requirement β, that is, the requirement that the probability be genuine, can still be satisfied by relation (4); hence, this relation can be interpreted as a conditional contrary to fact. Furthermore, the requirement that contrary implications must not both be interpreted as counterfactual conditionals is trivially satisfied for probability implications if the probabilities are genuine since in that case the probability values are unique. If both $P(A, B) \sim 1$ and $P(A, \bar{B}) \sim 1$ are asserted, we can infer from the latter value that $P(A, B) \sim 0$. But then the probability value of $P(A, B)$ is not unique, and at least one of these values is not genuine. For instance, when we say 'If higher life exists on one of the other planets (A), the atmospheric conditions must be similar to those on the earth (B)' we have $P(A, B) \sim 1$. This is a genuine probability because the statement is derived from a law of nature, although the reference class A may be empty. When we say 'If higher life exists on one of the other planets, the atmospheric conditions need not be similar to those on the earth' we have $P(A, \bar{B}) \sim 1$. If the reference class A is assumed to be empty, we can assign any value to $P(A, \bar{B})$; $P(A, \bar{B}) \sim 1$ is not genuine because it is not derivable from a nomological statement. From $P(A, B) \sim 1$ we can infer that $P(A, \bar{B}) \sim 0$. Thus $P(A, \bar{B})$ is not unique.

Let us apply these considerations to the problem of free action. Assume that a person wants to vote for the candidate of his political party. If we know his party affiliation (situation A), we can predict with a high probabability that he has the volition to vote for candidate $B(VB)$, and, therefore, that he will vote for candidate B. Now assume the person learns that candidate B has been bribed into being an advocate of monopolies. The situation A has now changed into a situation $A.K$, where 'K' stands for a certain kind of knowledge acquired by the person. Since he has decided not to vote for such a candidate, he is now willing to vote for candidate $\bar{B}$ of the opposite party. Consequently, in the situation $A.K$ there is a high probability that the person will do the action $\bar{B}$. This switching over from one volition to the opposite one is possible only if the foregoing relations between A and B and between A and VB were not formal implications; otherwise, $A.K$ would have the same consequent as A. A formal implication does not admit of exceptions. Now assume that in the situation $A.K$ our prospective voter receives the further information that candidate $\bar{B}$ is actively participating in a movement to promote war. Calling this additional information L, we can describe the new situation as $A.K.L$. Since our voter is even more strongly opposed to war than to monopolies, he will once again be willing to vote for candidate B. Thus, in the situation $A.K.L$ the probability of the action B is very high. Our actual decisions are characterized by such repeated switching back and forth, a phenomenon that can be expressed only in terms of probability implications. That such wavering may have grounds in additional knowledge, and could be predicted if the changes in the situation were known, does not make the action less free. Predictability without reference to volition is compatible with the assertion of freedom of action if the probability interpretation of causality is carried through.

In this context, we should like to add a remark concerning a point discussed previously, namely, the discrimination between the situation A and the volition VB. If both were combined into one situation A', defined as $A.VB$, it is obvious that in the situation A' a volition $V\bar{B}$ is impossible because VB

excludes $V\bar{B}$. The reference class $A'.V\bar{B}$ is therefore empty, and cannot even be used for the probability interpretation since it does not satisfy requirement β. With respect to the situation A', the freedom conditional is meaningless even for the probability interpretation. It is for this reason that we have introduced the rule that the situation A must be defined in such a way as not to overlap with the volition VB. Only if this rule is adhered to is the freedom conditional meaningful. But this result offers no difficulty to the definition of freedom. It merely compels us to restrict the assertion of freedom to cases where the distinction between a non-volitional situation A and a volition VB can be made; only in such cases can an action be called free.

The differences between the deterministic and the probabilistic interpretations of freedom have now been stated precisely enough to enable us to extend the analysis from the problem of free action to the problem of free will. These differences originate in the different logical structures of the implications utilized, since formal implications are subject to an invariance principle, whereas probability implications are not. As a consequence of this structural difference, only the probability interpretation of causality offers the logical possibilities for a general definition of the freedom of the will, a definition which we shall present in the following section.

IX

What has been defined so far is *freedom of action*; what remains to be investigated is whether *freedom of will* can also be defined.

This distinction can be clarified in the following manner: A person may be free to perform the action B, that is, he will be able to do B if he has the volition VB. But we may be able to predict that he will have the volition VB, and thus it would appear that this will is not free. This objection has often been raised against those who assert that there is freedom of will, and it has been argued that though there is freedom of action, there is no freedom of will.

In this formulation, the argument is based upon the assumption that predictability contradicts freedom. Since we have already pointed out that this assumption is false, we can show that the argument is inconclusive. As is to be expected, our investigation will show that complete freedom of will can be defined only by means of a probability interpretation of causality. Again, we shall give the analysis on the basis of both the probabilistic and the deterministic conceptions of causality.

Assuming the probability interpretation, we shall proceed by defining the statement 'a volition is free' in the same way as we have defined the statement 'an action is free', that is, by means of relations (3)–(4). However, an application of these relations to the problem of free will requires that the time order of events be indicated. This order will be expressed by the addition of subscripts. If the action B is performed at the time t_1, we shall write 'B_1'; if the volition V towards the action B occurs at the time t_0, we shall write 'V_0B_1'. The numerical order of the subscripts indicates the direction of the events in time. Relations (3)–(4) then assume the form

$$P(A_0 . V_0B_1, B_1) \sim 1 \tag{46}$$
$$P(A_0 . V_0\bar{B}_1, \bar{B}_1) \sim 1 \tag{47}$$
$$P(A_0 . I_0B_1, B_1) = u. \tag{48}$$

If the action B is predictable even when the volitional status of the person is unknown, we have by analogy with (13)

$$P(A_0, B_1) \sim 1. \tag{49}$$

We shall now consider cases where the occurrence of a volition towards the action B is predictable on the basis of the situation A alone. For instance, if we know that a person is a convinced Democrat, we can predict that he will decide to vote for the Democratic candidate. In this example, the volition V_0B_1 was predictable at an earlier time t_{-1}; thus we have

$$P(A_{-1}, V_0B_1) \sim 1. \tag{50}$$

Relation (50) does not contradict the statement that the volition V_0B_1 is free. The compatibility of (50) with the assertion that the will was free at the time t_0 can be demonstrated by the same method we employed previously in order to account for free action under similar circumstances. By analogy with the preceding definition of 'freedom of action', 'freedom of will' means that a volition at the time t_{-1} would have influenced a volition at the time t_0. Indeed, at the time t_{-1} a person may have the volition to have a volition at the time t_0 towards B or towards $\bar{B}$. In such cases we write '$V_{-1}V_0B_1$' or '$V_{-1}V_0\bar{B}_1$', respectively. The freedom of the volition V_0B_1 is therefore defined by the following conditions, which are analogous to (46)–(48):

$$P(A_{-1} . V_{-1}V_0B_1, V_0B_1) \sim 1 \tag{51}$$
$$P(A_{-1} . V_{-1}V_0\bar{B}_1, V_0\bar{B}_1) \sim 1 \tag{52}$$
$$P(A_{-1} . I_{-1}V_0B_1, V_0B_1) = u. \tag{53}$$

Since relations (51)–(53) do not contradict relation (50), we have shown that a volition towards B can be both predictable and free. The freedom of the will is thus formulated by an *iteration of the freedom conditional.* Such a treatment is justified because a volition is an event that can have a causal influence on other events, be they volitions or actions.

Obviously, the literation can be continued backward in time, and we could therefore assert that the volition $V_{-1} V_0 B_1$ was free although predictable at an earlier time. Hence, freedom of will can be defined as follows:

> *A volition is free if it would have been influenced by a preceding volition.*

As before, the phrase 'would have been influenced' is defined by a conditional contrary to fact, in this case expressed by the freedom condition (51)–(53). Similarly, the definition of 'efficient' is extended to volitions influencing other volitions. A volition is efficient if it is highly probable that it will be followed by a corresponding action or by a corresponding volition and if it is highly probable that a volition in the opposite direction would have been followed by an action or a volition in the opposite direction.

The objection might be raised that the iterated form '$V_{-1} V_0 B_1$' does not conform to actual psychological behavior. At the time t_{-1} we have a volition which is usually directed not towards $V_0 B_1$, but towards B_1; what we experience might therefore be better symbolized by '$V_{-1} B_1$'. Although, psychologically speaking, the objection is debatable, we shall take it seriously for the moment in order to show that it would not affect our explication of free will. It is easy to transcribe our formulations in accordance with the objection. We simply omit in relations (51)–(53) the intermediate term 'V_0' on the left side of the comma, but leave it on the right. The relations then state that if at the time t_{-1} there is a volition to do B at the time t_1, then there exists a high probability that at the intermediate time t_0 there is also a volition to do B at the time t_1. A person may not even be aware of having the volition V_0 although he might become conscious of having it if a specific occasion for its manifestation were to arise. For instance, a person might be asked how he intends to vote in a forthcoming election, and thus become aware of his intentions. Such situations can also be meaningfully formulated through probability relations. Written in the transcribed version, relations (51)–(53) could be used just as well for a definition of freedom of will. The iterated form employed above may be regarded as a rational reconstruction, suitable to show the structural identity of relations (51)–(53) with relations (46)–(48) or relations (3)–(4).

Incidentally, the longer the time distance between the volition and the action to be performed, the lower the probability that the action will occur because the opportunity for revising one's volition is greater; the causal chain from the volition to its execution can therefore be interpreted as a continuous probability sequence [1949f, Section 46].

X

Let us now investigate whether a definition of 'free will' analogous to that given by relations (51)–(53) can be given on the basis of the deterministic interpretation of causality. Introducing, as before, formal non-truth-functional implications in place of probability implications of a degree close to 1, we can transcribe relations (51)–(53) as follows:

$$A_{-1}.V_{-1}\,V_0B_1 \;\dot{\supset}\; V_0B_1 \qquad (54)$$

$$A_{-1}.V_{-1}\,V_0\bar{B}_1 \;\dot{\supset}\; V_0\bar{B}_1 \qquad (55)$$

$$A_{-1}.I_{-1}\,V_0B_1 \;\dot{\supset}\; ? \qquad (56)$$

These relations, like relations (17)–(19), can be interpreted as conditionals contrary to fact if the corresponding proper implications also hold, that is if the sign '$\dot{\supset}$' in (54)–(56) can be replaced by the sign '$\rightarrow$'. Relations (54)–(56) thus constitute a definition of 'freedom of will'. As in the probability interpretation, a volition is free if another volition at a preceding time would have influenced it.

If we accept this interpretation of relations (54)–(56), we must realize that the following relation

$$A_{-1} \;\dot{\supset}\; V_0B_1 \qquad (57)$$

cannot also hold. Relation (57) applies when a volition at the time t_0 to perform an action B at a later time t_1 is a physically necessary consequence of the situation A at the time t_{-1}. Applying the invariance principle of implication, we can derive from (57) the relation

$$A_{-1}.V_{-1}\,V_0\bar{B}_1 \;\dot{\supset}\; V_0B_1 \qquad (58)$$

which is the contrary implication of (55). This result shows that the combination of A_{-1} and $V_{-1}\,V_0\bar{B}_1$ in the antecedent of (58) and of (55) is physically impossible. Consequently, relations (54)–(56) cannot be interpreted as conditionals contrary to fact, and thus lose their meaning as freedom conditions. In connection with freedom of will, the determinist faces the same dilemma that he faced in connection with freedom of action. Either he has to give up the assertion that the volition V_0B_1 is free if relation (57) holds, since on the deterministic conception, in contrast to the probability conception, a volition which can be predicted without reference to a preceding volitional state is not free; or if he wants to maintain that the volition V_0B_1 is free, he must replace (57) by (50), in which case determinism no longer holds.

The fact that on the deterministic interpretation a free volition is not predictable without reference to a preceding volition leads the determinist into difficulties. Actually, instances of volition do not form an infinite series backward in time, but arise at a certain moment for the first time. The person who wills today to diet tomorrow may have willed yesterday to will today to diet tomorrow. But there has been a first occurrence of this volition. According to the deterministic interpretation, this first occurrence of the volition was predictable, and since it was not preceded by another instance of volition, it must have been predictable on the basis of a situation A_{-1} whose description does not include any reference to the volitional state of the person. If V_0B_1 is the first occurrence of a volition, relation (57) must hold. Consequently, the antecedent of relation (55) is physically impossible, and requirement ϵ is violated. Relations (54)–(55) can therefore no longer be interpreted as a freedom conditional, and the first instance of a volition can be called neither free nor unfree; it is neutral.

This result does not hold for the probability conception of causality. On this interpretation, too, the causal chain contains the first instance of a volition and this volition was predictable on the basis of a situation A whose description does not include any reference to the volitional state of the person. But this fact does not exclude the freedom of the first instance of a volition because relation (50) does not prevent relations (46)–(47) from being interpreted as conditionals contrary to fact. Let V_0B_1 be the first instance of a volition. Then it remains permissible to say: If V_0B_1 had been preceded by a volition $V_1\bar{B}_1$, then, instead of the volition V_0B_1, the volition $V_0\bar{B}_1$ would have occurred. This conditional contrary to fact is not assertable on the deterministic interpretation, and therefore the first instance of a volition cannot be called free by the determinist. Yet it cannot be called unfree either, as was pointed out above; it is neutral.

Some defenders of the thesis that determinism does not contradict free will have misunderstood this point. They are right when they argue that the existence of instances of volition which are not preceded by other instances of volition is not, in general, incompatible with freedom of the will. But they are wrong when they maintain that *on their theory* the existence of such instances is compatible with freedom. They believe that a first instance of a volition may be called free since it was possible to have a volition in the opposite direction. But on their theory such an assertion is false. They are not aware of the distinction between 'possible' and 'possible in the situation A'.

This is the reason why we explained above the difference between absolute and relative modalities (see p. 174). To say that a volition $V\bar{B}$ is physically possible is to say that there are situations which do not exclude such volitions. This statement requires only absolute modalities. But the state of affairs in which a volition arises for the first time is, on the deterministic conception of causality, characterized by relation (57). To say that a volition $V_0\bar{B}_1$ was possible requires the use of a relative modality because 'possible' in this context means 'possible at the time t_0' or 'possible relative to the situation A at the time t_{-1}', and the situation A excludes a volition $V_0\bar{B}_1$. In fact, this is the meaning of (52) if A_{-1} is a physically possible situation since then the contrary implication to (52) cannot be synthetic nomological; and the physical possibility of A_{-1} is, of course, assumed because A_{-1} is supposed to occur. When it is argued that a volition $V_0\bar{B}_1$ would have been possible if there had been a volition $V_{-1}V_0\bar{B}_1$, it must be answered that $V_{-1}V_0\bar{B}_1$ was impossible in the situation A_{-1} if relation (55) holds. The invariance principle of implication, applied to relation (57), leads to the contrary implication of relation (55), and therefore to the conclusion that the antecedent of (55) is physically impossible, that is, to the conclusion

$$A_{-1} \dot{\supset} \overline{V_{-1}V_0\bar{B}_1}. \tag{59}$$

It follows that the statement 'If there had been a volition $V_{-1}V_0\bar{B}_1$, it would have produced a volition $V_0\bar{B}_1$' is meaningless relative to the situation A_{-1}. A first instance of a volition cannot be free; it is neutral if determinism holds. Whoever wants to unite causality with freedom must construct an interpretation of causality which abandons relation (57), that is, he must give up determinism.

We shall call an action or a volition *completely free* if it is an element in a causal chain in which every element preceding it is free. It should be noted that this definition does not require that the chain have an infinite number of prior elements. The definition does not say that every efficient volition *is in fact* preceded by another effieicnt volition, but rather that it *is physically possible* for it to be preceded by an efficient volition. This physical possibility is expressed by a conditional contrary to fact. The chain may very well have a first element, and it may even consist of only one element. However, it follows from the definition of 'complete freedom' that, on the deterministic interpretation of causality, no action nor any instance of a volition is completely free since all actual chains of volitions have a first element, and this first instance of a volition is not free, but neutral. According to the probability interpretation, on the other hand, there can be completely free actions and volitions.

The following objection might be raised: If the first instance of a volition had been preceded by a volition in the opposite direction, we should have to ask whether the earlier volition had been free. Since this question can be answered only by reference to a further preceding instance of volition which, in turn, would have to be free, we would obtain a negatively infinite chain of possible instances of volition. We think these doubts can easily be dispelled. In the first place, the possibility of such a chain cannot be assumed; secondly, the objection misses the point. The first actual instance of volition is free if it is possible for it to have been preceded by an efficient volition. Whether this possible preceding volition would have been free is irrelevant; in fact, it might have been unfree. For instance, this preceding volition might have been the effect of a posthypnotic suggestion, and if it had occurred, the whole ensuing chain of volitions and actions would not have been completely free; but since it did not occur, no such conclusion can be drawn. On the probability interpretation, if V_0B_1 is the first instance of a volition, it is possible for it to have been preceded by another volition, and then V_0B_1 would not have been a first instance. The preceding volition could have been either free or unfree; both possibilities always exist. But a *possible* unfree predecessor does not make an *actual* volition unfree – does not make an *actual* volition lack complete freedom; only an *actual* unfree predecessor would do so.

XI

In traditional discussions of the freedom of the will, it has often been argued that our will cannot be free because volitions are the product of non-volitional conditions. The foregoing analysis shows that this argument refutes only the deterministic interpretation of free will, but does not apply to the probability interpretation. Although volitions emerge from non-volitional situations, and although they may be predictable on the basis of these situations, they can be called free on the probability interpretation since freedom is explicated by an admissible conditional contrary to fact. Only the deterministic interpretation must renounce the claim that emerging volitions are free since this interpretation cannot combine the conditional contrary to fact with the predictability of the emerging volition.

On the probability conception, we can explicate *free action*, *unfree action*, *free volition*, and *unfree volition*. Only if free action is combined with free volition do we have complete freedom; in all other combinations, we have freedom to a certain degree or no freedom. Let us summarize once more our explications.

An action B is free in the situation A if it is possible for the action B to have been preceded by a causally relevant volition and B is not physically impossible in the situation A. For example, I want to lift my arm, and I do so. If I had had the opposite volition, I could have refrained from lifting my arm. Relations (3)–(4) hold.

To the class of free actions belong the so-called actions under duress. For example, a prisoner of war is told that he must either sign a false statement or he will be shot. His action is free because if he wants to, he can refuse to sign; but then he will be shot. If B is the action of refusing to sign, relations (3)–(4) hold. However, the prisoner is not free both to refuse to sign and to remain alive. This lack of freedom is expressed by the fact that relation (3) does not hold if B is remaining alive while refusing to sign. Although we may not hold a person morally responsible for signing a false statement under duress, this is not because we consider his action unfree, but rather because we feel he is free only to choose between two evils; therefore we do not reproach him for deciding to commit the evil that saves his life.

Some of our actions are automatic – they need not be preceded by a volition. For example, the answer 'Just fine' to the question 'How are you?' is highly predictable, under ordinary circumstances, without any knowledge of the volitional state of the person addressed. Relation (13) applies in this case. Nonetheless, the action is free since a preceding volition could have evoked the same or a different response. Therefore relations (3)–(4) hold.

However, if the situation is not conventional, for example, if a person were asked how he was by his psychoanalyst, we should not venture to predict his answer unless we knew much more about him. Moral actions of normal people are free; under unusual circumstances, they differ from conventional behavior in not being predictable unless one knows the volitional state of the agent. In such circumstances, relation (13) does not apply. For example, we cannot predict whether a certain man will be a conscientious objector or fight for his country; but if we know what he wants to do, we can predict what he will do.

An action B is unfree if a preceding volition is irrelevant to it. For example, I may want to remain suspended in midair when I am falling from a tower, but my volition has no influence upon the act of falling. Relation (3) does not hold. The desired event is physically impossible because of the law of gravitation.

A volition V is free if it is possible for it to have been preceded by an earlier volition which is causally relevant to it. For example, I may will today to will to study tomorrow. Volitions directed towards changing our future intentions or motivations, such as New Year's resolutions, have this iterated form. Relations (51)–(52) apply in such cases.

A volition V is unfree if a preceding volition is causally irrelevant to it, or if the opposite volition is physically impossible. For example, under hypnotic suggestion the will is not free; the opposite volition is physically impossible, and one of the alternatives of relation (32) applies. Other examples are furnished by drug addiction and by the unconscious determination of volition studied in psychoanalysis; but in these cases, the degree of probability in relation (50) may be considerably lower than 1. Although in the given examples the volitions are unfree, the actions are free and relations (3)–(4) apply. However, we cannot say that the actions are completely free.

XII

Determinism assumes that the series of more and more exact descriptions of a momentary cross section of the universe has a last term from which the description of a later state of the universe follows with certainty. The preceding analysis shows that, because of this assumption, the deterministic interpretation of causality is incompatible with freedom of the will. The deterministic assumption, schematized in (1), leads to the use of formal implications; but the invariance principle of implication prevents us from maintaining both the freedom condition and the predictability, without reference to prior volitions, of either actions or volitions. However, if the series of descriptions has no last term, a formal implication cannot be asserted, and the freedom condition remains compatible with predictability.

Our conclusions are the result of logical analysis. Which of the two forms of causality holds for our physical world is an empirical question. Yet it seems that there is hardly any evidence for the deterministic hypothesis. Experience shows only that there are descriptions which admit of highly probable predictions, and that these descriptions can be improved by further analysis of the parameters of the situation. From this we inductively infer the hypothesis of causality; that is, we infer that the probability values of the predictions can be brought as close to 1 as we wish by means of further analysis of the physical conditions. To assert that there exists an ultimate description which forecasts the future with certainty goes beyond what could ever be inferred from observation. Even classical physics does not need such an additional assumption; all its theorems can be interpreted in accordance with the conception of causality expressed in schema (2).

It is well known that quantum mechanics has gone one step further in the rejection of determinism. Heisenberg's Principle of Indeterminacy asserts that the probability values of predictions concerning the microcosm cannot be brought as close to 1 as we wish, but must stop at a given boundary lower than 1.

The quantum mechanical conception can be represented by the following schema:

$$\begin{array}{lllll} D^{(1)} & D^{(2)} & D^{(3)} & \ldots & D \\ p^{(1)} & p^{(2)} & p^{(3)} & \ldots & p < 1. \end{array} \tag{60}$$

For quantum physics, as we see, there is an ultimate description D, the best that we can construct; it is given by the ψ-function of a momentary physical state of a particle. But this description permits us to predict the future state only with a probability p lower than 1, and no means are available ever to improve this probability. Schema (60) expresses the specific indeterminism asserted in quantum physics, namely, that there exists a limit of predictability [1944b], and [1956b], Section 25. Only if large numbers of elementary phenomena are involved can the probability values of predictions be increased. Thus causality is abandoned, and this form of indeterminism may therefore be called *statisticalism*. Statisticalism explains the apparent causalism of macrophenomena as resulting from the law of large numbers. Important as this version of indeterminism is for the theory of causality, it has no bearing upon the definitions of freedom of action and freedom of will given above. Since the freedom condition (3)–(5), at any stage of exactness, is compatible with probabilistic causalism, it is *a fortiori* compatible with statisticalism. The theory of free will which we have developed is compatible with the quantum mechanical conception of causality, but is not dependent upon it.

The solution of the problem of free will presented here may appear simple; in fact, the logical apparatus employed for it is none too complicated. Yet we were able to offer a solution only after we had given a precise explication of the relevant terms. So long as the term 'causality' is used in an imprecise manner, the difference between causality and determinism cannot be seen. And so long as no precise definition of 'freedom' is given, there is a temptation to employ such vague terms as 'internal causes' and 'external causes' and to make such vague statements as 'Freedom exists when the causes of the action are internal', or 'Freedom exists when the causal chain goes through the person as an agent'. The historical controversy up to the present time abounds in such expressions. We may read, for example, that lack of freedom is due to 'compulsion', to the 'constraining' volitions of other people or to restrictions imposed by the laws of the state. It is true that other persons may desire me to act in a certain way, but that does not deprive my actions of their freedom; otherwise, it would make no sense to call someone inconsiderate. The laws of the State forbid certain kinds of actions, but that does not prevent people from disobeying these laws on the basis of their own free decisions; otherwise, it would make no sense to hold people responsible for their actions. If we decide to defer to other persons' desires or to obey the laws of the State, certain volitions concerning our actions are entailed by these decisions. But these volitions are no less free for being entailed.

The thesis that causality does not exclude free will has often been maintained, but it could not be made consistent so long as causality was identified with determinism. The present investigation shows that determinism excludes complete freedom of will and action while probabilistic causality does not. And it shows, in addition, that the abandonment of causality as a consequence of quantum physics is also compatible with freedom provided that sufficiently high probabilities are left to permit us to regard volitions as causally relevant to our actions. It is astonishing to see that so old an issue as the freedom of the will admits of so simple a solution once the picture language of traditional philosophy is replaced by the precise language of the mathematical logician.

NOTES

[1] The restriction imposed upon this probability can be computed as follows. We use the notation

$$P(A, VB) = a_1 \quad P(A, V\bar{B}) = b_1 \quad P(A, IB) = c_1 \quad P(A.VB, B) = 1 - a_2$$
$$P(A.V\bar{B}, \bar{B}) = 1 - b_2 \quad P(A.IB, B) = 1 - c_2 \quad P(A, B) = 1 - d_2$$

The rule of elimination (1949f, p. 76) furnishes

$$1 - d_2 = a_1(1 - a_2) + b_1 b_2 + c_1(1 - c_2).$$

Eliminating c_1 by the use of the relation $a_1 + b_1 + c_1 = 1$ and solving for b_1 we find

$$b_1 = \frac{d_2 - c_2(1 - a_1) - a_1 a_2}{1 - b_2 - c_2} \text{ and thus } b_1 \leqq \frac{d_2}{1 - b_2 - c_2}$$

The inequality follows because the quantities occurring are all between 0 and 1. If the values b_2, c_2 and d_2 are smaller than a given small quantity δ (the value a_2 must then also be $< \delta$ since $a_2 \leqq d_2$ is then easily demonstrable), we have

$$b_1 \leqq \frac{\delta}{1 - 2\delta} \quad \text{or practically: } b_1 \leqq \delta$$

If $\delta = 0$, i.e., if the approximate equalities (3)–(5) and (13) are replaced by strict equalities, we have $b_1 = 0$. But otherwise b_1 need not be 0.

43. ON THE EXPLICATION OF ETHICAL UTTERANCES*

[1959a8]

[The following essay is based on rather extensive notes, dated December 31, 1952, by my late husband. As far as I remember, they constituted his reaction to certain ideas that were expressed at the Christmas meeting of the Pacific Division of the American Philosophical Association, held that year in Los Angeles at the University of Southern California. They may be regarded as supplementing the analysis of the nature of ethics set forth in the last two chapters of his book, *The Rise of Scientific Philosophy* [1951a].

The essay addresses itself to the familiar issue of a cognitive versus a non-cognitive explication of ethical discourse, and tries to show, by a careful logical analysis of the reconstructed language, not that a cognitive interpretation is false, but that it is incomplete – M.R.]

We can distinguish between two versions of the *volitional interpretation* of ethical judgments:

1. *The expressive version.* An ethical utterance includes a token-reflexive expression of volition by the speaker. In addition, it may be elliptic for certain implications relevant to the action. In the expressive version of the volitional interpretation, 'should' or 'ought' is an expression of volition by the speaker and, in correct wording, is given as an imperative. For example, 'The South should desegregate' can be translated into 'Abolish segregation'. This expression may be elliptic for such sentences as 'If segregation is to be abolished, then the schools will have to be integrated'.

The token-reflexivity of every imperative is borne out by the fact that the correlated indicative statement will always have to include a reference to the particular person who utters the imperative, so that there are as many different correlated statements as there are persons uttering the same imperative.

2. *The declarative version.* An ethical utterance is to be explicated as a declarative statement about the volitional state of the speaker. In addition, it may be elliptic for certain implications relevant to the action. In the declarative version of the volitional interpretation, 'should' or 'ought' is to be explicated by a declarative statement about the existence of a volitional state of the speaker, namely, by a correlated indicative statement of the kind just mentioned. Our example, 'The South should desegregate', would then have to be explicated by 'Mr. Eisenhower wants the South to desegregate' or 'Mr. Stevenson wants the South to desegregate'. In the declarative interpretation,

*From *Modern Philosophy of Science: Selected Essays,* ed. and tr. by Maria Reichenbach, Routledge & Kegan Paul, London, 1959, pp. 193–198. Copyright © Maria Reichenbach 1959 except in U.S. Copyright © in U.S. by Maria Reichenbach 1959.

the token-reflexivity of an imperative is made explicit.

In contrast to the volitional interpretation, there is a cognitive one, the *relational interpretation*. In the relational interpretation, 'should' or 'ought' expresses logical or physical entailment, and, in correct wording, the utterance containing these terms is neither an imperative nor a statement about the volitional state of a person. The utterance is elliptic for certain implications relevant to the action. In addition, it may include a token-reflexive expression of volition by the speaker, but this fact is irrelevant for the meaning of the utterance and can be omitted in the explication. In our example, the utterance 'The South should desegregate' would be implied by the decision of the Supreme Court. It is elliptic because the antecedent 'Since the Supreme Court decided that segregated schools are unconstitutional' is not explicitly stated. It is irrelevant in this case whether the speaker wishes the South to act in this way independently of the Supreme Court's decision.

When we ask which interpretation fits ordinary usage better, we find it difficult to give an answer because ordinary usage is ambiguous, and both interpretations, the volitional and the relational, may be used. It is possible to show, however, that the adequacy of the explication can be judged from another point of view, namely, that of *completeness*. If we accept the criterion completeness, a rational reconstruction of language based on the relational interpretation is inadequate because it is necessarily incomplete.

In order to justify this contention, let us give the relational interpretation in a more precise form. This interpretation assumes that a certain set V of valuations or preferences entails the action A that is under consideration, or, more precisely, that the action A *satisfies* the set V. This relation of satisfaction can be construed in a very wide sense. A may be said to satisfy the set V if A satisfies more urgent preferences, while dissatisfying others; or if A satisfies preferences for a longer time, not merely for the moment, even though certain momentary desires may be frustrated by A, etc. If conditions of this kind hold, we may say, briefly, that A supplies maximal satisfaction of V. The meaning of 'maximal' would have to be exactly specified, but may be regarded as understood for the present investigation. We see that the ethical utterance 'one should do A' is elliptic with respect to the set V; the complete statement would read 'For the set V, A supplies maximal satisfaction'. The question arises: what is the set V referred to in the usual ethical utterance?

We might say that V is the set of generally accepted valuations. But this interpretation can hardly be maintained because in ethical discussions we frequently want to criticize the existing valuations; otherwise, the utterance would merely mean that it is usually regarded as good to do A. Our example about segregation illustrates this point. The recent decision by the Supreme Court represents the opposite of the valuations hitherto widely accepted in the South. In order to escape this difficulty, we would have to add in ethical discussions a description of the set V. We would have to say: if these and these basic valuations are assumed, then A supplies maximal satisfaction. Though clumsy, such a mode of speech would certainly have its advantages because it would clarify ethical discussions. Let us call this mode of speech *referential* since it includes explicit reference to the set V. Obviously, if the relational interpretation is to be made non-elliptic, it requires the referential form.

Reference to V is necessary because, logically speaking, there exists for every given action A a set V for which A is good, i.e., for which A supplies maximal satisfaction, and there exists another set V for which A is bad, i.e., for which A lowers integral satisfaction considerably; in fact, there are always many sets of both kinds. Why do adherents of the relational interpretation, in particular the pragmatists, believe that it is not necessary to mention the set V explicitly? It cannot be taken for granted that everybody has the same set of valuations since various religions, forms of government, habits, and cultural patterns condition people for different sets of preferences.

Although such a cultural relativity of valuations has to be accepted as a matter of fact, some philosophers would argue that this observation is not the end, but the starting point, of ethical discussion. The adequacy of cultural conditioning, or of sets of valuations, should be judged with respect to the satisfaction of basic human needs. It seems that the pragmatists believe that there exists only one such set, i.e., they believe that human beings are sufficiently similar so that the same set of ultimate valuations can be derived. This assumption, if true, would still be empirical; we have no *a priori* grounds for it. But it appears highly questionable whether this assumption is true. Apart from the suspicion that this argument commits the old fallacy of going from what is desired to what is desirable, there seem to be a multitude of human needs, and as soon as good and bad needs are distinguished, or their number is arbitrarily restricted, value judgments are smuggled in. Sociological, psychological, and psychoanalytic investigations do not always agree on what these needs are. Therefore, the assumption of such a unique set appears unjustified for ethical discussions, and in fact the assumption is not necessary if the referential mode of speech is employed. In the following analysis, the referential mode of speech is always assumed for the relational interpretation.

There are many instances where ethical discussion would profit greatly if ethical utterances were given in the referential mode of speech, for example, in discussions about such general issues as political strategy, or a revision of the existing law code, or moral questions. The explicit formulation of the set V may be very helpful in clarifying these issues. In contrast, the volitional state of the speaker may be irrelevant; we would not lose very much if it were not indicated in the reconstructed language. In such utterances as 'Stealing is wicked', 'The divorce laws should be modified', etc., the volitional attitude of the speaker appears merely as an addition, interesting from a personal or psychological point of view, but dispensable for the purpose of the discussion.

There are other cases, however, where an expression of the volitional state of the speaker is less dispensable. For instance, when we say 'Paul should not be so nasty to his mother', or 'Peter should marry Jeanne after having lived with her for so many years', we wish to express our personal judgment. We want to make our judgment clear perhaps because we believe that a statement of our disapproval (or, in other cases, of our approval) will influence the person in question, or perhaps only because we want to avoid suspicion by a third person that we share certain views with which we should not like to be identified. But if an addition of this kind is made in the reconstructed language, even if the addition is formulated as a declarative statement and not in the expressive mode, the purely relational interpretation is abandoned, and the volitional interpretation is introduced.

While the relational interpretation may still be defended in these instances by the contention that the volitional state of the speaker is none too significant, there exists a realm of discourse in which an indication of the volitional state of the speaker is clearly indispensable. Such is the case when value judgments are used by an individual for directing his own behavior. If an action A is shown to be good with reference to a valuational set V, there remains the problem of deciding whether the individual accepts the set V; and it is in this context that statements about his volitional state are required.

Let us analyse this problem in more detail. In order to construct a fully cognitive theory of ethics, we must add to the relational interpretation of ethical rules a *cognitive ascertainment of subjective valuations*; that is, we must regard the finding of valuations within oneself as a sort of cognitive discovery. I find that I value freedom, interesting company, music, etc. We may then formulate the following argument:

(1) If V is the set of my valuations, then the action A will give me maximal satisfaction.
(2) V is the set of my valuations.
(3) The action A will give me maximal satisfaction.

Note that the conclusion is not an imperative but a cognitive prediction about my future experiences.

The question arises: can we derive the imperative 'Do *A*!' from the cognitive conclusion? Obviously not. Given the conclusion, I can still ask 'Should I do *A*?' Here the word 'should' is employed in the imperative sense.

The adherents of the relational interpretation may answer: these troubles arise only because you have now used 'should' in the imperative sense. If you interpret 'should' in your question as implicational, the utterance 'I should do *A*' means 'If I do *A*, I shall have maximal satisfaction'. This conclusion can be derived from our premises; in fact, it is identical with the one we derived above.

However, this interpretation provides a pseudo-solution. If the imperative 'should' is ruled out from our vocabulary, the question raised reappears in another formulation. It turns up as the question: how can I proceed from the cognitive conclusion 'The action *A* will give me maximal satisfaction' to the action *A*?

It is this transition, the transition from the cognitive conclusion to the action *A*, which is not covered by logic and which is not a cognitive affair. The step from knowledge to action cannot be cast into the form of a logical inference. In other words, I can accept the conclusion and yet decide against maximal satisfaction, i.e., I can refuse to do *A* without being logically inconsistent. The recognition of the non-logical relation between knowledge and action is the essence of the non-cognitive theory of ethics.

Due to its omission of any treatment of the relation between knowledge and action, the cognitive theory of ethics is incomplete. It gives formulations concerning ethics up to a certain point, and then is silent where something important remains to be said. It allows us to state all the knowledge required for moral action, but does not indicate the fact that something is still to be done. This omission is dangerous. Because of its silence about the transition from knowledge to action, the cognitive theory of ethics may mislead us into believing that this transition is also cognitive, whereas it is not. We do not wish to assert that the proponents of such a theory have *said* that this transition is cognitive but the reader of their publications cannot help feeling that they assume it is cognitive.

Let us show how the non-cognitive theory of ethics includes an explicit description of the transition from knowledge to action. For this purpose, linguistic utterances are used which express the volitional state of the speaker with regard to certain actions. This function is fulfilled by imperatives that, owing to their token-reflexivity, can always be construed as indicating a hidden 'I will'. [For a detailed analysis of token-reflexive terms see [1947c], Section 50, and [1951a], Ch. 17.] In this way we arrive at an exhaustive theory, i.e., at a rational reconstruction of language which is complete and which includes an indication of the relation between knowledge and action.

What is an imperative? Applied to ourselves, it is a signal for action. It functions as the first step in a chain of muscular actions ending in the action *A*; it releases the chain like a button which we press and which releases some physical mechanism. If the action *A* is to be postponed for some reason, we may still give the signal, using it, so to speak, as an anticipatory sign for delayed action. Applied to other persons, an imperative is likewise a signal, and is intended to release a chain of actions on their part leading to the action *A*. Though we do not always succeed with this method, there are many cases where it does have success. These signals occur in our language behavior and possess great significance. Without them, it would be awkward to apply language to action; there would remain a gap unbridged by linguistic utterances. Any rational reconstruction of language that omits these signals is necessarily incomplete.

The fact that imperatives cannot be omitted indicates that the question 'What should I do?' cannot be answered by drawing an inference, but only by making a volitional decision. Our valuations are always affirmed anew, or at least require volitional efforts to be sustained. The cognitive theory of ethics is incomplete in this sense. In its fear of indicating the non-cognitive relation between knowledge and action, it either cancels these terms, or construes the terms actually used as implications rather than as token-reflexive signals. The non-cognitive theory of ethics, in contrast, assigns a definite place to these terms within the reconstructed language. It regards them as utterances which are neither true nor false; casts them in the grammatical form of imperatives, or characterizes them by an added exclamation mark; uses them to express the non-cognitive transition from knowledge to action; and points out that these utterances are not derivable from declarative statements. It is for these reasons that only the non-cognitive theory of ethics supplies an adequate explication of ethical utterances.

BIBLIOGRAPHY OF WRITINGS OF HANS REICHENBACH

1912: (a) 'Die Studienreise englischer Studenten nach Deutschland', *Münchner Akademischer Rundschau* **6**, no. 2: 27 (Oct. 24, 1912)

(b) 'Die Gleichberechtigung der Reifezeugnisse höherer Lehransalten', *Münchner Akademische Rundschau* **6**, no. 3: 46 (Nov. 9, 1912)

(c) 'Studentenschaft und Katholizismus', *Das Monistische Jahrhundert*, W. Ostwald, ed., No. 16: 533–538 (Nov., 1912) (For English trans. see [1978 a-2])

(d) 'Die Neuorganisierung der Münchner Freistudentenschaft', *Dresdner Studentische Blätter*, no. 6: 1–4 (Nov. 20, 1912)

(e) 'Student und Schulproblem', *Münchner Akademische Rundschau* **6**, no. 6: 94–97 (Dec. 16, 1912)

(f) 'Der Student', *Münchner Studentisches Taschenbuch* (Munich: Max Steinebach, Wintersemester, 1912–13) pp. 42–44 (For English trans. see [1978 a-1])

1913: (a) 'Der Student und die pädagogischen Bestrebungen der Gegenwart', *Münchner Akademische Rundschau* **6**, no. 10: 178–179 (March 5, 1913)

(b) 'Die Militarisierung der deutschen Jugend. I. Der Tatbestund', *Die Freie Schulgemeinde* **3**, no. 4: 97–110 (July 1913)

(c) (*with Carl Landauer*) 'Die freistudentische Idee. Ihr Inhalt als Einheit', *Freistudententum: Versuch einer Synthese der freistudentischen Ideen*, by Hermann Kranold, Carl Landauer and Hans Reichenbach (Max Steinebach, Munich, 1913) pp. 25–40 (For English trans. see [1978 a-3])

(d) 'Der Wandervogel und die Juden', Beilage des *Berliner Börsen-Courier* no. 539 (Nov. 16, 1913)

(e) 'Tagebuch', *Der Student* [Neue Folge der *Berliner Freistudentischen Blätter*] **6**, no. 6: 80–81 (Nov. 28, 1913)

(f) 'Freischar oder Freistudentschaft', *Der Student* [Neue Folge der *Berliner Freistudentischen Blätter*] **6**, no. 7: 88–89 (Dec. 5, 1913)

(g) 'Warum treiben wir Körperkultur?, *Berliner Freistudentischen Blätter*, no. 3: 1–4 (1913) (For English trans. see [1978 a-4])

(h) 'Verhandlungen' (Zweiter Deutscher Kongress für Jugendbildung und Jugendkunde, Munich, Oct. 3–5, 1912) *Arbeiten des Bundes für Schulreform* no. 6 (B. G. Teubner, Leipzig and Berlin, 1913) pp. 165–166

1914: (a) 'Militarismus und Jugend', *Die Tat* **5**, no. 12: 1234–1238 (March 1914)

(b) 'Von der Georgia Augusta. Die Hochschule', *Göttinger Akademische Wochenschau* **10**: 21 (1914)

(c) 'Der Sinn der Hochschulreform', *Studentenschaft und Jugendbewegung* (Max Steinebach, Munich, 1914) pp. 7–11 (For English trans. see [1978 a-5])

(d) 'Die Jungdeutschlandbewegung. Die Jugendbewegung der Gegenwart und ihre Bedeutung für die Hochschule', *Studentenschaft und Jugendbewegung* (Max Steinebach, Munich, 1914) pp. 12–33

(e) 'Zum Lietzschen Vortragsabend', *Göttinger Akademische Wochenschau* **10** no. 5: 38 (June 12, 1914)

(f) 'Jugendbewegung und Freie Studentenschaft', *Münchner Akademische Rundschau*, no. 13: 224 (Nov. 13, 1914)

(g) 'Zwei Fichtefeiern', *Münchner Akademische Rundschau*, no. 4: 29 (Dec. 21, 1914)

1915: (a) 'Hans Wegener, *Wir jungen Männer*', *Die Tat* **6**, no. 2: 218–220 (May 1915)

(b) *Der Begriff der Wahrscheinlichkeit für die mathematische Darstellung der Wirklichkeit*. Inaugural Dissertation, University of Erlangen (Barth, Leipzig, 1915) 79 pp.

1916: (a) 'Der Begriff der Wahrscheinlichkeit für die mathematische Darstellung der Wirklichkeit', *Zeitschrift für Philosophie und Philosophische Kritik* **161**, 210–239 (1916); **162**, 9–112, 223–253 (1916). Reprint of Reichenbach's dissertation, [1915b].

1918: (a) 'Die Sozialisierung der Hochschule', unpublished manuscript, 1918 (For English trans. see [1978 a-7])

(b) 'Programm der sozialistischen Studentenpartei', (orig. source and place of publication have not been established) (For English trans. see [1978 a-6])

(c) 'Bericht der sozialistischen Studentenpartei Berlin' and 'Erläuterungen zum Programm', Winter semester 1918/19 (orig. source and place of publication have not been established; published in Staatsarchiv München, Staatsanwaltschaft München I, no. 2973, vol. 2 (For English trans. see [1978 a-8])

1919: (a) 'Der Begriff der Wahrscheinlichkeit für die mathematische Darstellung der Wirklichkeit', Autoreferat. *Die Naturwissenschaften* **7**, no. 27: 482–483 (1919). Reichenbach's summary of his dissertation, [1915b].

(b) 'Student und Sozialismus', *Der Aufbau*, no. 5 [Flugblätter der Jugend, Berlin]

(c) 'Sterzinger, *Zur Psychologie und Naturphilosophie der Geschicklichkeitsspiele*' (Review), *Naturwiss.* **7**, 644 (1919).

1920: (a) 'Die Einsteinsche Raumlehre', *Die Umschau* **24**, no. 25: 402–405 (1920)

(b) Über die physikalischen Voraussetzungen der Wahrscheinlichkeitsrechnung', *Zeitschrift für Physik* **2**, no. 2: 150–171 (1920). See also [1921a]

(c) 'Die physikalischen Voraussetzungen der Wahrscheinlichkeitsrechnung', *Die Naturwissenschaften* **8**, no. 3: 46–55 (1920) (For English trans. see [1978 b–52a])

(d) 'Nachtrag', *Die Naturwissenschaften* **8**, no. 19: 349 (1920) (For English trans. see [1978 b–52b])

(e) 'Philosophische Kritik der Wahrscheinlichkeitsrechnung', *Die Naturwissenschaften* **8**, no. 8: 146–153 (1920) (For English trans. see [1978 b-53])

(f) *Relativitätstheorie und Erkenntnis Apriori* (Berlin: Springer, 1920) (For English trans. see [1965a])

1921: (a) 'Nachtrag' (Suppl. to [1920b]), *Zeitschrift für Physik* **4**, no. 3: 448–450 (1921)

(b) 'Fürth, Reinhold, *Schwankungserscheinungen in der Physik*' (review), *Die Naturwissenschaften* **9**, no. 7: 111 (Feb. 18, 1921)

(c) 'Erwiderung auf H. Dinglers Kritik an der Relativitätstheorie', *Physikalische Zeitschrift* **22**, 379–380 (1921)
(d) 'Bericht über eine Axiomatik der Einsteinschen Raum-Zeit Lehre' (Address at the German Congress of Physics, Jena 1921), *Physikalische Zeitschrift* **22**, 683–687 (1921)
(e) 'Erwiderung auf Herrn Theodor Wulfs Einwände gegen die allgemeine Relativitätstheorie', *Astronomische Nachrichten* **213**, 307–310 (1921)
(f) 'Die Einsteinsche Bewegungslehre', *Die Umschau* (Frankfurt) **25**, no. 35: 501–505 (1921)
(g) 'Entgegnung' (answer to Oskar Kraus), *Die Umschau* (Frankfurt) **25**, no. 46: 684–685 (1921)

1922: (a) 'Der Nobelpreis für Einstein', *Neue Zürcher Zeitung* (Nov. 22, 1922) (For English trans. see [1978 a-9])
(b) 'Die Relativitätstheorie in der Streichholzschachtel', *Neue Zürcher Zeitung* (April 29, 1922) (For English trans. see [1978 a-10])
(c) 'La signification philosophique de la théorie de la relativité', *Revue Philosophique de la France et de l'Étranger*, **94**, 5–61 (1922)
(d) 'Relativitätstheorie und absolute Transportzeit', *Zeitschrift für Physik* **9**, nos. 1–2: 111–117 (1922)
(e) 'Erwiderung auf Herrn Andersons Einwände gegen die allgemeine Relativitätstheorie', *Astronomische Nachrichten* **213**, 373–376 (1922)
(f) 'Der gegenwärtige Stand der Relativitätsdiskussion', *Logos* **10**, no. 3: 316–378 (1922) (For English trans. see [1959a]; reprinted in [1978 b-44])

1923: (a) 'Müller, Aloys, *Die Philosophischen Probleme der Einsteinschen Relativatätstheorie*' (Review), *Die Naturwissenschaften* **11**, no. 2: 30–31 (Jan. 12, 1923)

1924: (a) 'Eine neue Erfindung in der astronomischen Messtechnik', *Neue Zürcher Zeitung* (Aug. 6, 1924)
(b) 'Radiotechnik und Kultur', *Radio–Umschau* (Frankfurt), no. 14 (May 18, 1924)
(c) *Was Ist Radio*?, vol. 1 of the series *Die Radio-Reihe* (Berlin: Richard C. Schmidt, and Stuttgart: Verlag die Zeit-Max Kahn (paperbound), 1924; 2nd Schmidt ed. 1929)
(d) 'Die Bewegungslehre bei Newton, Leibniz und Huyghens', *Kantstudien* **29**, 416–438 (1924) (For English trans. see [1959a]; reprinted in [1978 b-45])
(e) 'Entgegnung' (answer to O. Brühlmann), *Annalen der Philosophie* **4**, nos. 4–5: 195–198 (1924)
(f) 'Die relativistische Zeitlehre', *Scientia*, 361–374 (Dec. 1924) (For English trans. see [1978 b-46])
(g) Articles in *Physikalisches Handwörterbuch*, Berliner-Scheel, eds. (Springer, Berlin, 1924):
Bernouillisches Theorem;
Diskret;
Erwartung, mathematische;
Ergodenhypothese;
Fehlertheorie;

Gesetz der grossen Zahlen;
Intramolekularbewegung;
Liouvillescher Satz;
Loschmidtsche Zahl pro Mol:
Maxwellsche Geschwindigkeitsverteilung;
Mittel;
Molare Unordnung;
Statistisches Gleichgewicht;
Stirlingsche Formel;
Stosszahlansatz;
Umkehreinwand;
Unabhängige Ereignisse;
Wahrscheinlichkeit;
Wahrscheinlichkeitsnachwirkung;
Wahrscheinlichkeitsrechnung;
Zufall.

(h) *Axiomatik der relativistischen Raum-Zeit-Lehre* [Die Wissenschaft, vol. 72] (Vieweg, Braunschweig, 1924; reprinted Braunschweig, 1965) (For English trans. see [1969a])

1925: (a) 'Metaphysik und Naturwissenschaft', *Symposion* **1**, no. 2: 158–176 (1925) (For English trans. see [1978 a-31])

(b) 'Planetenuhr und Einsteinsche Gleichzeitigkeit', *Zeitschrift für Physik* **33**, no. 8: 628–634 (1925)

(c) 'Über die physikalischen Konsequenzen der relativistischen Axiomatik', *Zeitschrift für Physik* **34**, no. 1: 32–48 (1925) Supplement to [1924-h]

(d) 'Die Kausalstruktur der Welt und der Unterschied zwischen Vergangenheit und Zukunft', *Sitzungsberichte*, *Bayerische Akademie der Wissenschaft* (Nov. 1925) pp. 133–175 (For English trans. see [1978 b-47])

(e) 'Wahrscheinlichkeitsgesetze und Kausalgesetze', *Die Umschau* (Frankfurt) **29**, no. 40: 789–792 (1925)

1926: (a) 'Tycho Brahes Sextanten', *Hamburger Fremdenblatt* (Dec. 18, 1926) (For English trans. see [1978 a-11])

(b) 'Ist die Relativitätstheorie widerlegt?', *Die Umschau* (Frankfurt) **30**, no. 17 (April 24, 1926)

(c) 'Die Probleme der modernen Physik', *Die Neue Rundschau*, 414–425 (April 1926)

(d) 'Die Auswirkung der Einsteinschen Lehre', *Kunstwart*, 35–39 (Oct. 1926) (For English trans. see [1978 a-12])

(e) 'Ein offener Brief an die Funkstunde A.G., Berlin', *Radio-Umschau* (Frankfurt) **3**, no. 4 (Jan. 24, 1926) (For English trans. see [1978 a-13])

(f) 'Erwiderung auf eine Veröffentlichung von Herrn Hj. Mellin', *Zeitschrift für Physik* **39**, nos. 2–3: 106–112 (1926)

1927: (a) 'Grundsteinlegung für das Haus der Chemie. Marcellin Berthelots Werk', *Berliner Tageblatt* (Oct. 22, 1927) (For English trans. see [1978 a-14])

(b) 'Erinnerungen an Svante Arrhenius', *Berliner Tageblatt* (Oct. 5, 1927) (For English trans. see [1978 a-15])

(c) 'Ein neues Atommodell', *Die Umschau* (Frankfurt) **31**, no. 15 (April 9, 1927) (For English trans. see [1978 a-16])

(d) 'Die Umgestaltung des naturwissenschaftlichen Weltbildes: Atomtheorie-Relativitätstheorie', *Exakte Naturwissenschaften* [Sammelband der Lessinghochschule] **2**, no. 17: 247–272 (1927)

(e) 'Lichtgeschwindigkeit und Gleichzeitigkeit', *Annalen der Philosophie* **6**, no. 4: 128–144 (1927)

(f) *Von Kopernikus bis Einstein* (Ullstein, Berlin, 1927) (Czech trans. Prague, 1928; for English trans. see [1942a])

1928: (a) 'Zum Tode von H. A. Lorentz', *Berliner Tageblatt* (Feb. 6, 1928) (For English trans. see [1978 a-17])

(b) 'Denker der Zeit-Bertrand Russell', *Vossische Zeitung* (Feb. 12, 1928) (Reichenbach's contribution to a series of articles entitled 'Great Thinkers of our Time'; for English trans. see [1967a])

(c) 'Philosophie der Naturwissenschaften', *Vossische Zeitung* (Jan. 3, 1928) (For English trans. see [1978 a-18])

(d) 'Raum und Zeit: Von Kant zu Einstein', *Vossische Zeitung* (Mar. 4, 1928) (For English trans. see [1978 a-19])

(e) 'Kausalität oder Wahrscheinlichkeit?', *Vossische Zeitung* (July 18, 1928) (For English trans. see [1978 a-20])

(f) 'Die Weltanschauung der exakten Wissenschaften', *Die Böttcherstrasse* (Bremen), 44–46 (Nov. 1928) (For English trans. see [1978 a-21])

(g) 'Wandlungen im physikalischen Welbild', *Zeitschrift für angewandte Chemie* **41**, no. 14: 347–352 (1928)

(h) *Philosophie der Raum-Zeit-Lehre* (Walter de Gruyter, Berlin and Leipzig, 1928) 330 pp. (For English trans. see [1958a]; Italian trans., Milan, 1977)

1929: (a) 'Einstein's neue Theorie', *Vossische Zeitung* (Jan. 25, 1929) (For English trans. see [1978 a-26])

(b) 'Neue Wege der Wissenschaft. Physikalische Forschung', *Vossische Zeitung* (March 31, 1929) (For English trans. see [1978 a-22])

(c) 'Neue Wege der Wissenschaft. Philosophische Forschung', *Vossische Zeitung* (June 16, 1929) (For English trans. see [1978 a-23])

(d) 'Neue Wege der Wissenschaft. Mathematische Forschung', *Vossische Zeitung* (Aug. 18, 1929) (For English trans. see [1978 a-24])

(e) 'Die neue Naturphilosophie', *Deutsche Allgemeine Zeitung* (Oct. 13, 1929) (For English trans. see [1978 a-25])

(f) 'Crise de la Causalité', *Documents* (Paris), 105–108 (May 1929)

(g) 'Ziele und Wege der physikalischen Erkenntnis', *Handbuch der Physik*, vol. 4: *Allgemeine Grundlagen der Physik* (Springer, Berlin, 1929) pp. 1–80 (For English trans. see [1978 b-48])

(h) 'Neuere Forschungsergebnisse in der Naturphilosophie', *Forschungen und Fortschritte* [Nachrichtenblatt der deutschen Wissenschaft und Technik] vol. 5, no. 16: 185 (1929)

(i) 'Bertrand Russell', *Obelisk Almanach* (Drei-Masken Verlag, Berlin and Munich, 1929) pp. 82–92 (For English trans. see [1978 a-32])

(j) 'Zur Einordnung des neuen Einsteinschen Ansatzes über Gravitation und Elektrizität', *Zeitschrift für Physik* **59**, nos. 9–10: 683–689 (1929)

(k) 'Die neue Theorie Einsteins über die Verschmelzung von Gravitation und Elektrizität', *Zeitschrift für angewandte Chemie* **42**, no. 5: 121–123 (1929)

(l) 'Stetige Wahrscheinlichkeitsfolgen', *Zeitschrift für Physik* **53**, nos. 3–4: 274–307 (1929)

(m) 'Das Kausalproblem in der gegenwärtigen Physik', *Zeitschrift für angewandte Chemie* **42**, no. 19: 457–459 (1929)

1930: (a) 'Johannes Kepler. Zur dreihundertsten Wiederkehr seines Todestages', *Die Woche*, 1329–1330 (Nov. 1930) (For English trans. see [1978 a-27])

(b) 'Der heutige Stand der Wissenschaften. Die exakten Naturwissenschaften', *Die Literarische Welt*, no. 38: 3 (1930) (For English trans. see [1978 a-28])

(c) 'Probleme und Denkweisen der gegenwärtigen Physik', *Deutsche Rundschau*, 37–44; 131–141 (July–Aug. 1930)

(d) 'Zur Einführung', Introduction to *Erkenntnis* **1**, no. 1: 1–3 (Felix Meiner, Leipzig, 1930)

(e) 'Die philosophische Bedeutung der modernen Physik', *Erkenntnis* **1**, no. 1: 49–71 (1930) (For English trans. see [1978 a-33])

(f) 'Tagung für Erkenntnislehre der exakten Wissenschaften in Königsberg', *Die Naturwissenschaften* **18**, no. 50: 1093–1094 (1930) (For English trans. see [1978 a-34])

(g) 'Kausalität und Wahrscheinlichkeit', *Erkenntnis* **1**, nos. 2–4: 158–188 (1930) (For English trans. of Part III see [1959a]; reprinted in [1978 b-55])

(h) *Atom und Kosmos. Das Physikalische Weltbild der Gegenwart* (Berlin: Deutsche Buch-Gemeinschaft, 1930) (For English trans. see [1932h]; Spanish trans. Madrid, 1931; French trans. Paris, 1934; Hungarian trans. Budapest, 1937)

1931: (a) 'Hundert gegen Einstein', *Vossische Zeitung* (Feb. 24, 1931) (For English trans. see [1978 a-29])

(b) 'Naturwissenschaft und Philosophie', *Frankfurter Zeitung* (April 29, 1931)

(c) 'Schlussbemerkung (zur Diskussion v. Aster-Vogel-Dingler)', *Erkenntnis* **2**, 39–41 (1931)

(d) 'Heinrich Scholz, *Geschichte der Logik*', (Review) *Erkenntnis* **2**, 471–472 (1931) (For English trans. see [1978 a-37])

(e) 'Bernhard Bavink, *Ergebnisse und Probleme der Naturwissenschaften*' (Review), *Erkenntnis* **2**, 468–471 (1931)

(f) 'Zum Anschaulichkeitsproblem der Geometrie', *Erkenntnis* **2**, no. 1: 61–72 (1931)

(g) 'Der physikalische Wahrheitsbegriff', *Erkenntnis* **2**, nos. 2–3: 156–171 (1931) (For English trans. see [1978 a-36])

(h) 'Bemerkungen zum Wahrscheinlichkeitsproblem', *Erkenntnis* **2**, nos 5–6: 365–368 (1931)

(i) 'Das Kausalproblem in der Physik', *Die Naturwissenschaften* **19**, no. 34: 713–722 (1931) (For English trans. see [1978 a-35])

(j) *Ziele und Wege der heutigen Naturphilosophie* (Leipzig: Felix Meiner, 1931) 64 pp. (For English trans. see [1959a], reprinted in [1978 a-38]; French trans. Hermann, Paris, 1932)

1932: (a) 'Ist der menschliche Geist wandelbar?', *Die Woche* (Berlin) **34**, no. 2: 39–40 (Jan. 9, 1932) (For English trans. see [1978 a-30])
(b) 'Der endliche Weltenraum', *Die Woche* (Berlin) **34**, no. 38 (Sept. 17, 1932)
(c) 'Bemerkung (Karl Popper, 'Ein Kriterium des empirischen Charakters theoretischer Systeme')', *Erkenntnis* **3**, 427–428 (1932/33)
(d) 'Die Kausalbehauptung und die Möglichkeit ihrer empirischen Nachprüfung', *Erkenntnis* **3**, no. 1: 32–64 (1932) (For English trans. see [1959a], reprinted in [1978 b-56])
'Schlussbemerkung', pp. 71–72 (included in [1978 b-56] reprinting above)
(e) 'Kausalität und Wahrscheinlichkeit in der Biologie', *Klinische Wochenschrift* **2**, no. 6: 251–253 (1932)
(f) 'Axiomatik der Wahrscheinlichkeitsrechnung', *Mathematische Zeitschrift* **34**, no. 4: 568–619 (1932)
(g) 'Wahrscheinlichkeitslogik', *Sitzungsberichte*, *Preussische Akademie der Wissenschaften*, *Phys.-Math. Klasse* **29**, 476–490 (1932)
(h) *Atom and Cosmos. The World of Modern Physics*, revised and updated by H.R., trans. by Edward S. Allen (George Allen & Unwin, London, 1932; MacMillan & Co., New York, 1933; Braziller, New York, 1957) (English trans. of [1930h]; Chap. 18 reprinted in *Basic Problems in Philosophy. Selected Readings*, D. J. Bronstein, Y. H. Krikorian, and P. P. Wiener, eds., Prentice-Hall, New York, 1947, pp. 343–353; 2nd ed. 1955, pp. 270–276)
(i) 'Kant und die moderne Naturwissenschaft. Naturwissenschaftliche Berichte', *Frankfurter Zeitung* 77, nos. 626–627: 2–3 (Aug. 23, 1932)

1933: (a) 'Vom Bau der Welt', *Die Neue Rundschau*, 39–60; 235–250 (July–August 1933)
(b) 'Kant und die Naturwissenschaft', *Die Naturwissenschaften* **21**, no. 33: 601–606; no. 34: 624–626 (1933) (For English trans. see [1978 a-39])
(c) 'Rudolf Carnap, *Der logische Aufbau der Welt*', (Review) *Kantstudien* **38**, 199–201 (1933) (For English trans. see [1978 a-40])
(d) 'Kausalität und Wahrscheinlichkeit in der gegenwärtigen Physik', *Unterrichtsblätter für Mathematik und Naturwissenschaften* **39**, no. 3: 65–69 (1933)
(e) 'Die logischen Grundlagen des Wahrscheinlichkeitsbegriffs', *Erkenntnis* **3**, nos. 4–6: 401–425 (1933) (For English trans. see [1949e])

1934: (a) 'Friedrich Schilling, *Projektive und nichteuklidische Geometrie*' (Review), *Erkenntnis* **4**, 378 (1934)
(b) 'In eigener Sache', *Erkenntnis* **4**, 75–78 (1934)
(c) 'Wahrscheinlichkeitslogik', *Erkenntnis* **5**, nos. 1–3: 37–43 (1934)
(d) 'Sur les fondements logiques de la probabilité', *Recherches philosophiques* **4**, 361–370 (1934/35)

1935: (a) 'Metaphysik bei Jordan?', *Erkenntnis* **5**, 178–179 (1935)
(b) 'Bemerkung zu H. Blumes finiter Wahrscheinlichkeitsrechnung', *Zeitschrift für Physik* **93**, 792–794 (1935)
(c) 'Zur Induktionsmaschine', *Erkenntnis* **5**, 172–173 (1935)
(d) 'Bemerkungen zu Carl Hempels Versuch einer finitistischen Deutung des Wahrscheinlichkeitsbegriffs', *Erkenntnis* **5**, no. 4: 261–266 (1935)

(e) 'Über Induktion und Wahrscheinlichkeit. Bemerkungen zu Karl Poppers *Logik der Forschung*', *Erkenntnis* **5**, no. 4: 267–284 (1935) (For English trans. see [1978 b-57])
(f) 'Bemerkungen zu Karl Marbes statistischen Untersuchungen zur Wahrscheinlichkeitsrechnung', *Erkenntnis* **5**, no. 5: 305–322 (1935)
(g) 'Wahrscheinlichkeitslogik und Alternativlogik', *Einheit der Wissenschaft* (Prager Vorkonferenz der Internationalen Kongresse für Einheit der Wissenschaft, Leipzig, 1935)
(h) *Wahrscheinlichkeitslehre. Eine Untersuchung über die Logischen und Mathematischen Grundlagen der Wahrscheinlichkeitsrechnung* (A. W. Sijthoff's Uitgeversmaatschappij, Leyden, 1935) 451 pp. (For English trans. see [1949f]; French trans. of 8 sections by H. Savonnet, Hermann, Paris, (1939) For 2nd German ed., see [1977a].

1936: (a) 'Ansprache bei der Begrüssungssitzung des Pariser Kongresses', *Actes du Congrès International de Philosophie Scientifique*, Paris 1935, vol. 1 (Hermann, Paris, 1936) pp. 16–18
(b) 'L'Empirisme logistique et la désagrégation de l'*a priori*', *Actes du Congrès International de Philosophie Scientifique*, Paris 1935, vol. 1 (Hermann, Paris, 1936) pp. 28–35
(c) 'Die Induktion als Methode der wissenschaftlichen Erkenntnis', *Actes du Congrès International de Philosophie Scientifique*, Paris 1935, vol. 4 (Hermann, Paris, 1936) pp. 1–7
(d) 'Wahrscheinlichkeitslogik als Form des wissenschaftlichen Denkens', *Actes du Congrès International de Philosophie Scientifique*, Paris 1935, vol. 4 (Hermann, Paris, 1936) pp. 24–30
(e) 'Logistic Empiricism in Germany and the Present State of its Problems', *The Journal of Philosophy* **33**, no. 6: 141–160 (March 12, 1936)
(f) 'Moritz Schlick†', *Erkenntnis* **6**, 141–142 (1936)
(g) 'Clemens Thaer, *Die Elemente von Euklid*', (Review) *Erkenntnis* **6**, 71–72 (1936)
(h) 'Die Bedeutung des Wahrscheinlichkeitsbegriffes für die Erkenntnis', *Actes du huitième Congrès International de Philosophie*, Prague, Sept. 2–7, 1934 (1936) pp. 163–169
(i) 'Warum ist die Anwendung der Induktionsregel für uns notwendige Bedingung von Voraussagen?', *Erkenntnis* **6**, no. 1: 32–40 (1936)
(j) 'Induction and Probability' (Discussion of H. Feigl), *Philosophy of Science* **3**, 124–126 (1936)

1937: (a) 'La Philosophie scientifique: une esquisse de ses traits principaux', *Travaux du IX Congrès International de Philosophie*, Paris 1937, vol. 4 (Paris, 1937) pp. 86–91
(b) 'Les fondements logiques du calcul des probabilités, *Annales de l'Institut Henri Poincaré* (Paris) vol. 7, part 5: 267–348 (1937)
(c) 'Causalité et induction', *Bulletin de la Société française de Philosophie* **37**, no. 4: 127–159 (1937)

1938: (a) 'On Probability and Induction', *Philosophy of Science* **5**, no. 1: 21–45 (1938)
(b) 'Reply to Everett J. Nelson's Criticism', *The Journal of Philosophy* **35**, no. 5: 127–130 (1938)

(c) *Experience and Prediction. An Analysis of the Foundations and the Structure of Knowledge* (Univ. of Chicago Press, Chicago, 1938; 1st Phoenix paper edition, 1961, Midway paper reprint, 1976) 410 pp. For German trans. see [1977a], vol. 4

1939: (a) 'Dewey's Theory of Science', in *The Philosophy of John Dewey*, P. A. Schilpp, ed. (The Library of Living Philosophers, Evanston, Ill., vol. 1, 1939) pp. 159–192.

(b) 'Über die semantische und die Objektauffassung von Wahrscheinlichkeitsausdrücken', *The Journal of Unified Science* (*Erkenntnis*) **8**, 50–68 (1939) (For English trans. see [1978 b-58])

(c) 'Bemerkungen zur Hypothesenwahrscheinlichkeit', *The Journal of Unified Science* (*Erkenntnis*) **8**, no. 4: 256–260 (1939)

1940: (a) 'On the Justification of Induction', *The Journal of Philosophy* **37**, no. 4: 97–103 (1939); reprinted in *Readings in Philosophical Analysis*, H. Feigl and W. Sellars, eds. (Appleton-Century-Crofts, New York, 1949) pp. 324–329.

(b) 'On Meaning', *The Journal of Unified Science* (*Erkenntnis*) **9**, 134–135 (1940)

1941: (a) 'Note on Probability Implication', *Bulletin of the American Mathematical Society* **47**, no. 4: 265–267 (1941)

1942: (a) *From Copernicus to Einstein*, tr. by R. B. Winn (Philosophical Library, New York, 1942) 123 pp. (English trans. of [1927f] with paperback ed., 1957)

1944: (a) 'Bertrand Russell's Logic', in *The Philosophy of Bertrand Russell*, P. A. Schilpp, ed. (The Library of Living Philosophers, Evanston, Ill., vol. 5, 1944) pp. 21–54

(b) *Philosophic Foundations of Quantum Mechanics* (Univ. of California Press, Berkeley and Los Angeles, 1944; paper edition, 1965) 182 pp.; German trans. Basel, 1949 and [1977a], vol. 5; Italian trans. Torino, 1954. §§ 29–37 reprinted as 'Three-valued Logic and the Interpretation of Quantum Mechanics' in *The Logico-Algebraic Approach to Quantum Mechanics*, vol. I: *Historical Evolution*, ed. C. A. Hooker, (Reidel, Dordrecht and Boston, 1975) pp. 53–97.

1945: (a) 'Reply to Donald C. Williams' Criticism of the Frequency Theory of Probability', *Philosophy and Phenomenological Research* **5**, no. 4: 508–512 (1945)

1946: (a) 'Reply to V. F. Lenzen's Critique', *Philosophy and Phenomenological Research* **6**, no. 3: 487–492 (1946)

(b) 'Reply to Ernest Nagel's Criticism of my Views on Quantum Mechanics', *The Journal of Philosophy* **43**, no. 9: 239–247 (1946)

1947: (a) 'Philosophy: Speculation or Science?', *The Nation* **164**, no. 1: 19–22 (Jan. 4, 1947); reprinted under the title 'The Nature of a Question' in *The Language of Wisdom and Folly*, I. I. Lee, ed. (Harper and Brothers, New York, 1949)

(b) 'The Scientist and Society' (review of Philipp Frank, *Einstein*: *His Life and Times*), *The Nation* **164**, 306–307 (March 15, 1947)

(c) *Elements of Symbolic Logic* (Macmillan Co., New York, 1947). For German trans. see [1977a], vol. 6, 444 pp.

1948: (a) 'Rationalism and Empiricism: An Inquiry into the Roots of Philosophical Error' [Presidential address, Pacific Division meeting, APA, Dec. 30, 1947], *The Philosophical Review* **57**, no. 4: 330–346 (1948); reprinted in [1959a] For German tr., see [1977a]

(b) 'Philosophy and Physics' [Faculty Research Lecture delivered March 25, 1946] (Univ. of California Press, Berkeley and Los Angeles, 1948) 13 pp.

(c) 'A Reply to a Review', *The Journal of Philosophy* **45**, no. 17: 464–467 (1948)

(d) 'The Principle of Anomaly in Quantum Mechanics', *Dialectica* **2**, nos. 3–4: 337–350 (1948); reprinted in *Readings in the Philosophy of Science*, H. Feigl and M. Brodbeck, eds. (Appleton-Century-Crofts, New York, 1953) pp. 509–520.

(e) 'Theory of Series and Gödel's Theorems' (unpublished mimeographed manuscript; first publication in [1978 a-41])

1949: (a) 'The Philosophical Analysis of Quantum Mechanics', *Library of the 10th International Congress of Philosophy*, August 11–18, 1948, Amsterdam, vol. 1 (North-Holland, Amsterdam, 1949) pp. 921–922

(b) 'The Philosophical Significance of the Theory of Relativity', in *Albert Einstein: Philosopher-Scientist*, P. A. Schilpp, ed. (The Library of Living Philosophers, Evanston, Ill., vol. 7, 1949) pp. 287–311; reprinted in *Readings in the Philosophy of Science*, H. Feigl and M. Brodbeck, eds. (Appleton-Century-Crofts, New York, 1953) pp. 195–211, and *Readings in Philosophy of Science*, P. P. Wiener, ed. (Scribners, New York, 1953) pp. 59–76; German trans. Stuttgart, 1955.

(c) 'Philosophical Foundations of Probability', *Proceedings of the Berkeley Symposium on Mathematical Statistics and Probability* (Univ. of California Press, Berkeley and Los Angeles, 1949) pp. 1–20

(d) 'A Conversation between Bertrand Russell and David Hume', *The Journal of Philosophy* **46**, no. 17: 545–549 (1949). For German trans., see [1977a], vol. 5.

(e) 'The Logical Foundations of the Concept of Probability', tr. by M. Reichenbach in *Readings in Philosophical Analysis*, H. Feigl and W. Sellars, eds. (Appleton-Century-Crofts, New York, 1949) pp. 305–323; reprinted in *Readings in the Philosophy of Science*, H. Feigl and M. Brodbeck, eds. (Appleton-Century-Crofts, New York, 1953) pp. 456–474. (English trans. with added footnotes of [1933e])

(f) *The Theory of Probability. An Inquiry into the Logical and Mathematical Foundations of the Calculus of Probability*, tr. by E. H. Hutten and M. Reichenbach (Univ. of California Press, Berkeley and Los Angeles, 1949; California Library Reprint Series No. 23, 1971) (English version and 2nd edition of [1935h])

1950: (a) 'On the Theory of Probability', *Felix Kaufmann: A Memorial* [a special issue of *12th Street*, vol. 3, no. 2, 1950], pp. 11–12

1951: (a) *The Rise of Scientific Philosophy* (Univ. of California Press, Berkeley and Los Angeles, 1951) 333 pp.; German trans. Berlin, 1953, Braunschweig 1968 and [1977a], vol. 1; French trans. Paris, 1955; Spanish trans. Buenos Aires and Mexico City, 1953; Swedish trans. Stockholm, 1957; Italian trans. Bologna, 1958; Japanese trans. Tokyo, 1958; Polish trans.

Ksiazka i Wiedza, Warsaw, 1960; Yugoslav trans. Belgrade, 1964; Korean trans., Seoul, 1960; Marathi Trans., Bombay, 1974. Chapter 14 reprinted in *Contemporary Philosophy, A Book of Readings*, J. L. Jarrett and S. M. McMurrin, eds. (Holt, New York, 1954) pp. 366–376.

(b) 'Why I wrote *The Rise of Scientific Philosophy*', *Book Find News*, no. 102 (Braziller, New York, 1951)

(c) 'The Verifiability Theory of Meaning', *Proceedings of the American Academy of Arts and Sciences* **80**, no. 1: 46–60 (1951); reprinted in *Readings in the Philosophy of Science*, H. Feigl and M. Brodbeck, eds. (Appleton-Century-Crofts, New York, 1953) pp. 93–102

(d) 'On Observing and Perceiving', *Philosophical Studies* **2**, no. 6 (1951) pp. 29–39

(e) 'Foreword', Catalog No. 130, Zeitlin and Ver Brugge [Booksellers], Los Angeles, California (May 1951)

(f) 'The Value of Old Books', *Antiquarian Bookman* (July 7, 1951) p. 2

(g) 'Über die erkenntnistheoretische Problemlage und den Gebrauch einer dreiwertigen Logik in der Quantenmechanik', *Zeitschrift für Naturforschung* (Tübingen) 6a, no. 11: 569–575 (1951) (For English trans. see [1978 b-49]

(h) 'Probability Methods in Social Science', in *The Policy Sciences. Recent Developments in Scope and Method*, D. Lerner and H. D. Lasswell, eds. (Stanford Univ. Press, Stanford, 1951) pp. 121–128

(i) 'The Freedom of the Will' (unpublished manuscript; first publication in [1959 a-7])

1952: (a) 'Are Phenomenal Reports absolutely Certain?', *The Philosophical Review* **61**, no. 2: 147–159 (1952)

(b) 'The Syllogism Revised', *Philosophy of Science* **19**, no. 1: 1–16 (1952)

(c) 'Logical Empiricism. Philosophy: Summaries of a Series of Meetings', *The Humanist*, 7–10 (1952)

(d) 'Les fondements logiques de la mécanique des quanta', *Annales de l'Institut Henri Poincaré* **13**, part 2: 109–158 (1952/53) [Four lectures given at the Institut Henri Poincaré, June 4–7, 1952] (For English trans. see [1978 b-50])

(e) 'On the Explication of Ethical Utterances' (unpublished manuscript; first publication in [1959 a-8])

1953: (a) 'La signification philosophique du dualisme ondes-corpuscules', tr. by O. Costa de Beauregard, in *Louis de Broglie, Physicien et Penseur* (Albin Michel, Paris, 1953) pp. 117–134 (For original English text see [1978 b-51])

1954: (a) 'The Emotive Significance of Time', *Idea and Experiment* **4**, no. 1: 3–9 (1954)

(b) 'Les fondements logiques de la théorie des quanta. Utilisation d'une logique à trois valeurs', *Collection de logique mathématique*, Série A, vol. 5, *Applications scientifiques de la logique mathématique* [Actes du 2e Colloque International de Logique Mathématique, Paris, Aug. 25–30, 1952] (Institut Henri Poincaré, Paris, 1954) pp. 103–114, incl. discussion

(c) (Discussion of) 'Jean-Louis Destouches, La logique et les théories physiques',

Collection de logique mathématique, Série A, vol. 5, *Applications scientifiques de la logique mathématique* [Actes de 2e Colloque International de Logique Mathématique, Paris, Aug. 25–30, 1952] (Institut Henri Poincaré, Paris, 1954) p. 126.

(d) 'Exposé introductif: Remarques sur l'application de la méthode inductive dans la physique'. *Collection de Logique mathématique*, Série A, vol. 5, *Applications scientifiques de la logique mathématique* [Actes du 2e Colloque International de Logique Mathématique, Paris, Aug. 25–30, 1952] (Institut Henri Poincaré, Paris, 1954) pp. 163–172, incl. discussion.

(e) *Nomological Statements and Admissible Operations* (North-Holland, Amsterdam, 1954) 140 pp. Reissued with new foreword as *Laws, Modalities and Counterfactuals*, see [1976a]

1956: (a) 'Can Operators reach through Quotes?', *Philosophical Studies* 7, no. 3: 33–36 (1956)

(b) *The Direction of Time*, M. Reichenbach, ed. (Univ. of California Press, Berkeley and Los Angeles, 1956; Paper edition, 1971) 280 pp; Spanish trans. Mexico City, 1958; German trans., [1977a], vol. 8

1958: (a) *The Philosophy of Space and Time*, tr. by M. Reichenbach and J. Freund (Dover, New York, 1958) 295 pp. (English trans. of [1928h])

1959: (a) *Modern Philosophy of Science*: *Selected Essays*, ed. and tr. by M. Reichenbach (Routledge & Kegan Paul, London, 1959) 214 pp. Italian trans. Milan, 1966 (1st. 4 articles); Spanish trans. Madrid, 1965. Among the contents are:

1. 'The Present State of the Discussion on Relativity', pp. 1–45, (English trans. of [1922f]; reprinted in [1978 b-44])
2. 'The Theory of Motion According to Newton, Leibniz, and Huyghens', pp. 46–66 (English trans. of [1924d]; reprinted in [1978 b-45])
3. 'Causality and Probability', pp. 67–78 (English trans. of Part III of [1930g]; reprinted in [1978 b-55])
4. 'Aims and Methods of Modern Philosophy of Nature', pp. 79–108 (English trans. of [1931j]; reprinted in [1978 a-38])
5. 'The Principle of Causality and the Possibility of its Empirical Confirmation', pp. 109–134 (English trans. of [1932d]; reprinted in [1978 b-56])
6. 'Rationalism and Empiricism', pp. 135–150 (reprint of [1948a])
7. 'The Freedom of the Will', pp. 151–192 (first publication; reprinted in [1978 a-42])
8. 'On the Explication of Ethical Utterance', pp. 193–198 (first publication; reprinted in [1978 a-43])

1965: (a) *The Theory of Relativity and A Priori Knowledge*, tr. with an introduction by M. Reichenbach (Berkeley and Los Angeles: Univ. of California Press, 1965) 116 pp. (English trans. of [1920f])

1967: (a) 'An Early Appreciation', tr. by M. Reichenbach in *Bertrand Russell – Philosopher of the Century* (George Allen & Unwin, London, 1967) pp. 129–133 (English trans. of [1928b])

1969: (a) *Axiomatization of the Theory of Relativity*, tr. by M. Reichenbach with an introduction by W. C. Salmon (Univ. of California Press, Berkeley and Los Angeles, 1969) 208 pp. (English trans. of [1924h])

1976: (a) *Laws, Modalities, and Counterfactuals*, with a foreword by W. C. Salmon (Univ. of California Press, Berkeley and Los Angeles, 1976) xliii + 140 pp. (expanded reissue of [1954])

1977: (a) Reichenbach's Collected Works are under publication in German, edited by Andreas Kamlah and Maria Reichenbach (Vieweg, Braunschweig, subsequently Wiesbaden, 1977ff.:

Vol. 1 *Der Aufstieg der Wissenschaftlichen Philosophie* [German trans. of *The Rise of Scientific Philosophy* (1951a)], with an introductory essay to the *Gesamtausgabe* by W. C. Salmon and the essay 'Rationalism and Empiricism' [1948a]. (1977)

Vol. 2 *Philosophie der Raum-Zeit-Lehre*, with a brief foreword by R. Carnap to the 1st English trans. [1928h], and the essay 'Zur Einordnung des neuen Einsteinschen Ansatzes über Gravitation und Elektrizität' [1929j]. (1977)

Vol. 3 *Die philosophische Bedeutung der Relativitätstheorie*.
1. *Relativitätstheorie und Erkenntnis Apriori* [1920f]
2. *Axiomatik der relativistischen Raum-Zeit-Lehre* [1924h]
3. 'Die relativistiche Zeitlehre' [1924f]
4. 'Über die physikalischen Konsequenzen der relativistischen Axiomatik' [1925c]
5. 'Planetenuhr und Einsteinsche Gleichzeitigkeit' [1925b]
6. 'Die philosophische Bedeutung der Relativitätstheorie' [German trans. of 1949b]
7. 'Der gegenwärtige Stand der Relativitätsdiskussion' [1922f]
8. 'Die Bewegungslehre bei Newton, Leibniz und Huygens' [1924d]

Vol. 4 *Erfahrung und Prognose* [German trans. of *Experience and Prediction* (1938c) by M. Reichenbach] with notes by Alberto Coffa.

Vol. 5 *Philosophische Grundlagen der Quantenmechanik und Wahrscheinlichkeit*
1. *Philosophische Grundlagen der Quantenmechanik* [German trans. of *Philosophic Foundations of Quantum Mechanics* (1944b)]
2. 'Über die erkenntnistheoretische Problemlage und den Gebrauch einer dreiwertigen Logik in der Quantenmechanik' [1951g]
3. 'Der Begriff der Wahrscheinlichkeit für die mathematische Darstellung der Wirklichkeit' [1915b]
4. 'Die physikalischen Voraussetzungen der Wahrscheinlichkeitstheorie' [1920c]
5. 'Philosophische Kritik der Wahrscheinlichkeitsrechnung' [1920e]
6. 'Die logischen Grundlagen des Wahrscheinlichkeitsbegriffs' [1933e]
7. 'Eine Unterhaltung zwischen B. Russell und D. Hume' [German trans. of 'A Conversation between Bertrand Russell and David Hume' (1949d)]

Vol. 6 *Grundzüge der symbolischen Logik* [German trans. of *Elements of Symbolic Logic* (1947c)]

Vol. 7 *Wahrscheinlichkeitslehre* (*Theory of Probability*) [1935h; as revised in 1949f]

Vol. 8 *Kausalität und Zeitrichtung*

1. *Die Richtung der Zeit* [German trans. of *The Direction of Time* (1957b)]
2. 'Die Kausalstruktur der Welt und der Unterschied zwischen Vergangenheit und Zukunft' [1925d]
3. 'Die Kausalbehauptung und die Möglichkeit ihrer empirischen Nachprüfung' [1932d]
4. 'Das Kausalproblem in der Physik' [1931i]
5. 'Kausalität und Wahrscheinlichkeit', Part III [1930g]

Vol. 9 *Wissenschaft und logischer Empirismus*

1. *Gesetzesaussagen und sinnvolle logische Verknüpfungen* [German trans. of *Nomological Statements and Admissible Operations* (1954e)]
2. 'Eine erneute Untersuchung des Syllogismus' [German trans. of 'The Syllogism Revised' (1952b)]
3. 'Bertrand Russell's Logik' [German trans. of 1944a]
4. 'Deweys Wissenschaftstheorie' [German trans. of 'Dewey's Theory of Science' (1939a)]
5. 'Die Verifizierbarkeitstheorie der Bedeutung' [German trans. of 'The Verifiability Theory of Meaning' (1951c)]
6. 'Sind Wahrnehmungsberichte absolut sicher?' [German trans. of 'Are Phenomenal Reports Absolutely Certain?' (1952a)]
7. 'Die Willensfreiheit' [German trans. of 'The Freedom of the Will' (1951i)]
8. 'Über die Explikation ethischer Äusserungen' [German trans. of 'On the Explication of Ethical Utterances' (1952e)]
9. 'Kant und die [moderne] Naturwissenschaft' [1932i]
10. 'Die philosophische Bedeutung der modernen Physik' [1930e]
11. *Ziele und Wege der heutigen Naturphilosophie* [1931j]
12. 'Metaphysik und Naturwissenschaft' [1925a]
13. 'Der physikalische Wahreitsbegriff' [1931g]
14. 'Logistic Empiricism in Germany and the Present State of its Problems' [1936e]
15. Complete Bibliography of the Works of Reichenbach
16. Table of Contents for the Full Collected Works

1978: *Hans Reichenbach, Selected Essays: 1909–1953,* Maria Reichenbach and Robert S. Cohen, eds., principal translations by Elizabeth Hughes Schneewind, further translations by Laurent Beauregard, Sheldon Gilman, Maria Reichenbach, and Gisela Lincoln.

(a): Vol. I

(1) 'The Student', pp. 102–103; English trans. of [1912f]
(2) 'The Student Body and Catholicism', pp. 104–107; English trans. of. [1912c]
(3) 'The Free Student Idea: Its Unified Contents', pp. 108–123; English trans. of [1913c]
(4) 'Why do we Advocate Physical Culture?', pp. 124–128; English trans. of [1913g]
(5) 'The Meaning of University Reform', pp. 129–131; English trans. of [1914c]
(6) 'Platform of the Socialist Students' Party', pp. 132–135; English trans. of [1918b]
(7) 'Socializing the University', pp. 136–180; English trans. of [1918a]
(8) 'Report of the Socialist Student Party, Berlin and Notes on the Program', pp. 181–185; English trans. of [1918c]
(9) 'The Nobel Prize for Einstein', pp. 189–191; English trans. of [1922a]
(10) 'Relativity Theory in a Matchbox: A Philosophical Dialogue', pp. 192–195; English trans. of [1922b]
(11) 'Tycho Brahe's Sextants', pp. 196–200; English trans. of [1926a]
(12) 'The Effects of Einstein's Theory', pp. 201–206; English trans. of [1926d]
(13) 'An Open Letter to the Berlin Funkstunde Corporation', pp. 207–211; English trans. of [1926e]
(14) 'Laying the Foundations of Chemistry: The Work of Marcellin Berthelot', pp. 212–215; English trans. of [1927a]
(15) 'Memories of Svante Arrhenius', pp. 216–218; English trans. of [1927b]
(16) 'A New Model of the Atom', pp. 219–225; English trans. of [1927c]
(17) 'On the Death of H. A. Lorentz', pp. 226–227; English trans. of [1928a]
(18) 'Philosophy of the Natural Sciences', pp. 228–231; English trans. of [1928c]
(19) 'Space and Time: From Kant to Einstein', pp. 232–235; English trans. of [1928d]
(20) 'Causality or Probability?, pp. 236–240; English trans. of [1928e]
(21) 'The World View of the Exact Sciences', pp. 241–244; English trans. of [1928f]
(22) 'New Approaches in Science: Physical Research', pp. 245–248; English trans. of [1929b]
(23) 'New Approaches in Science: Philosophical Research', pp. 249–253; English trans. of [1929c]
(24) 'New Approaches in Science: Mathematical Research', pp. 254–257; English trans. of [1929d]
(25) 'The New Philosophy of Science', pp. 258–260; English trans. of [1929e]
(26) 'Einstein's New Theory', pp. 261–262; English trans. of [1929a]
(27) 'Johannes Kepler', pp. 263–269; English trans. of [1930a]
(28) 'The Present State of the Sciences: The Exact Natural Sciences', pp. 270–272; English trans. of [1930b]
(29) 'One Hundred Against Einstein', pp. 273–274; English trans. of [1931a]
(30) 'Is the Human Mind Capable of Change?, pp. 275–282; English trans. of [1932a]
(31) 'Metaphysics and Natural Science', pp. 283–297; English trans. of [1925a]
(32) 'Bertrand Russell', pp. 298–303; English trans. of [1929i]

(33) 'The Philosophical Significance of Modern Physics', pp. 304–323; English trans. of [1930e]
(34) 'The Königsberg Conference on the Epistemology of the Exact Sciences', pp. 324–325; English trans. of [1930f]
(35) 'The Problem of Causality in Physics', pp. 326–342; English trans. of [1931i]
(36) 'The Physical Concept of Truth', pp. 343–356; English trans. of [1931g]
(37) 'Heinrich Scholz' *History of Logic*', pp. 356–358; English trans. of [1931d]
(38) *Aims and Methods of Modern Philosophy of Nature*, pp. 359–388; reprinted from [1959a]
(39) 'Kant and Natural Science', pp. 389–404; English trans. of [1933b]
(40) 'Carnap's *Logical Structure of the World*', pp. 405–408; English trans. of [1933c]
(41) 'Theory of Series and Gödel's Theorems' (Sections 17–22), pp. 409–428; first publication of the mimeographed text
(42) 'The Freedom of the Will', pp. 431–473; reprinted from [1959a]
(43) 'On the Explication of Ethical Utterances', pp. 474–479; reprinted from [1959a]

(b): vol. II

(44) 'The Present State of the Discussion on Relativity', pp. 3–47; reprinted from [1959a]
(45) 'The Theory of Motion According to Newton, Leibniz and Huyghens', pp. 48–68; reprinted from [1959a]
(46) 'The Relativistic Theory of Time', pp. 69–80; English trans. of [1924f]
(47) 'The Causal Structure of the World and the Difference between Past and Future', pp. 81–119; English trans. of [1925d].
(48) 'The Aims and Methods of Physical Knowledge', pp. 120–225; English trans. of [1929g]
(49) 'Current Epistemological Problems and the Use of a Three-Valued Logic in Quantum Mechanics', pp. 226–236; English trans. of [1951g]
(50) 'The Logical Foundations of Quantum Mechanics', pp. 237–278; English trans. of [1952d]
(51) 'The Philosophical Significance of Wave-Particle Dualism', pp. 279–289; original English text of [1953a]
(52a) 'The Physical Presuppositions of Probability Calculus', pp. 293–309; English trans. of [1920c]
(52b) '*Appendix*: A Letter to the Editor', pp. 310–311; English trans. of [1920d]
(53) 'A Philosophical Critique of the Probability Calculus', pp. 312–327; English trans. of [1920e]
(54) 'Notes on the Problem of Causality' [A Letter of January 25, 1924 from Erwin Schrödinger to Hans Reichenbach], pp. 328–332; English trans. of a letter printed in *Erkenntnis* **3**, 65–70 (1932).
(55) 'Causality and Probability', pp. 333–344; reprinted from [1959a]
(56) 'The Principle of Causality and the Possibility of Its Empirical Confirmation', pp. 345–371; reprinted from [1959a] with the brief 'Schlussbemerkung' from [1932d]
(57) 'Induction and Probability: Remarks on Karl Popper's *The Logic of Scientific Discovery*', pp. 372–387; English trans. of [1935e]

(58) 'The Semantic and the Object Conceptions of Probability Expressions', pp. 388–404; English trans. of [1939b]
(59) 'A Letter of March 28, 1949 from Hans Reichenbach to Bertrand Russell', pp. 405–411.

INDEX OF NAMES*

*'n' refers to note

VIENNA CIRCLE COLLECTION

1. OTTO NEURATH, *Empiricism and Sociology*. Edited by Marie Neurath and Robert S. Cohen. With a Section of Biographical and Autobiographical Sketches. Translations by Paul Foulkes and Marie Neurath. 1973, xvi + 473 pp., with illustrations. ISBN 90-277-0258-6 (cloth), ISBN 90-277-0259-4 (paper).

2. JOSEF SCHÄCHTER, *Prolegomena to a Critical Grammar*. With a Foreword by J. F. Staal and the Introduction to the original German edition by M. Schlick. Translated by Paul Foulkes. 1973, xxi + 161 pp. ISBN 90-277-0296-9 (cloth), ISBN 90-277-0301-9 (paper).

3. ERNST MACH, *Knowledge and Error. Sketches on the Psychology of Enquiry*. Translated by Paul Foulkes. 1976, xxxviii + 393 pp. ISBN 90-277-0281-0 (cloth), ISBN 90-277-0282-9 (paper).

4. MARIA REICHENBACH and ROBERT S. COHEN, *Hans Reichenbach: Selected Writings, 1909-1953 (Volume One)*. 1978, in press. ISBN 90-277-0291-8 (cloth), ISBN 90-277-0292-6 (paper). *Hans Reichenbach: Selected Writings, 1909-1953 (Volume Two)*. 1978, in press. ISBN 90-277-0909-2 (cloth), ISBN 90-277-0910-6 (paper). Sets: ISBN 90-277-0892-4 (cloth), ISBN 90-277-0893-2 (paper).

5. LUDWIG BOLTZMANN, *Theoretical Physics and Philosophical Problems. Selected Writings*. With a Foreword by S. R. de Groot. Edited by Brian McGuinness. Translated by Paul Foulkes. 1974, xvi + 280 pp. ISBN 90-277-0249-7 (cloth), ISBN 90-277-0250-0 (paper).

6. KARL MENGER, *Morality, Decision, and Social Organization. Toward a Logic of Ethics*. With a Postscript to the English Edition by the Author. Based on a translation by E. van der Schalie. 1974, xvi + 115 pp. ISBN 90-277-0318-3 (cloth), ISBN 90-277-0319-1 (paper).

7. BÉLA JUHOS, *Selected Papers on Epistemology and Physics*. Edited and with an Introduction by Gerhard Frey. Translated by Paul Foulkes. 1976, xxi + 350 pp. ISBN 90-277-0686-7 (cloth), ISBN 90-277-0687-5 (paper).

8. FRIEDRICH WAISMANN, *Philosophical Papers*. Edited by Brian McGuinness with an Introduction by Anthony Quinton. Translated by Hans Kaal (Chapters I, II, III, V, VI and VIII and by Arnold Burms and Philippe van Parys. 1977, xxii + 190 pp. ISBN 90-277-0712-X (cloth), ISBN 90-277-0713-8 (paper).

VIENNA CIRCLE COLLECTION

9. FELIX KAUFMANN, *The Infinite in Mathematics. Logico-mathematical writings.* Edited by Brian McGuinness, with an Introduction by Ernest Nagel. Translated from the German by Paul Foulkes. 1978, xviii + 236 pp. ISBN 90-277-0847-9 (cloth), ISBN 90-277-0848-7 (paper).

10. KARL MENGER, *Selected Papers in Logic and Foundations, Didactics, Economics.* 1978, in press. ISBN 90-277-0320-5 (cloth), ISBN 90-277-0321-3 (paper).

11. HENK L. MULDER and BARBARA F. B. VAN DE VELDE-SCHLICK, *Moritz Schlick: Philosophical Papers Volume I (1909-1922).* Translated by Peter Heath. 1978, xxxviii + 370 pp. ISBN 90-277-0314-0 (cloth), ISBN 90-277-0315-9 (paper).

12. EINO SAKARI KAILA, *Reality and Experience. Four Philosophical Essays.* Edited by Robert S. Cohen. 1978, in press. ISBN 90-277-0915-7 (cloth), ISBN 90-277-0919-X (paper).

THE VIENNA CIRCLE was a discussion group of philosophically interested scholars who met weekly from 1925 to 1936 at the University of Vienna. The Circle was formed around the physicist and philosopher, Moritz Schlick, and its members included Carnap, Feigl, Menger, Neurath, Waismann, and others. Reichenbach and Hempel belonged to associated groups and all recognised Mach, Boltzmann and Russell as their precursors.
However, the Circle was not a clique with a rigid organisation, nor had it any conscious aim of radically revising traditional views on the task and place of philosophy. Among the many subjects discussed by the Circle were Wittgenstein's *Tractatus*, the possibility of reducing all concepts of science to what is directly given in experience, the setting-up of a criterion of meaningfulness for non-logical utterances, the character of the basic propositions of empirical science, and the devising of a meta-language for the syntactic analysis of scientific language systems. Because of the great range of activities, the Vienna Circle's influence extended beyond philosophy. Its effects, directly or indirectly, can be seen in all those formal and empirical sciences (arithmetic, physics, geometry, biology and psychology, and the social services) whose structure and methodology and the problems of whose foundations were discussed by the Circle.
The group gradually began to break up in the 1930s with many of its members departing for new countries. The ending came with the assassination of Schlick in 1936 by a student, and the German annexation of Austria.
The Vienna Circle Collection as a publishing enterprise, is basically a series of anthologies translated into English of the most important work of individual members. Moreover, the series includes certain work by members of the Circle, or those who had a particular influence in it, which are of special scientific or historical importance. Many of the works are translated into English for the first time, and those which have been, most are out of print. Some material has hitherto remained unpublished, especially the letters which will be presented in this series.

D. Reidel Publishing Company

Dordrecht : Holland / Boston : U.S.A. / London : England

Hans Reichenbach

Selected Writings, 1909-1953

[Volume One]

Edited by Maria Reichenbach and Robert S. Cohen

D. Reidel Publishing Company, Dordrecht : Holland / Boston : U.S.A. / London : England